Tailored Polymers & Applications

TAILORED POLYMERS & APPLICATIONS

Edited by:
Y. Yagci, M.K. Mishra, O. Nuyken,
K. Ito, G. Wnek

///VSP///

VSP BV
P.O. Box 346
3700 AH Zeist
The Netherlands

Tel: +31 30 692 5790
Fax: +31 30 693 2081
vsppub@compuserve.com
www.vsppub.com

© VSP BV 2000

First published in 2000

ISBN 90-6764-326-2

All rights reserved. No part of this publication may be reproduced, stored in a retrieval system, or transmitted in any form or by any means, electronic, mechanical, photocopying, recording or otherwise, without the prior permission of the copyright owner.

Printed in The Netherlands by Ridderprint bv, Ridderkerk

Contents

Preface — vii

Amphiphilic polymers for micellar networks
R. Weberskirch, O. Nuyken — 1

Amphiphilic poly(ethylene oxide) macromonomers, polymerization and copolymerization
K. Ito, M. Maniruzzaman, H. Nishimura, T. Hattori, K. Tano, S. Kawaguchi — 15

Synthetic strategies to access linear and branched amphiphilic copolymers based on polystyrene and poly(ethylene oxide)
D. Taton, Y. Gnanou — 25

Arborescent polymers: Designed macromolecules with a dendritic structure
M. Gauthier — 43

Towards functional metallo-supramolecular assemblies and polymers
U.S. Schubert — 63

Syntheses and functions of glycopeptide surface-modified dendrimers
M. Okada, K. Aoi, K. Tsutsumiuchi — 85

"Perfectly branched" hyperbranched polymers
G. Maier, C. Zech, B. Voit, H. Komber — 97

Preparation and properties of heteroarm polymers and graft copolymers of poly(ethylene oxide) and polystyrene by macromonomer method
Y. Tsukahara, M. Takatsuka, K. Hashimoto, K. Kaeriyama — 111

Simultaneous block copolymerization via macromolecular engineering. One-step, one-pot initiation of living free radical and cationic polymerization
D.Y. Sogah, R.D. Puts, O.A. Scherman, M.W. Weimer — 123

Selection of polymeric materials for biomedical applications
M. El Fray — 133

Synthesis and thermo-responsive properties of poly(N-vinyl caprolactam)/
polyether segmented networks
N.A. Yanul, Y.E. Kirsh, S. Verbrugghe, E.J. Goethals, F.E. Du Prez   139

Poly(rotaxane)s as building blocks for the preparation
of cyclodextrin-containing membranes
L. Duvignac, A. Deratani   151

Design, synthesis, and uses of phosphazene high polymers
H.R. Allcock, J.M. Nelson, C.R. deDenus, I. Manners   165

Design, synthesis and thermal behavior of liquid crystalline block
copolymers by using transformation reactions
I.E. Serhatli, Y. Hepuzer, Y. Yagci, E. Chiellini, A. Rosati, G. Galli   175

Application of chiral polybinaphthyl-based Lewis acid catalysts
to the asymmetric organozinc additions to aldehydes
L. Pu   197

Polymer-protected metal nanocatalysts
A.B.R. Mayer, J.E. Mark   207

ATRP "living"/controlled radical grafting of solid particles to create new
properties
H. Böttcher, M.L. Hallensleben, R. Janke, M. Klüppel, M. Müller, S. Nuss,
R.H. Schuster, S. Tamsen, H. Wurm   219

Functionalization of polymers prepared by "living" free radical
polymerization
E. Beyou, P. Chaumont, C. Devaux, N. Jarroux, N. Zydowicz   233

Functional beaded polymers *via* metathesis polymerization:
Concepts and applications
M.R. Buchmeiser, F. Sinner, M. Mupa   253

Electron-beam initiated cationic polymerization
J.V. Crivello, T.C. Walton, R. Malik   265

Complexation behavior of diazosulfonate polymers
K.E. Geckeler, O. Nuyken, U. Schnöller, A. Thünemann, B. Voit   287

Surface modification of silica particles and glass beads
R.M. Ottenbrite, H. Bin, J. Wall, J.A. Siddiqui   297

Molecular design of polymers containing metallo-porphyrin moieties
as photoactive function
M. Kamachi   309

Preface

This book contains a selection of papers presented at the APME'99 Conference, the Third International Symposium on Advanced Polymers via Macromolecular Engineering held in Colonial Williamsburg, VA, USA during July 31–August 4, 1999.

Our aim has been to provide researchers in both academia and industry with central concepts and recent developments in the field of Tailored Polymers & Applications. This book should also prove useful for graduate students having some background in polymer chemistry.

The book focusses on the synthesis of targetted polymers with specific properties using macromolecular architecture. Various controlled polymerization techniques such as ionic, metathesis and recently developed 'living'/controlled polymerization and some of their combinations were covered. Polymers with different topologies such as block and graft copolymers, hyper branched polymers, heteroarm polymers, dentrimers, segmented networks and supramolecular systems were presented with emphasis on the special characteristics that influence their amphiphilic, liquid crystalline, complexation, thermo- and photosensitive and catalytic properties. The uses of such macromolecular structures in various fields such as surface modifiers, catalysts and coating and biomedical applications were discussed.

We would like to thank VSP (publisher) for the publication of this book.

Y. Yagci, M.K. Mishra, O. Nuyken, K. Ito, G. Wnek

Amphiphilic polymers for micellar networks

RALF WEBERSKIRCH and OSKAR NUYKEN

Technische Universität München, Lehrstuhl für Makromolekulare Stoffe, Lichtenbergstr. 4, D-85747 Garching b., München, Germany

Abstract—New amphiphilic, watersoluble triblock polymers containing a lipophilic and a fluorophilic segment, bridged by a hydrophilic block, were synthesized and their phase behavior was studied in detail. From viscosity studies, fluorescenic and ^{19}F-NMR spectroscopy it could be shown that those compounds aggregate. Depending on the composition of the triblock copolymers, micelle formation can either start with the lipophilic or with the fluorophilic part of the compound and continue until a micellar network of fluorophilic and lipophilic micelles, linked via hydrophilic segments, is formed.

Keywords: Amphiphilic triblock copolymers; micellar networks; oxazoline polymerization.

1. INTRODUCTION

The motivation for our studies on micellar networks was the idea to mimic nature. From specifically designed triblock copolymers, containing a lipophilic and fluorophilic segment bridged by a hydrophilic block, we expected that they would be able to build micelles with poor lipophilic and fluorophilic cores side by side. Thus, these triblock copolymers could function as simple models for compartimentation in nature, as for example in eukaryotic cells or domain formation in polypeptides (Fig. 1) [1, 2].

The favoured class of compounds for our studies were amphiphilic molecules, which are known for their tendency of self assembly above a certain concentration. Typical examples are

anionic surfactants $\quad C_{12}H_{25}SO_3^-\ Na^+$
cationic surfactants $\quad C_{16}H_{33}N(CH_3)_3^+\ Br^-$
nonionic surfactants $\quad C_{12}H_{25}O(CH_2CH_2O)_nH$.

A field of growing interest are the amphiphilic polymers [3–5]. This is not only due to the increasing number of synthetic routes for those polymers, but also becouse of their capability of self assemby as shown in Fig. 2.

Typical polymers with amphiphilic properties can have different architectures. Some selected examples are presented in Fig. 3.

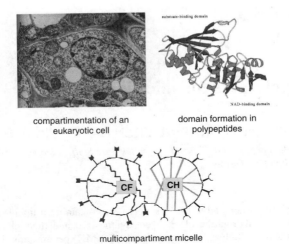

Figure 1. Examples for multicompartment systems.

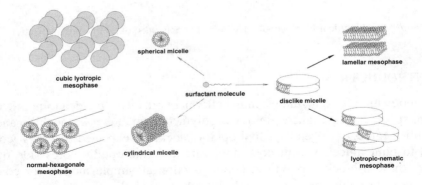

Figure 2. Self assembly of amphiphilic surfactants.

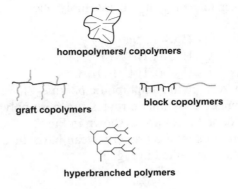

Figure 3. Architecture of amphiphilic polymers.

Typical examples of block copolymers with amphiphilic properties are shown in Fig. 4.

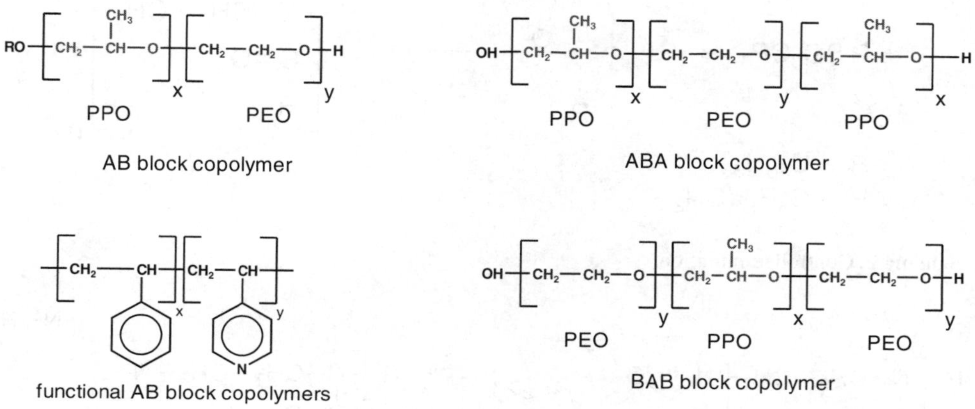

Figure 4. Block copolymers with amphiphilic properties.

The aim of this work was the synthesis of phase separated micelles of different polarity (lipophilic and fluorophilic core), which coexist in water side by side. Those compartments should allow selective mass transport and different chemical reactions, depending on their polarity. Therefore, we consider those multiphilic compounds to be suitable models for more complex biological systems, which allow a better understanding of multicompartment systems in nature.

2. CHOICE OF MONOMER AND POLYMER

Living cationic polymerization of 2-oxazoline derivatives opens many possibilities of structure variation [7–15].

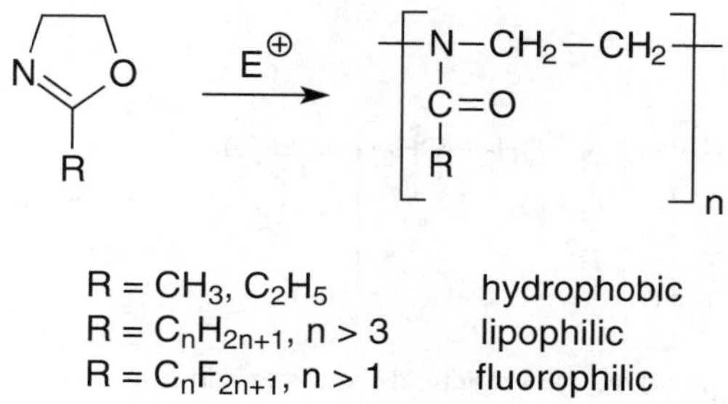

$R = CH_3, C_2H_5$ hydrophobic
$R = C_nH_{2n+1}, n > 3$ lipophilic
$R = C_nF_{2n+1}, n > 1$ fluorophilic

Scheme 1. Polymerization of 2-oxazoline.

Moreover, controlled initiation and controlled termination allow further modification.

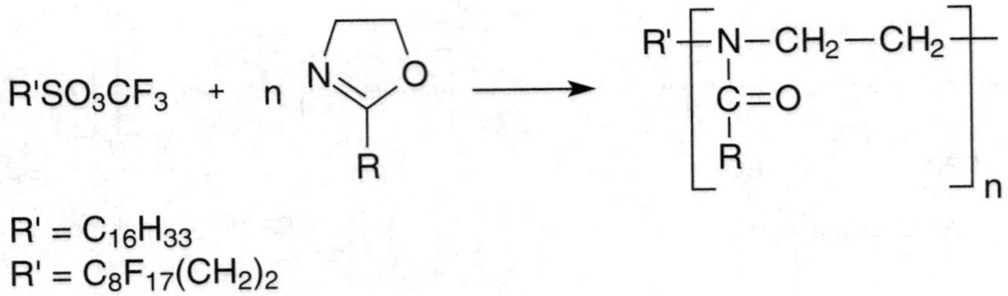

Scheme 2. Controlled initiation.

$R' = C_{16}H_{33}$
$R' = C_8F_{17}(CH_2)_2$

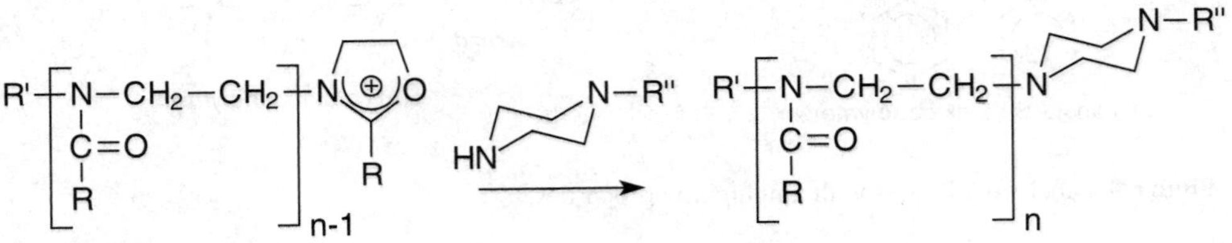

Scheme 3. Controlled termination (and controlled initiation).

Furthermore, it is also possible to design controlled structures via sequential monomer addition.

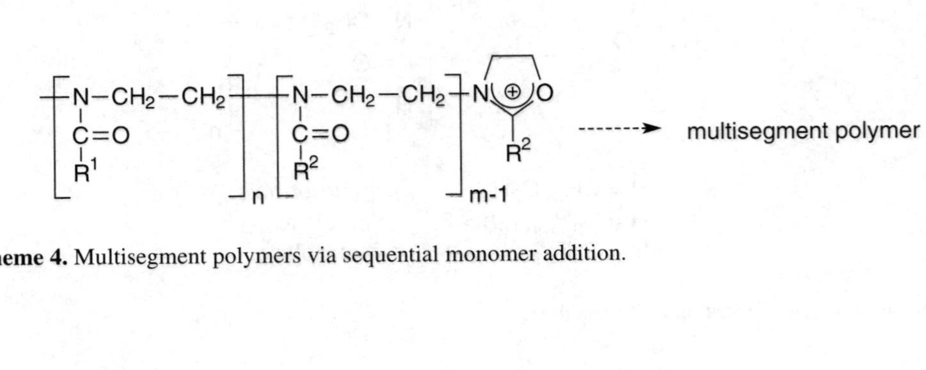

Scheme 4. Multisegment polymers via sequential monomer addition.

Some of the characteristic data of the AB- and ABA-type blockcopolymers are summerized in Table 1.

Table 1.
Structure, number of repeating units and dispersity of selected AB- and ABA blockcopolymers based on 2-methyl-2-oxazoline [15]

Polymer	R	n calculated	n by ^{1}H-NMR	M_w/M_n	degree of modification in %
FP	$C_8F_{17}-(CH_2)_2$	35	36	1.14	–
LP	$C_{16}H_{33}$	25	27	1.13	–
FPL 1	C_6H_{13}	30	31	1.16	100
FPL 2	C_8H_{17}	30	30	1.13	97
FPL 3	$C_{10}H_{21}$	30	31	1.16	100
FPL 5	$C_{12}H_{25}$	50	52	1.18	87
FPL 7	$C_{16}H_{33}$	30	32	1.21	86
FPL 8	$C_{16}H_{33}$	50	51	1.27	93

FP: fluorophilic-hydrophilic diblockcopolymer.
LP: lipophilic-hydrophilic diblockcopolymer.
FLP: fluorophilic-hydrophilic-lipophilic triblockcopolymer.

Calculated and experimental size of the hydrophilc segments are in excellent agreement with each other. The size of these segments is controlled by the ratio of monomer to initiator. The fluorophilic segments are introduced by controlled initiation (yielding FP and LP) and by combined controlled initiation and termination with a monoalkyl substituted piperazine (yielding FPL 1 through FPL 8). The overall dispersity is reasonably narrow in all cases.

3. COMPOSITION AND PROPERTIES OF THE BLOCKCOPOLYMERS

The most suitable method for the determination of the molar masses of the blockcopolymers LP, FP and FPL is ^{1}H-NMR spectroscopy. Typical examples are presented in Figs. 5 and 6.

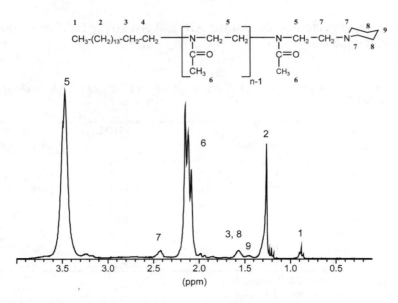

Figure 5. ^{1}H-NMR spectrum of a LP-blockcopolymer in CDCl$_3$.

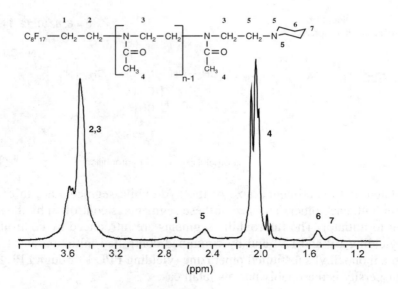

Figure 6. ^{1}H-NMR spectrum of FP-blockcopolymer in CDCl$_3$.

Table 2 shows the excellent agreement between theoretical (calculated) and experimental DPs, determined via endgroup analysis (piperazine endgroup). However, it can also be seen that GPC needs a specific calibration, as the determined values (based on polystyrene standards) deviate from reality by a factor of approximately 0.5.

Table 2.
Number average of the molar masses of selected polymers (LP and FP-type) by endgroup analysis and GPC

polymer	$\overline{DP_n}$ [a)] calculated	$\overline{DP_n}$ [b)] by ^{1}H-NMR	$\overline{M_n}$ [b)] in g/mol	$\overline{M_n}$ [c)] in g/mol	$\overline{M_w}/\overline{M_n}$ [c)]
LP 1	10	11	1245	1100	1.22
LP 2	25	27	2605	1350	1.15
LP 3	30	31	2945	1850	1.14
LP 4	50	52	4730	2346	1.26
FP 1	10	11	1466	1130	1.14
FP 2	20	19	2146	1020	1.17
FP 3	35	36	3592	1335	1.13
FP 4	50	52	4953	2030	1.25

a) $\overline{DP_n} = \dfrac{[M]}{[I]} \cdot$ conversion; b) ^{1}H-NMR endgroup analysis; c) GPC, calibrated with polystyrene standards.

Light scattering measurements provide information about size, shape and degree of aggregation of colloids, hence, we carried out static and dynamic light scattering experiments. Typical static light scattering experiments of a FP block copolymer in water are shown in Fig. 4. Polymer solutions in the range of 0.2 g L^{-1} were measured at angles between 40° and 135°. Extrapolation to c = 0 and ϑ = 0 results in a molar mass of $M_w \approx 80000$ g mol^{-1} which is a strong indication for aggregation.

$$80000 \cdot \frac{1}{1.16} \cdot \frac{1}{3590} \cong 19$$

$$M_w \cdot \frac{1}{M_w/M_n} \cdot \frac{1}{M_{n,\text{single chain}}} = DP_{\text{aggr.}}$$

Moreover, the independency of the scattering function Kc/R of the observation angle in the range of 0.2 to 10 g L^{-1} is also a strong indication for aggregation. Further evidence results from the fact that the second virial coefficient A is equal to zero, which means there is little or no interaction between the species in solution.

Hydrodynamic radii of the aggregates were determined by dynamic light scattering. A typical result is presented in Fig. 8.

The hydrodynamic radius of r_h = 5.5 nm is an additional strong indication for aggregates. It is interesting to note that only above a concentration of c $\approx$ 50 g L^{-1} the scattering varies with the observation angle (Fig. 9).

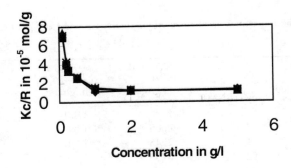

Figure 7. Kc/R at 45°, 90° and 135° as a function of the surfactant concentrations (LP 3) in water at 20 °C.

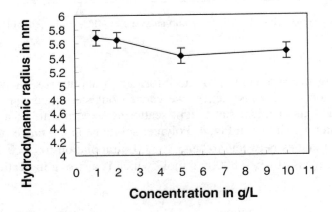

Figure 8. Hydrodynamic radii as a function of the polymer concentration (FP 3) in water at 20 °C.

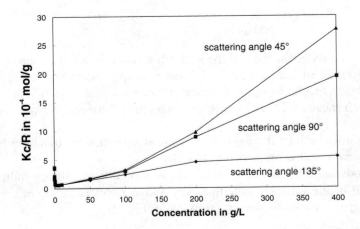

Figure 9. Kc/R as a function of the surfactant concentration and observation angle.

The experimental results can easiliy explained with the following Fig. 10.

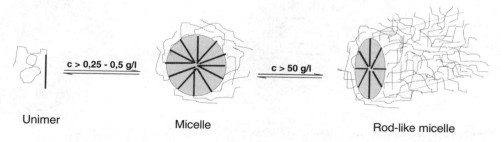

Figure 10. From unimers via micelles to higher aggregates for LP and FP-type polymers.

Above the critical micelle concentration (cmc) the unimers start to aggregate. With increasing concentration the number of aggregates but not the number of molecules per aggregate increase, and only above a rather high concentration (here: $c > 50$ g L^{-1}) the aggregates change their shape which might be explained by the formation of extended micelles (i. e. fibrilles etc.).

Further support comes from viscosity measurements, which show low values up to 50 g L^{-1} for FPL-type polymers in water. Above this concentration a dramatic increase of the η_{spec}/c-values is observed (Fig. 11). This can be explained by an interaction of aggregates, i. e. by crosslinking (Fig. 12).

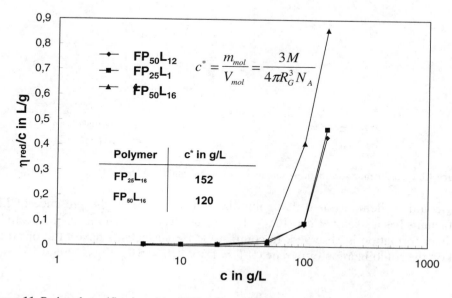

Figure 11. Reduced specific viscosity of FPL-type polymers in water at 20 °C.

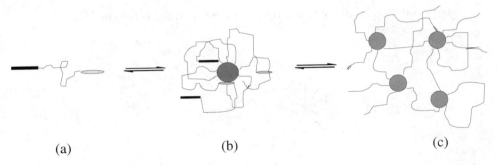

(a) (b) (c)

Figure 12. From unimers (a) via micelles (b) to network (c) for FLP-type polymers.

By means of fluorescence spectroscopy micelles with fluorophilic and hydrophilic core can be identified. Therefore we have studied the solubilizations of pyrene in water in the presence of different but well characterized surfactants of the FP, LP and FLP-type. Figs. 13a and 13b show typical differences of the fluorescence spectra of pyrene in different solvents (n-heptane and water). These characteristic differences are known as the Ham-effect. The ratio I_1/I_3 is an indicator for the micropolarity ($I_1/I_3 = 1.82$ for pyrene in water and $I_1/I_3 = 0.81$ for n-heptane).

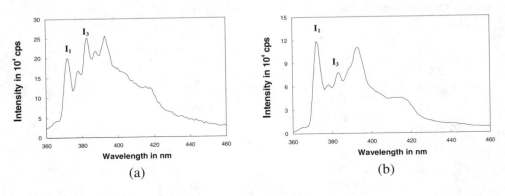

(a) (b)

Figure 13. Fluorescence spectra of pyrene in (a) n-heptane and (b) water.

It could be demonstrated that solubilization of pyrene in the presence of LP-surfactants has a strong effect ($I_1/I_3 = 1.28$; $I_1/I_E = 1.47$) which means that a strong excimer formation is observed. In contrast the FP-surfactants have little or no effect on the solubilization of pyrene ($I_1/I_3 = 1.56$; $I_1/I_E = 6.7$) (Figs. 14 and 15).

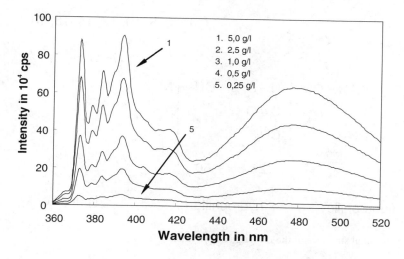

Figure 14. Fluorescence spectra of pyrene in water in the presence of LP-diblockcopolymer, $\lambda_E = 336$ nm, T = 25 °C, $I_1/I_E = 1.47$.

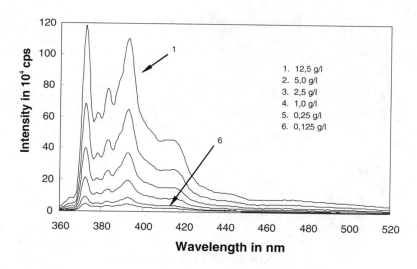

Figure 15. Fluorescence spectra of pyrene in water in the presence of FP-diblock copolymer, $\lambda_E = 336$ nm, T = 25 °C, $I_1/I_E = 6.7$.

The results support earlier observations that micelles with fluorophilic core are almost unable to solubilize pyrene [16]; therefore, no excimers are formed, and consequently high values for I_1/I_E are observed. Additional support came frome experiments with FPL-type polymers of different compositions. Polymers of the following structure were synthesized:

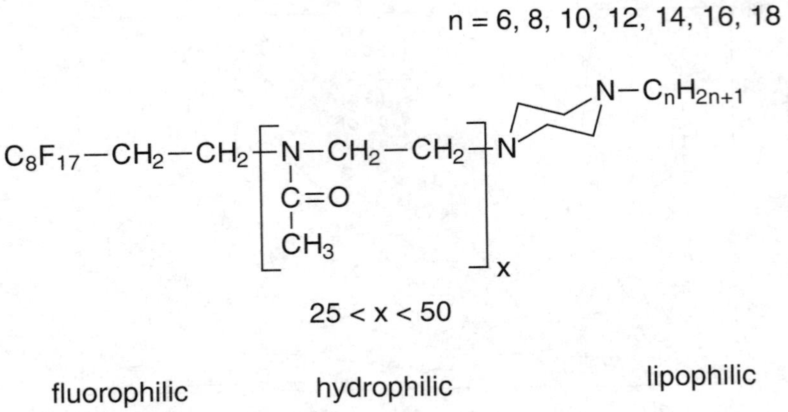

$n = 6, 8, 10, 12, 14, 16, 18$

$C_8F_{17}-CH_2-CH_2-[N-CH_2-CH_2-]_x-N\text{-}C_nH_{2n+1}$

with side chain $C=O$, CH_3 on repeat unit

$25 < x < 50$

fluorophilic   hydrophilic   lipophilic

Scheme 5. General structure of the investigated polymers.

Depending on the length of the alkyl chain the lipophilic part will contribute different hydrophobic energies to the molecule (Fig. 16).

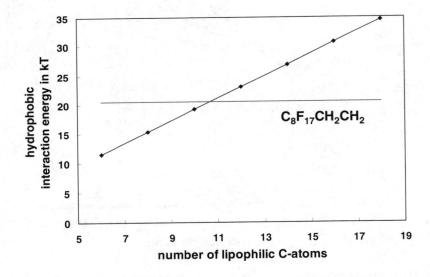

Figure 16. Hydrophobic energy of the fluorinated and of the alkyl part of the FPL-polymer.

This means, that above a certain chain length the properties of FPL-triblocks are determined by the alkyl part, consequently the formation of micelles with an alkyl core are favoured. Below this concentration micelles with fluorophilic core were formed first. Therefore, $FP_{25/50}L_n$ with $n \geq 10$ shows the characteristics of pure LP ($I_1/I_3 = 1.28$, $I_1/I_E = 1.47$) and with $n < 10$ the characteristics of pure FP ($I_1/I_3 = 1.56$, $I_1/I_E = 6.7$). From these results we can draw the following picture (Fig. 17):

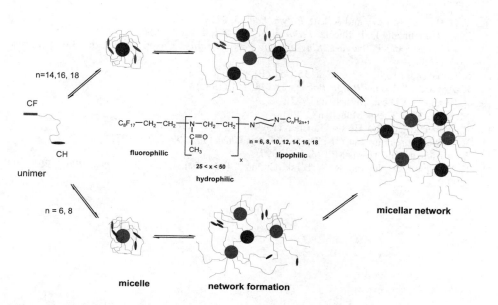

Figure 17. Total view of the network formation.

These investigations were continued with ^{19}F-NMR studies and T_2 relaxation analysis. The transversale relaxation of a ^{19}F-NMR of FPL-polymers can be described by a biexponential curve [17, 18].

$$A_{short} \cdot \exp\left(-\frac{\tau}{T_{short}}\right) + B_{long} \cdot \exp\left(-\frac{\tau}{T_{long}}\right) \quad \text{with} \quad A + B = 1$$

From solid state ^{19}F-NMR measurements one can determine the T_2 relaxation to be 210 μs ≈ T_{short}. This value is very similar to $T_{2,short}$ in solution experiments. Furthermore, since the properties of the part which shows short relaxation increases with the concentration, we conclude that the aggregated fluorinated chains are responsible for T_{short} due to their limited mobility.

REFERENCES

1. P. Lehmann, Diplom Thesis Mainz (1995).
2. R. Weberskirch, Diplom Thesis Mainz (1995).
3. B. Chu, *Langmuir*, **11**, 414 (1995).
4. M. R. Munch and P. H. Gast, *Macromolecules*, **21**, 1360 (1988).
5. R. Nagarajan and K. Ganesh, *Macromolecules*, **22**, 4312 (1989).
6. C. L. Zhou, M. A. Winnik, G. Riess and M. D. Groucher, *Langmuir*, **6**, 514 (1990).
7. W. Seeliger, E. Aufderhaar, W. Diepers, R. Feinhauer, R. Nehring, W. Thiermann and H. Hellmann, *Angew. Chem.*, **78**, 613 (1966).

8. T. G. Bassiri, A. Levy and A. Litt, *Polym. Lett.*, **5**, 871 (1967).
9. D. A. Tomalia and D. P. Sheetz, *J. Polym. Sci. Part A*, **4**, 2253 (1966).
10. S. Kobayashi and T. Saegusa, *Ring-opening Polymerization*, Chapt. 11, Elsevier, London, New York (1984).
11. S. Kobayashi *Progr. Polym. Sci.*, **15**, 751 (1990).
12. K. Aoi and M. Okada, *Progr. Polym. Sci.*, **21**, 151 (1996).
13. P. Kuhn, PhD Thesis München (1997).
14. M. Ruile, PhD Thesis München (1998).
15. R. Weberskirch, PhD Thesis München (1999).
16. K. Kalyanasundaram and J. K. Thomas, *J. Am. Chem. Soc.*, **99**, 2039 (1977).
17. J. Preuschen, PhD Thesis MPI Mainz (1999).
18. H. W. Spieß, O. Nuyken, R. Weberskirch and J. Preuschen, *Macromol. Chem. Phys.*, **201**, 995 (2000).

Amphiphilic poly(ethylene oxide) macromonomers, polymerization and copolymerization

KOICHI ITO,* MOHD. MANIRUZZAMAN, HIROKAZU NISHIMURA, TERUYUKI HATTORI, KEISUKE TANO and SEIGOU KAWAGUCHI

Department of Materials Science, Toyohashi University of Technology, Tempaku-cho, Toyohashi 441-8580, Japan

Abstract—Amphiphilic poly(ethylene oxide) (PEO) macromonomers such as **1-3** homopolymerize very rapidly, via micellar organization, in water to afford regular comb polymers. Poly(macromonomer)s obtained exhibit either star-like or bottlebrush conformation depending on the relative size of backbone to arm. Micellar copolymerization between PEO macromonomers apparently enhanced the incorporation of those with a more hydrophobic polymerizing end-group. Micellar copolymerization of solubilized styrene with **1** (1/1 mol/mol) afforded an apparently unimolecular, highly branched graft copolymer of *ca* 30 nm size with MW of ca 10^7. Emulsion (in water) and dispersion copolymerization (in methanol/water mixture of 9/1 v/v) of styrene with **1** afforded nearly monodisperse polymeric microspheres of submicron to micron size.

Keywords: Poly(ethylene oxide) macromonomer; micellar organized polymerization and copolymerization; bottlebrush; emulsion copolymerization; dispersion copolymerization.

1. INTRODUCTION

In the last two decades, macromonomer technique has been established as one of most convenient approaches to well-defined branched polymers such as star- and comb-shaped polymers and copolymers as well as to core-shell polymeric microspheres [1, 2]. In particular, amphiphilic macromonomers are interesting because their organization affects dramatically their polymerization and copolymerization behavior and may afford a variety of amphiphilic polymer architecture. We have been interested in poly(ethylene oxide) (PEO) macromonomers such as **1** and **2** or **3**, which have distinct amphiphilic character in that the PEO chains are hydrophilic while the polymerizing end groups are hydrophobic. This paper will review our recent results of polymerization and copolymerization of these macromonomers in water or an alcoholic medium and discuss control in structure of the amphiphilic polymer systems ranging from nm to μm in size.

---

* To whom correspondence should be addressed. E-mail: itoh@tutms.tut.ac.jp

CH₃O−[CH₂CH₂O]$_n$−(CH₂)$_m$−⟨C₆H₄⟩−CH=CH₂

(**1**) C1-PEO-Cm-S-n

CH₃O−[CH₂CH₂O]$_n$−(CH₂)$_m$−OC(=O)C(CH₃)=CH₂

(**2**) C1-PEO-Cm-MA-n

CH₃O−[CH₂CH₂O]$_n$−C(=O)C(CH₃)=CH₂

(**3**) C1-PEO-MA-n

2. HOMOPOLYMERIZATION IN WATER (MICELLAR POLYMERIZATION)

1 polymerized unusually rapidly in water with 4,4′-azobis(4-cyanovaleric acid) (AVA) to very high DPs, apparently because of their micellar organization with their hydrophobic, polymerizing end-groups locally concentrated into the cores [3]. The kinetics apparently followed normal ones ($R_p \sim [M][I]^{1/2}$, $DP \sim [M]/[I]^{1/2}$) but with an apparently high propagation rate constant (k_p) and a very low termination rate constant (k_t) as compared to styrene, as given in Table 1. To be noted, the k_p values of the macromonomers in water are remarkably higher than those in benzene, as a result of micellar organization in the former solvent. The initiator efficiency was also found to be higher in water than in benzene. With increase in alkylene chain length, m in **1** [3, 4] and also in **2** [5], the polymerization rate is apparently higher, as expected because of increasing micellar organization. All these facts support the micellar polymerization mechanism of the PEO macromonomers in water, together with the direct evidence of aggregation of the macromonomers and their models [6, 7].

The polymers obtained are regular comb polymers. Thus poly(macromonomer) from **1** is a polystyrene as a backbone but with PEO as the side chains each attached to every phenyl ring, as in Fig. 1 (a). They should serve as a simple model of amphiphilic branched polymers, and indeed a poly(C1-PEO-C4-S-50) in water exhibited chain conformations such as star-like with a z-average square root radius of gyration, $<S^2>_z^{1/2} \sim 5$ nm, and semiflexible bottlebrush, with $<S^2>_z^{1/2} \sim$ 10–500 nm, depending on their low and high DPs in the backbone, respectively, as in Fig. 1 (b) [2, 8]. It appears that the crossover of the conformation from starlike to bottlebrush occurs when the extended length of the main chain exceeds the diameter

Table 1.
Kinetic Parameters of Radical Polymerization of PEO Macromonomers (**1**) under Irradiation at 20°C[a] [3]

Monomer/ solvent	Initiator[b]	$R_p \times 10^6$ mol L^{-1} s^{-1}	$[P\bullet] \times 10^6$ mol L^{-1}	k_p L mol^{-1} s^{-1}	k_t L mol^{-1} s^{-1}
C1-PEO-C1-S-48/ benzene	tBPO	3.9	2	40	1800
C1-PEO-C1-S-48/ water	AVA	210	4	1100	5400
C1-PEO-C4-S-48/ benzene	tBPO	15	5	45	4500
C1-PEO-C4-S-48/ water	AVA	510	5	2100	9000
Styrene/bulk[c]	tBPO	940	0.6	35	1.9×10^7

[a] [M] = 50 × 10^{-3} mol L^{-1}, [I] = 2.5 × 10^{-3} mol L^{-1} for the macromonomers, and [I] = 0.2 mol L^{-1} for styrene. Values of R_p, [P•], k_p, and k_t of the macromonomers in this work may include errors of 10–30%.

[b] tBPO = *tert*-butyl peroxide, AVA = 4,4′-azobis(4-cyanovaleric acid).

[c] Data at 25°C from B. M. Burnett, G. G. Cameron, and S. N. Joiner, *Trans. Faraday Soc. I*, **69**, 322 (1973).

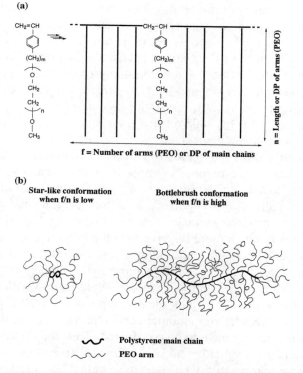

Figure 1. Schematic drawing of (a) polymerization of a PEO macromonomer (**1**) to a comb polymer and (b) star-like and bottlebrush conformations of the comb polymer (poly-**1**) in water.

of the brush cylinder or rod (double of the size of the random PEO arms). The polystyrene main chain is highly restricted or rigid in its conformational mobility (persistence length ~17 nm as compared to 1.1 nm for a standard linear polystyrene) as a result of so extensive steric congestion of the PEO arms.

3. MICELLAR COPOLYMERIZATION

(a) Copolymerization of PEO Macromonomers with Styrene. A limited amount of hydrophobic monomer can be solubilized into micelles of an amphiphilic macromonomer in water, and they are expected to copolymerize ("micellar copolymerization") to afford a highly branched graft copolymer with a very compact conformational size. An attempt was made using styrene and the PEO macromonomers, **1** [9].

The copolymerization between equimolar styrene and **1** (m=7, n=18) with AVA at 60°C was found to proceed very rapidly, apparently transparently and azeotropically to give ca 30 nm sized microparticles (as estimated by DLS in water) in almost quantitative yields. The apparent MW and $<S^2>_z^{1/2}$ of the copolymers were estimated by SEC-MALLS and DLS measurements in DMF to be as high as 10^7 and 100 nm, respectively. This suggests interestingly that the microparticle in water consists of one graft copolymer, i. e., supposedly "unimolecular particle". This microparticle should have a densely crowded and folded main chain composed of a random copolymer of equimolar styrene and p-heptylstyrene surrounded by a number of PEO arms. The size is just between the star-like and the bottlebrush conformations of the homopoly(macromonomers) in Fig. 1 (b), reasonable because the very high MW (DP~10^4 for the main chain) will expand the overall size but copolymerization with styrene in the main chain as well as shorter PEO arms (n = 18) will release their steric congestion to allow dense folding of the hydrophobic main chain.

(b) Copolymerization between PEO Macromonomers. A copolymerization between two kinds of macromonomers appears to manifest any polymer chain effects, if any, due to solubility, organization, or conformational difference. In fact, PEO macromonomers with p-vinylbenzyl and methacrylate end groups, **1** (m = 1) and **3**, were copolymerized in water with a result in Fig. 2 [10] showing that the former is apparently much more reactive than the latter, most probably as a result of higher incorporation of the more hydrophobic, former monomer into the micelles. Also in water, **1** (m = 4) with a more hydrophobic styryl end was more readily incorporated into a copolymer in copolymerization with a less hydrophobic analog, **1** (m = 1). The results clearly support a micellar copolymerization mechanism where the relative monomer concentrations around the actual reaction sites, i. e., in the micelles have to be considered. Thus the micellar copolymerization has to be analyzed by taking account of the relative monomer partition in addition to the monomer reactivity ratios as in so-called bootstrap effect [11].

In contrast to the micellar copolymerization in water, copolymerizations in benzene or methanol apparently proceed azeotropically ($r_A \sim r_B \sim 1$), indicating apparently similar reactivity of the macromonomers involved. This result in case of **1** (m = 1) and **3** is surprising in view of the well-known alternating tendency of their low molecular weight monomers like styrene and MMA ($r_A \sim r_B \sim 0.5$). Thus the same PEO chains appear to make the intrinsic reactivity difference of their end groups almost insignificant. On the other hand, different polymer chains such as polystyrene vs PEO with a same polymerizing end group (*p*-vinylbenzyl) were found to show apparently different reactivities, probably reflecting some incompatibility or conformational effects.

These results suggest importance of the polymer chain effects on the apparent reactivity of the chain ends to be considered in addition to their intrinsic chemical reactivities.

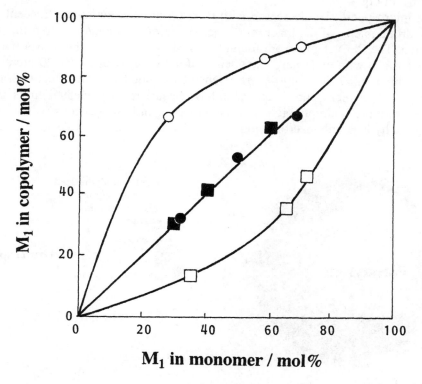

Figure 2. Copolymer composition curves for copolymerization between PEO macromonomers at 60°C with a total monomer concentration, $[M]_0 = 50$ mmol L^{-1}. (○) **1** (m=1, n=16) as M_1 and **3** (n=16) as M_2 in D_2O with AVA, (□) **1** (m=1, n=48) as M_1 and **1** (m=4, n=48) as M_2 in D_2O with AVA, (■) **1** (m=1, n=48) as M_1 and **3** (n=48) as M_2 in C_6D_6 with AIBN, (●) **1** (m=1, n=48) as M_1 and **1** (m=4, n=48) as M_2 in CD_3OD with AIBN.

4. EMULSION COPOLYMERIZATION WITH STYRENE

An excessive amount of hydrophobic monomer can be emulsified with a water-soluble macromonomer and copolymerized to give well-stabilized latex with polymer particles of submicron size [1, 2]. Here the graft copolymers produced in situ serve as effective steric stabilizers for the resulting (co)polymer particles of core-corona structure. The surface or corona is composed of the grafted macromonomer side chains extending into the media while the main chains anchor into the particle cores consisted of the substrate polymers, as in Fig. 3.

Application of **1** to emulsion copolymerization with styrene using KPS or AIBN as an initiator at 65°C afforded stable latex composed of monodisperse microspheres of submicron size [12]. The copolymerization was found to proceed azeotropically, with the microsphere volume increasing linearly with conversion, as shown in Fig. 4.

The microsphere size decreased with increase in macromonomer concentration and hydrophobicity of the macromonomer (increasing m and/or decreasing n), as expected. The PEO chains were found by TEM observation to be densely grafted on the surface with extremely extended conformation, with ca 10–20 nm as h in Fig. 3 with **1** (m=7, n=45), to exert extensive excluded volume effect (steric stabilization). The microspheres appear to have been largely nucleated either in micelles with more hydrophobic macromonomers or in homogeneous (water) phase with more hydrophilic macromonomers.

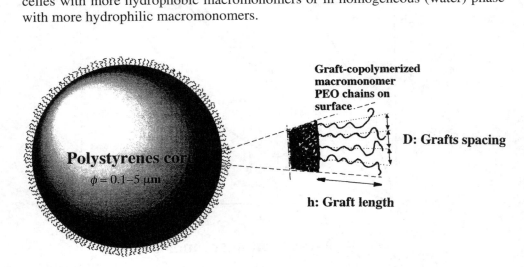

Figure 3. Schematic representation of a polymeric microsphere obtained by emulsion or dispersion copolymerization of a substrate monomer with a macromonomer; heere styrene with a PEO macromonomer.

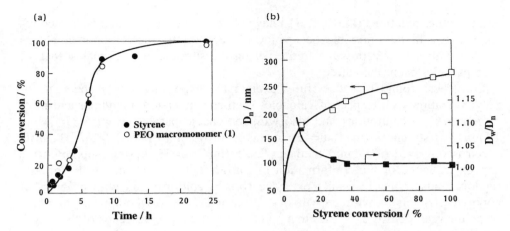

Figure 4. Plots of (a) conversion vs time and (b) particle diameter and distribution vs styrene conversion for the emulsion copolymerization of styrene with **1** (m=7, n=16) at 65°C. [**1**] = 8 g L^{-1}, [KPS] = 1g L^{-1}.

5. DISPERSION COPOLYMERIZATION WITH STYRENE

Copolymerization of an excessive amount of a soluble monomer (but its polymer being insoluble to precipitate out) with a very soluble macromonomer provides a convenient technique to afford well-stabilized dispersions with usually monodisperse polymeric particles of submicron to micron size [1, 2]. Here also, the graft copolymers produced in situ serve as effective steric stabilizers for the resulting polymer particles of core-corona structure, just like that in Fig. 3.

Copolymerization of **1** with styrene [13] (or with *n*-butyl methacrylate [14]) in methanol-water (9/1 or 8/2 v/v) comprises a successful example. The particle size could be controlled according to the power law as follows:

$$R = \theta^{\frac{1}{3}} \left(\frac{3W_o}{\rho N_A} \right)^{\frac{2}{3}} \left(\frac{M_D r_1}{W_{Do} S_{crit}} \right)^{\frac{1}{2}} \left(\frac{0.386 k_2}{4\pi k_p} \right)^{\frac{1}{6}} \left(\frac{k_t}{2 k_d f [I]_o} \right)^{\frac{1}{12}} \quad (1)$$

where R = the radius of the particle core, which consists of homopolymers from the substrate monomer plus trunk chains of the graft copolymers, W_o = weight of the substrate monomer in feed, W_{Do} = weight of the macromonomer (dispersant) in feed, θ = conversion of the substrate monomer, M_D = molecular weight of the macromonomer, ρ = density of the polymers in the particle core, N_A = the Avogadro's number, S = cross-sectional area per copolymerized macromonomer chain on the particle core surface (core-corona interface), S_{crit} = S at the critical point when the formation of sterically stabilized particles is established, k_2 = rate constant of coalescence of similar-sized microparticles (assumed to be 10^9 L mol^{-1} s^{-1}), r_A = the monomer reactivity ratio of the substrate monomer before the critical point, and the other notations (k_p, k_d, f, $[I]_o$) have the conventional meanings. S_{crit} corresponds to a highest possible occupancy of each of the macromonomer chains grafted with a lowest possible frequency on the surface, and no coalescence of such particles occur thereafter.

Eq. 1 was found to successfully explain the experiments of dispersion copolymerization with respect of valuables such as conversion, monomer and macromonomer concentrations and also with reasonable parameters of S_{crit}/r_A, e.g., 10 nm^2 for styrene/**1** (m=1, n=43) in methanol-water (9:1 v/v) [13]. ^{1}H NMR spectroscopy [15] of *n*-butyl methacrylate/**1** (m=7, n=53) system supported the expected change of the conformation of the grafted PEO chains, from initially random (mushroom) to finally extended (brush) conformation with progress of copolymerization or with growth of the particles. Crossover appears to occur when the graft density increases to a point of significant coalescence of the neighbouring grafts with a mean spacing, D~1.5 nm.

Methacrylate-ended macromonomers, **2** (m=6 or 10), were also found to be very effective in preparing polystyrene microsheres [5]. On the other hand, ω-hydrophobic alkyl group caused a disappointing effect in that an ω-dodecyloxy PEO methacrylate with a similar HLB gave extensive coagulation, indicating possible corona looping or inter-particle bridging of some PEO chains through the hydrophobic groups on both ends. With MMA as a substrate monomer, a simple macromonomer, **3**, was rather more suitable in producing monodisperse particles than that with a spacer, **2** (m=6) [5]. A dependence, different from Eq. 1, on macromonomer concentration such as $R \sim W_{Do}^{-1.1}$ was observed in MMA polymerization with **1** (m=1) in methanol-water mixture (7/3 v/v) but normal ($R \sim W_{Do}^{-0.52}$) in a more polar mixture (5/5 v/v) [16]. It appears that some compatibility among the substrate polymer (PMMA), the macromonomer chains (PEO) and the medium (methanol) play a role in such an abnormal power law as observed in the former.

Proper balance and location of the hydrophilic and hydrophobic segments relative to the polymerizing group appear to be a factor of key importance. Eq. 1 serves as a reference expression in designing the particle size in dispersion (co)polymerization with macromonomers, particularly in case of producing microspheres of clear-cut core-corona structure like those from styrene or *n*-butyl methacrylate / PEO macromonomer / methanol-water. A similar power law appears to be applicable to emulsion copolymerization of styrene with a hydrophobic macromonomer where nucleation in the continuous phase (water) may be dominant.

6. CONCLUSION

Amphiphilic PEO macromonomers such as **1-3**, in particular those with hydrophobically enhanced (co)polymerizing end groups, provide a convenient approach to design a variety of well-defined, amphiphilic, branched polymer architecture, including star, comb (bottlebrush), and microsheres of controlled sizes.

REFERENCES

1. K. Ito, *Prog. Polym. Sci.*, **23**, 581 (1998).
2. K. Ito and S. Kawaguchi, *Adv. Polym. Sci.*, **142**, 129 (1999).
3. E. Nomura, K. Ito, A. Kajiwara and M. Kamachi, *Macromolecules*, **30**, 2811 (1997).
4. D. Chao, S. Itsuno and K. Ito, *Polym. J.*, **23**, 1045 (1991).
5. H. Furuhashi, S. Kawaguchi, S. Itsuno and K. Ito, *Colloid Polym. Sci.*, **275**, 227 (1997).
6. K. Ito, K. Hashimura, H. Tanaka, G. Imai, S. Kawaguchi and S. Itsuno, *Macromolecules*, **24**, 2348 (1991).
7. S. Kawaguchi, A. Yekta, J. Duhamel, M. A. Winnik and K. Ito, *J. Phys. Chem.*, **98**, 7891 (1994).
8. (a) S. Kawaguchi, K. Akaike, Z.-M. Zhang, H. Matsumoto and K. Ito, *Polym. J.*, **30**, 1004 (1998); (b) S. Kawaguchi and K. Ito, *Colloids & Surfaces, A, Physicochem. & Eng. Aspects*, **153**, 173 (1999).
9. M. Maniruzzaman, S. Kawaguchi and K. Ito, presented at APME'99, Williamsburg, VA, Jul. 31-Aug. 5 (1999): *Macromolecules*, **33**, 1583 (2000).
10. H. Nishimura, T. Hadama and K. Ito, *Polym. Prepr., Jpn.*, **44**, 1013 (1995).
11. H. J. Harwood, *Makromol. Chem., Macromol. Symp.*, **10/11**, 331 (1987).
12. K. Tano, S. Kawaguchi and K. Ito, presented at APME'99, Williamsburg, VA, Jul. 31-Aug. 5, (1999); S. Kawaguchi, K. Tano, M. Maniruzzaman and K. Ito, presented at 4th Internat. Symp. Polymers in Dispersed Media, Lyon, France, Apr. 11–15, 1999.
13. M. B. Nugroho, S. Kawaguchi, K. Ito and M. A. Winnik, *Macromol. Reports*, **A32**, Suppl. 5&6, 593 (1995).
14. S. Kawaguchi, M. A. Winnik and K. Ito, *Macromolecules*, **28**, 1159 (1995).
15. S. Kawaguchi, M. A. Winnik and K. Ito, *Macromolecules*, **29**, 4465 (1996).
16. T. Hattori, M. B. Nugroho, S. Kawaguchi and K. Ito, 9th Polymeric Microspheres Symposium, Nov. 11–13, 1996, Tsukuba, Preprints, p. 35–36.

Synthetic strategies to access linear and branched amphiphilic copolymers based on polystyrene and poly(ethylene oxide)

DANIEL TATON* and YVES GNANOU

Laboratoire de Chimie des Polymères Organiques, UMR 5629-ENSCPB-CNRS-Université Bordeaux-1, Avenue Pey-Berland, B.P. 108, 33402 Talence Cedex, France

Abstract—The synthetic methodologies developed to prepare both linear-type and branched amphiphilic copolymers based on polystyrene (PS) and poly(ethylene oxide) (PEO) are reviewed. These nonionic macromolecular amphiphiles include linear block, graft, star-block, miktoarm star, star-graft, dumbbell and arborescent copolymers. Emphasis is placed on the possibilities to obtain well defined architectures free of any contaminant through the control of the polymerization procedures.

Keywords: Amphiphilic; linear; branched; copolymers; poly(ethylene oxide); polystyrene.

1. INTRODUCTION

Amphiphilic copolymers consisting of hydrophilic and hydrophobic parts are macromolecular compounds that are able of self-aggregating into micro domains in the presence of a selective solvent of one of the two constituents. This self-association that occurs *via* intra- or intermolecular interactions can promote the formation of three-dimensional structures of enhanced viscosity such as stable micelles or more organized assemblies such as vesicles; this aggregation phenomenon is reversible depending upon the concentration of the solution, its temperature, etc... [1].

For instance, Bates and coll. have described new types of structures formed by self-assembling of poly(ethylene oxide)-b-poly(butadiene) diblock copolymers [1e]. The latter were found to organize in water at low concentration from membrane-like bilayers to cylinders to spheres with increasing the hydrophilic PEO part. Then, these amphiphilic copolymers could be chemically cross-linked after dispersion in water affording the so-called giant wormlike rubber micelles.

Most of the nonionic amphiphilic copolymers are based on poly(ethylene oxide) (PEO) the hydrophobic components being constituted of either polystyrene,

---

*To whom correspondence should be addressed. E-mail: taton@iagp.enscpb.u-bordeaux.fr

poly(propylene oxide), poly(butadiene) or alkyl chains Because they are little sensitive to the ionic strength of the solution, macromolecular amphiphiles can thus be preferred over micromolecular ionic surfactants for certain specific applications. They are used as drug delivery systems (a hydrophobic core of a micelle can serve as host for a guest drug), adhesives, ion-conducting components or to stabilize colloidal suspensions or control the rheology of aqueous formulations [1].

Even if most of the studies aiming at investigating the properties of polymeric amphiphiles have been conducted in solution or in dispersion media, amphiphilic PEO-based copolymers have also been utilized in the solid state to template the generation of highly ordered silica structure. New nano– and mesoscopic architectures have been synthesized through the self-assembly process of oligomeric surfactants [2]. By controlling the hydrophilic-lipophilic balance (HLB), the nature of the mesophase formed (lamellar, cubic, …) can be adjusted.

Among micelle-like polymeric compounds, copolymers based on polystyrene (PS) and PEO have been extensively investigated. Generally, these copolymers exhibit either a linear block (PEO-b-PS) or a graft (PS-g-PEO) copolymer structure and suitable methodologies have been developed towards this end; the characterization of their solution behavior is also well documented. In 1995, Piirma [1a] on the one hand, Velichkova et al. [1b], on the other, reviewed the synthesis and characterization of linear-type and graft copolymers based on PEO or other hydrophilic polymers (polyacrylic acid, polyoxazoline, …).

Attempts at associating both ethylene oxide (EO) and styrene (St) units in other non-linear morphologies are more recent. In such systems (like star or arborescent copolymers) the number of branching points can be precisely controlled unlike the case of graft homologues [1]. Such amphiphilic star or arborescent species are also expected to form stable 'monomolecular micelles' [3]: Newkome has coined this term to describe a polymeric system consisting of a hydrophobic inner dendritic core carrying hydrophilic functions at its surface. In contrast to conventional micelles for which the micellar behavior only appear above the critical micellar concentration (cmc), the formation of a unimolecular micelle can be observed for a single molecule and is independent of the concentration.

On the other hand, water soluble branched polymers carrying short hydrophobic moieties (the so-called associative polymers) are known to enhance the viscosity of aqueous formulations [1d] better than do their linear homologues.

The present paper is aimed at reviewing the routes that have been recently considered to prepare original amphiphilic branched assemblies based on PS and PEO, each of these structures differing from the other by the topological arrangement of its subchains. The methodologies used to produce linear-type PS-b-PEO and graft PS-g-PEO amphiphilic copolymers are first discussed.

2. LINEAR BLOCK COPOLYMERS BASED ON PS AND PEO

2.1. By coupling reactions

Linear PS/PEO block copolymers can be obtained by condensation reactions involving linear precursors fitted with antagonist functional groups (Scheme 1, route a- and b-) [1]. Following this approach, α,ω-(bis hydroxy)-PEO chains were reacted with linear PS chains carrying terminal acyl chloride functions in order to generate PS-b-PEO-b-PS copolymers [4]. The easiness of this methodology notwithstanding samples obtained in this way generally suffer from compositional heterogeneity: unreacted chains usually contaminate the expected copolymers and a last step of fractionation is required to isolate the latter [1].

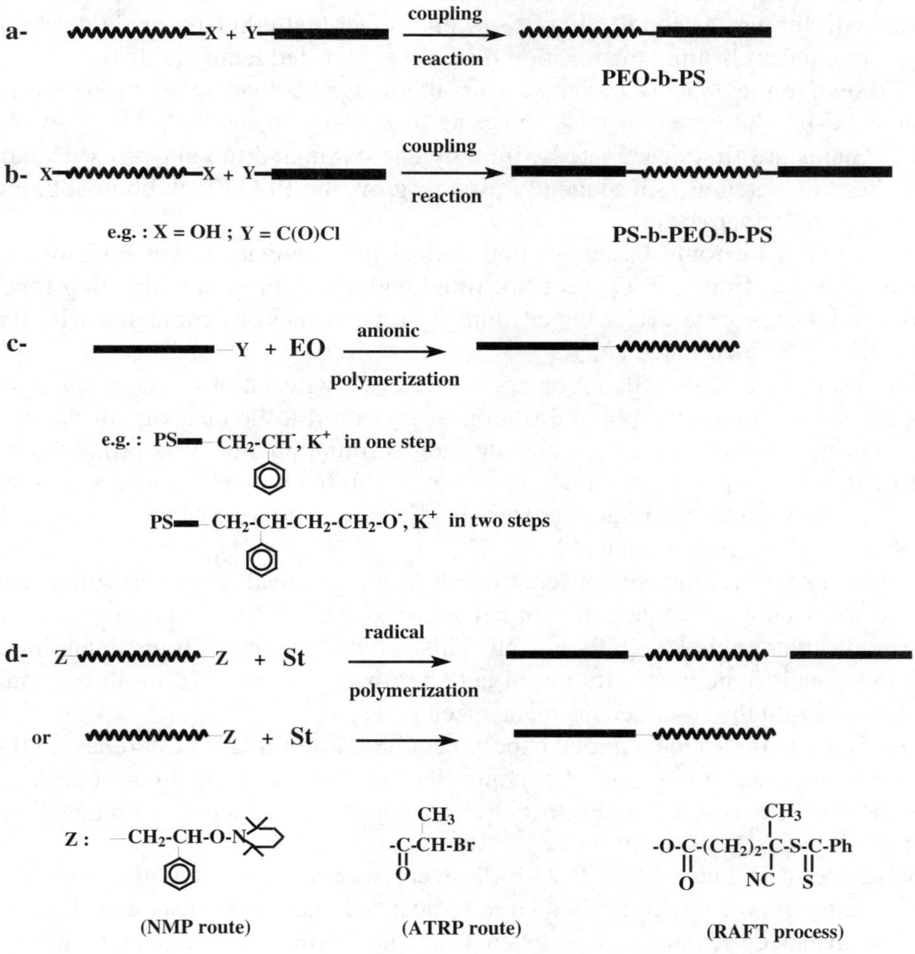

Scheme 1. Access to linear block PS-b-PEO or PS-b-PEO-b-PS copolymers.

2.2. Sequential polymerization of the monomers

The second route to access block PS/PEO copolymers consists in the sequential anionic polymerization of styrene and ethylene oxide, using for instance cumyl potassium as initiator [5]. This affords PS-b-PEO copolymers in one step without isolating the intermediate PS (Scheme 1, route c-) but the samples isolated may be contaminated with a substantial amount of PEO homopolymers. The presence of the latter originates from the initiation of EO polymerization by CH_3OK that is formed as a side product from the cleavage of cumyl methyl ether by potassium to make cumyl potassium.

A way to overcome this inconvenience was proposed by Quirk and coll. who sequentially polymerized the two monomers and obtained a minimal amount of contaminants [6]. After polymerizing the first PS block by lithiated species, these authors added potassium 2,6-di-*t*-butylphenoxide to allow the growth of the PEO block (with lithium as counter-ion, it is well known that ethylene oxide does not polymerize unless heating the reaction medium at elevated temperatures).

As shown in the Scheme 1 (route c-), an alternative pathway to synthesize well-defined PS-b-PEO copolymers is to resort to a two-step method. The growing PS^-Li^+ chains are first deactivated with ethylene oxide and the ω-hydroxyl functions thus generated are subsequently used to grow the PEO block upon addition of potassium derivatives [7].

In the synthetic route based on free radical polymerization, the PS block is grown in second from a PEO precursor fitted with the appropriate initiating functions: the latter process offers the advantage of its easiness as compared with the ionic techniques mentioned above.

For instance, PEO-based oligomers end-capped with chloroacetate functions have served to initiate the polymerization of styrene [8]; the cleavage of the carbon-chlorine bond necessary to generate radical initiating sites was promoted by heating these PEO precursors in the presence of $[Mn(CO)_{10}]$. However, due to the lack of control of the free radical process in this case, a mixture of tri- and diblock PEO/PS copolymers was obtained.

The recent development of controlled/'living' radical polymerization has opened up a new and potentially broad route to well defined polymers and to macromolecular engineering [9] as well. This could be achieved by maintaining a very low steady state concentration of active growing chains that are in dynamic equilibrium with the so-called dormant species.

The first attempt to make block copolymers based on PS and PEO relies on the use of photoiniferters to trigger the 'controlled' radical polymerization of styrene. This, however, resulted in only poor yields; moreover, the samples obtained exhibited a broad polydispersity [10].

Better defined linear PS-b-PEO-b-PS were recently reported, the outer PS blocks being grown by 'controlled' free radical polymerization (Scheme 1, route d-). For instance, advantage was taken from the controlled character of the nitroxide mediated polymerization (NMP) of styrene to prepare such PEO-b-PS copolymers [11]. However, as the polymerization was carried out at 130°C for 14 hours, the authors could not avoid the presence of PS chains formed by the thermal self-initiation of styrene. A last step of selective extraction with cyclohexane was therefore required to rid these contaminants from the expected amphiphilic compounds. In order to produce the same copolymers with a still better defined structure, the same authors resorted to the Atom Transfer Radical Polymerization (ATRP) of styrene [12]. Using a PEO-based macroinitiator fitted with two bromopropionate functions at its end groups, they grew two PS blocks upon heating the reaction medium at 130°C in the presence of styrene and CuBr complexed with bipyridyl.

ATRP involves the successive transfer of halide from the dormant polymer chain end to transition metal compounds (CuX with X = Br, Cl) activated by bipyridyl or some other ligands and vice-versa, the chain growth occurring in between this back and forth transfer when the growing chain is in the free radical form [9d].

With respect to the synthesis of PS-b-PEO block copolymers, ATRP was found more efficient than the NMP route since a lower amount of homopolystyrene chains was formed by the thermal process in the former case.

Rizzardo and his team have recently developed the so-called reversible addition-fragmentation chain transfer (RAFT) process which involves the use of dithiobenzoate compounds to control the free radical polymerization of a wide range of monomers, those containing acid functionality (e.g. methacrylic acid) included [9e]. Block copolymers of controlled structure including amphiphilic PEO-b-PS were readily synthesized by applying the RAFT process [13]. After generating PEO ω-dithioester precursors upon coupling PEO monomethyl ether with an acid functional dithioester, PS blocks could be grown from these macroinitiators by RAFT. The authors indicated that their PEO-b-PS copolymers were of narrow polydispersity and did not contain any homoPEO impurity.

3. AMPHIPHILIC GRAFT COPOLYMERS BASED ON PS AND PEO (SCHEME 2)

Graft copolymers contain branching points that may be distributed along the polymer backbone either randomly or under a controlled manner; they can be generated by three distinct routes named 'grafting onto', 'grafting from' or *via* the copolymerization of macromonomers with smaller species. These strategies have been used to produce copolymers constituted of a PS backbone and PEO grafts (Scheme 2).

3.1. 'Grafting onto' method

Advantage of the 'grafting onto' method is associated with the fact that the graft size can be strictly controlled, although the deactivation of the 'living' chains or the grafting of ω-functional chains onto a backbone carrying antagonist functions are generally not quantitative. For instance, PS-g-PEO copolymers were synthesized by

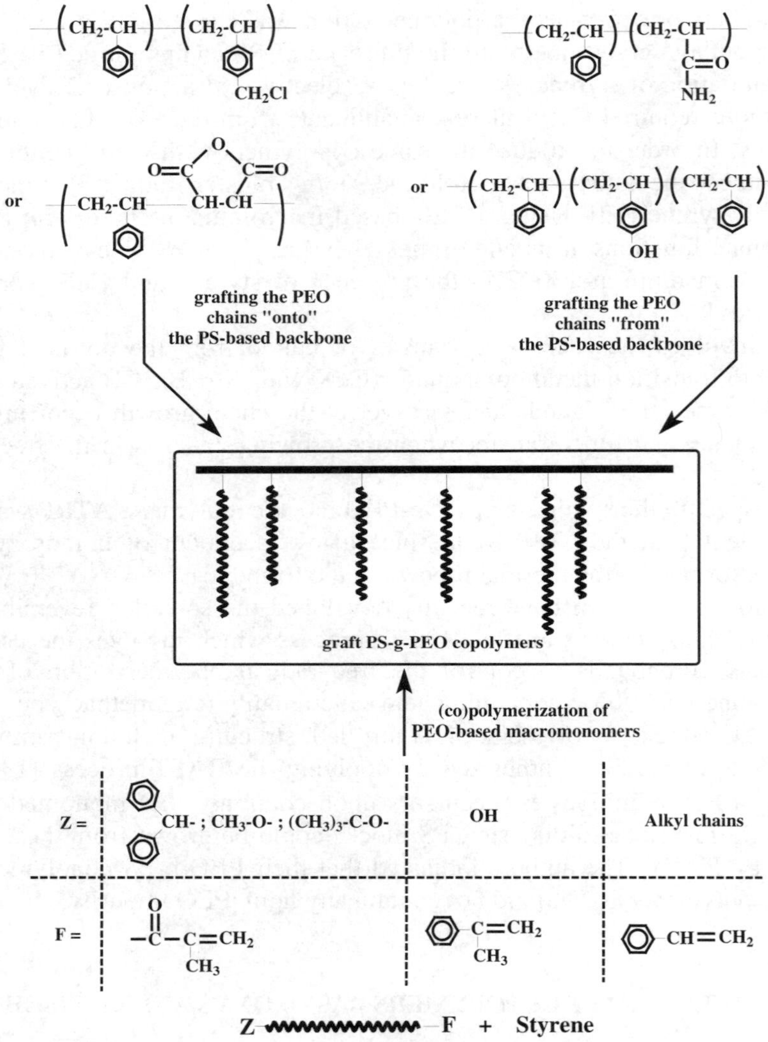

Scheme 2. Access to graft PS-g-PEO copolymers.

Rempp and coll. [14] and George and coll. [15] by adding randomly chloromethylated PS chains into a medium containing 'living' PEO's. In such cases, the branching points were randomly placed along the copolymer backbone. The same feature was observed when reacting N-bromosuccinimide modified poly(p-methylstyrene) with 'living' PEO chains [16].

Wang *et al.* recently showed that a high grafting efficiency could be obtained when grafting PEO chains fitted with a Schiff base as end group onto chloromethylated PS *via* the Decker-Forster reaction [17].

Another option to produce amphiphilic PS-g-PEO copolymers was to use a terpolymeric backbone based on styrene, maleic anhydride and methyl methacrylate units, obtained by free radical copolymerization of the corresponding monomers. The ω-hydroxylated PEO chains were further reacted with pendant anhydride functions [18]. Hydrolysis of the residual anhydride functions afforded graft copolymers carrying carboxylic acid groups along the backbone. The grafting reactions were, however, of limited yields (~60%) and gelation was observed when attempting to reach high conversion because of the presence of dihydroxy-PEO. Moreover, the conventional radical process used to synthesize the terpolymer suggested that the latter were polydisperse.

A similar method was used to graft ω-amino PEO chains onto an alternating poly(styrene-*alt*-maleic anhydride) copolymer [19].

3.2. 'Grafting from' method

Only a few reports have described the possibility of producing PS-g-PEO copolymers by the 'grafting from' method. From a random copolymer made of styrene and acrylamide, PEO grafts could be grown after partially ionizing the NH_2 functions of acrylamide and adding EO [20]. Not only did the resulting graft copolymers exhibit a broad molar mass distribution because of the use of the free radical process to prepare the backbone but they were also found to contain homoPEO's.

Graft copolymers based on PS and PEO were also obtained from well-defined poly(styrene-b-*tert*butoxystyrene-b-styrene) triblock copolymers synthesized by sequential 'living' anionic polymerization of the corresponding monomers [21]. Treating these copolymers under mild acidic conditions resulted in the formation of poly(styrene-b-hydroxystyrene-b-styrene) triblock copolymers. The subsequent metallation of the pendent phenolic units allowed one to grow the PEO grafts upon addition of EO. In this case as well, the authors acknowledged the contamination of their block-graft copolymers by linear PEO's.

3.3. (Co)polymerization of macromonomers

The approach consisting in homo- and copolymerizing macromonomers is a powerful tool to synthesize tailor-made graft copolymers [22]. The main difficulty is to obtain a complete conversion of macromonomers when copolymerizing the latter with a smaller comonomer: the incompatibility between the macromonomers and the growing polymer may be the cause of the lack reactivity of the latter species.

PEO-based macromonomers were copolymerized with styrene (or alkyl methacrylates) by conventional (uncontrolled) free radical route [23]. As a result, the copolymers obtained exhibited a rather large fluctuation in composition, not to mention the presence in substantial amount of unreacted PEO chains. These observations are drawn from the work of Rempp and coll. who reported the synthesis of PS/PEO graft copolymers by the macromonomer route [23].

With the advent of controlled radical polymerization mediated by nitroxides, the possibilities are now open to prepare relatively well defined amphiphilic graft copolymers (polydispersity indices are in the range 1.2-1.3). For instance, α-methacryloyl PEO macromonomers were copolymerized with styrene at 125°C [24] using 4-hydroxyl-2,2,6,6-tetramethyl pyperidine-1-oxy as counter-radical. Because chain growth occurs in between cycles of the reversible deactivation-activation, much better defined graft copolymers could be obtained.

In an original contribution to this field, Ito and coll. showed that macromonomers of PEO can undergo fast and complete homopolymerization, provided water is chosen as the continuous medium [25]. The authors indeed took advantage of the presence of hydrophobic polymerizing groups at PEO chains to organize the latter in micelles, their polymerizing unsaturation residing inside the particle formed. The corresponding PEO-based polymacromonomers were found to be shape-persistent in solution, behaving as 'bottle-brushes' or spheres as a function of their degree of polymerization.

4. STAR, ARBORESCENT, AND STAR-GRAFT COPOLYMERS BASED ON PS AND PEO

Apart from linear-type and miscellaneous graft structures based on PS and PEO, only a few reports deal with the preparation of other architectures such as stars or arborescent copolymers consisting of these monomer units. The synthetic methodologies to produce such branched amphiphilic structures are described below.

Among branched structures, star polymers represent the most elementary way to arrange subchains since each star contains only one branching point [26]. According to the terminology established by Hadjichristidis [27], one has to distinguish between two types of star copolymers whether their arms are themselves di- or triblock copolymers or of different chemical nature. In the former case, such species would be called star block copolymer and for the latter case Hadjichristidis proposed the term of miktoarm stars.

There are essentially two methods to get access to star polymers: either by linking a given number of linear chains to a central core ('arm-first' method) [28] or by growing branches from an active core ('core-first' method) [29]. The central core may be formed also upon reaction of short 'living' chains with a difunctional monomer such as divinylbenzene.

Following the latter approach, Tsitsilianis *et al.* reported the synthesis of amphiphilic miktoarm star polymers constituted of PS and PEO chains [30]. First, living polystyryl potassium were added to divinylbenzene leading to the formation of star polymers with poly(divinylbenzene) cores, the latter being further used to grow the PEO arms upon introduction of EO (Scheme 3). This synthetic pathway afforded miktoarm stars with a total and average number of arms close to 20. The main drawback of this approach, indeed, is its lack of reproducibility and predictability; moreover, the samples obtained did exhibit a certain heterogeneity in composition that reflected the fluctuation of their functionality.

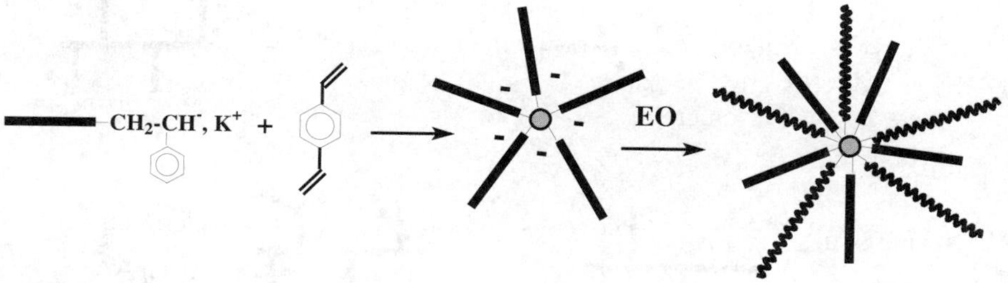

Scheme 3. Synthesis of miktoarm PS_nPEO_n copolymers.

An earlier study made by Xie *et al.* described the synthesis of well-defined PS_2PEO_2 (and PS_2PEO) miktoarm star polymers obtained by the 'arm-first method' [31]. The reaction of two molar amounts of living PS chains with one molar amount of $SiCl_4$ (or CH_3SiCl_3) used as electrophilic agent first yielded the two arm PS Si-based species which were further reacted, through their remaining Si-Cl bonds, with two (or one) molar amounts of living PEOs'. Although attractive, this approach afforded miktoarm samples that contain fragile Si-O-C bond at their central core.

Gauthier and coll. have applied an elegant 'graft-on-graft approach' (combining both the 'grafting onto' and the 'grafting from' techniques) for the synthesis of their arborescent-graft copolymers that can also be viewed as dendrimers with polymeric generations [32]. These dendrimers were constituted of a highly branched hydrophobic PS core surrounded by a hydrophilic PEO shell. As depicted in Scheme 4, highly branched PS samples were first synthesized by repeated grafting of 'living' PS linear chains onto chloromethylated PS substrates giving rise to the PS with an increasing arborescent structure. The fact that the chloromethyl groups were randomly introduced means that the overall degree of branching is not strictly controlled. The living PS chains used for the grafting of the last generation were obtained from a lithiated initiator carrying a protected hydroxyl functionality. Its subsequent hydrolysis performed under acidic conditions afforded arborescent PS with terminal hydroxyl functions from which the PEO corona could be generated by polymerizing ethylene oxide. The authors reported that such highly branched macromolecular amphiphiles were subject to aggregation in THF. A mixture of 4% of acetic acid in THF was indeed required to prevent the formation of aggregates for copolymers containing less than 66% of PEO, while dimethylformamide was found the best suited for copolymers with a smaller amount of PS (not more than 19%). The thorough characterization of these dendritic amphiphiles revealed their capability to form stable core-shell structures.

Polymacromonomers can be viewed as star macromolecules rather than comb copolymers, provided the degree of polymerization of their backbone is sufficiently low, as suggested by Tsukahara and coll. [33]. Exploiting the possibilities

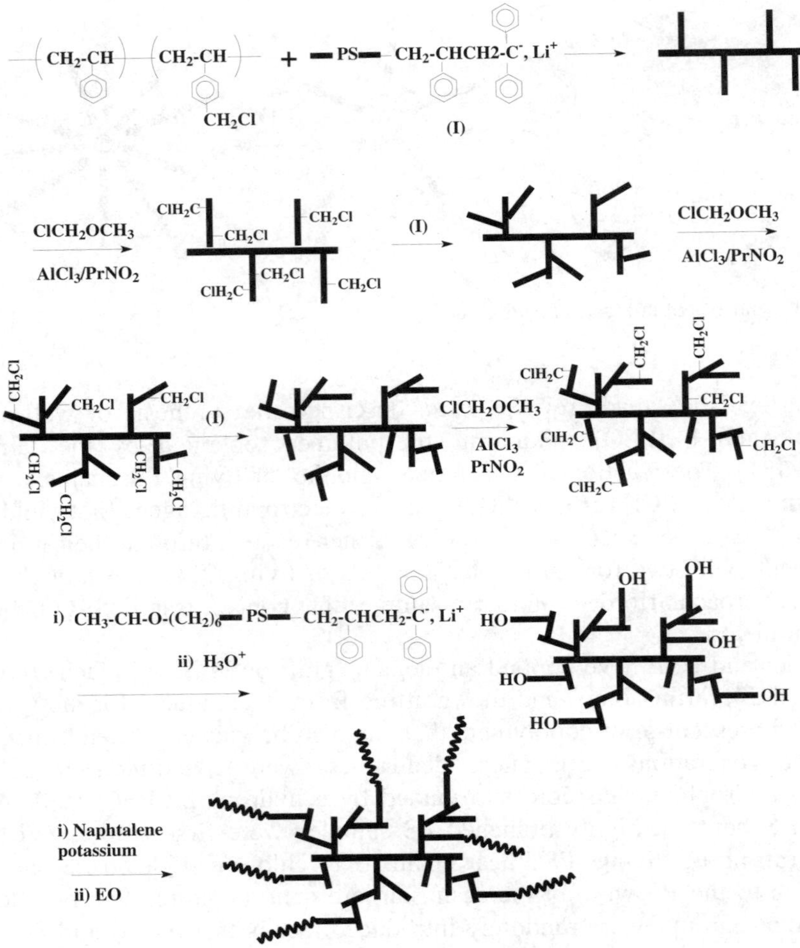

Scheme 4. Synthesis of highly branched (dendritic) amphiphilic copolymers.

offered by the 'living' ring-opening methathesis polymerization (ROMP) of ω-norbornenyl macromonomers, Héroguez et al. prepared novel well-defined macromolecular architectures made of PS/PEO, as those depicted in Scheme 5 [34]. For instance, 'Janus-type' amphiphilic structures were obtained from the sequential ROMP of PS and PEO macromonomers, 100% crossover efficiencies being achieved during the diblock synthesis.

The PS-PEO miktoarm copolymeric homologues were generated by statistical copolymerization of PEO and PS macromonomers, their copolymerization being triggered by addition of the requested amount of the Schrock initiator into a solution containing the two macromonomers. The values of molar masses were found in rather good agreement with the targeted values.

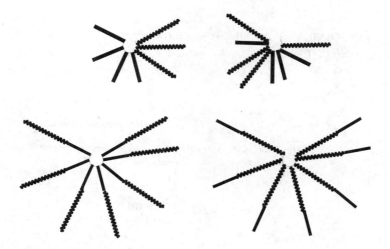

Scheme 5. Star polymers from PEO/PS-based polymacromonomers.

Polymacromonomers with a hydrophilic outer layer behaving like spheres in solution were also prepared by polymerization of α-norbornenyl-(PS-b-PEO) macromonomers. The remarkable selectivity of ROMP and its straightforward character were monitored by SEC coupled with a light scattering detector: 100% conversion was accomplished and a good control of the polymacromonomer molar mass was achieved as well.

Gitsov and coll. have described the synthesis of 'star-graft' copolymers in which hydrophilic triarm PEO stars are attached as side moieties onto a hydrophobic copolymer backbone made of (phtalimido acrylate-co-styrene) chains [35]. For this purpose, the phtalimido acrylate units were partially modified with tris(hydroxymethyl)aminomethane (Scheme 6). Then, isocyanato terminated methoxy-PEO chains were condensed with the side hydroxyl functions leading to star-graft architectures for which the number of branching points could be varied although in an uncontrolled manner. The behavior of these branched amphiphilic copolymers, in the solid state and in solution, was found to be affected by the degree of branching and the hydrogen bonding occurring *via* the urethane groups linking the PEO grafts to the backbone.

As illustrated in Scheme 7, the so-called dumbbell-type architecture consisting of two 'umbrella' connected each other through a linear polymer has been described by Bayer and Stadler [36]. In this case, the branches are PEO chains and the connector is a linear PS chain. The strategy envisaged to synthesize such a branched structure is the following. First, a well-defined poly(butadiene-b-styrene-b-butadiene) triblock copolymer was prepared under anionic conditions. Then, the pendent double bonds of the 1,2-butadiene units were partially modified into hydroxyl groups *via* the hydroboration-oxidation reaction. After deprotonation of these OH groups by cumylpotassium, the polymerization of ethylene oxide

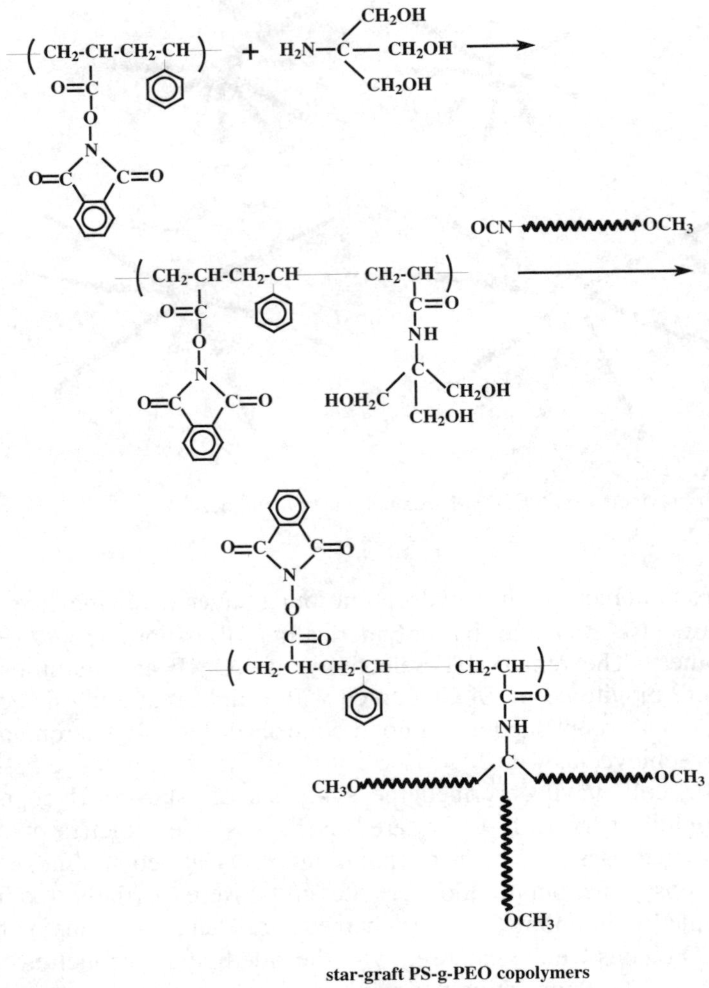

Scheme 6. Synthesis of star-graft PS-g-PEO copolymers.

was carried out (in the presence of a cryptand in order to prevent aggregation of the corresponding alkoxides), producing the dumbbell-type species. The main limitation of this approach was related to the hydroboration-oxidation step that was not complete; therefore, the number of PEO arms was not strictly controlled and the hydroxylated precursor was also subject to cross-linking.

We have recently reported the synthesis of well-defined star-block PS_6-b-PEO_6 copolymers obtained by the 'core-first' method [37]. Starting from an initiator containing six phenylethyl chloride groups, hexaarm polystyrene stars were first produced by 'living' cationic polymerization of styrene. In a second step, the six secondary chlorine atoms end-fitting each arm of the PS stars were transformed

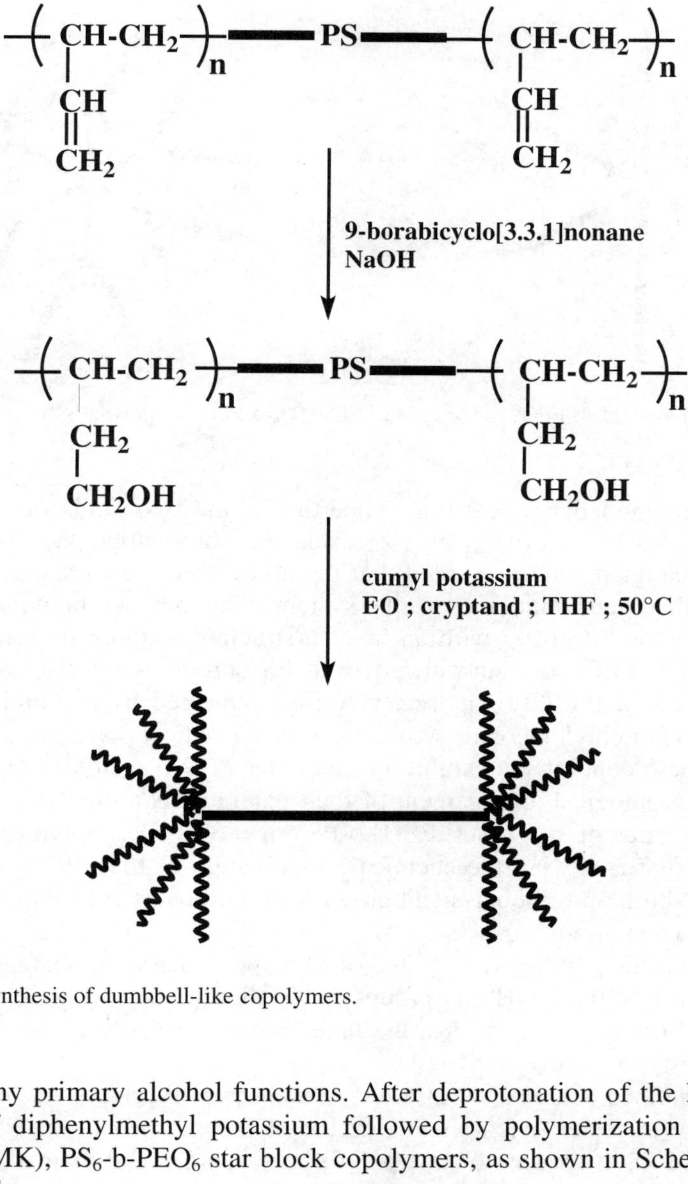

Scheme 7. Synthesis of dumbbell-like copolymers.

into as many primary alcohol functions. After deprotonation of the latter with a solution of diphenylmethyl potassium followed by polymerization of ethylene oxide (DPMK), PS_6-b-PEO_6 star block copolymers, as shown in Scheme 8, could be obtained.

An extension of this class of branched PS/PEO copolymers are the arborescent PS_6-b-PEO_{12} macromolecules (Scheme 8) consisting of the above inner hydrophobic hexaarm PS star surrounded by a generation of 12 hydrophilic PEO chains [38]. In this case, it is necessary to introduce at each PS arm end a branching agent containing two protected hydroxyl functions. After deprotection of the 12 hydroxyls followed by their titration with DPMK, PEO arms could be grown by anionic polymerization of ethylene oxide.

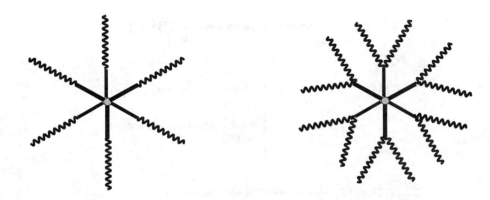

Scheme 8. Star block PS$_6$-b-PEO$_6$ and arborescent-like PS$_6$-b-PEO$_{12}$ copolymers.

The investigation of the solution properties of the two kinds of copolymers (PS$_6$-b-PEO$_6$ and PS$_6$-b-PEO$_{12}$) by size exclusion chromatography both in THF and in an aqueous medium revealed that the arborescent copolymers were more inclined to self-organize into unimicellar systems than their star homologues.

Amphiphilic architectures with an inverse structure to those described above, that is an inner PEO core surrounded by a PS corona, were also synthesized. These novel block PS/PEO copolymers were engineered by combining anionic polymerization of ethylene oxide with atom transfer radical polymerization of styrene [38]. These copolymers exhibit a same inner PEO part but differ from each other by the topological arrangement of their external PS arms. As illustrated in Scheme 9, a series of star block PEO$_n$-b-PS$_n$ (n = 3 or 4) copolymers were obtained, after first achieving the chemical modification of the PEO hydroxyl end groups into 2-bromoproprionates and growing PS corona by ATRP of styrene under controlled conditions.

Arborescent PEO$_n$-b-PS$_{2n}$ (n = 1, 2, 3 or 4) homologues were obtained from the PEO samples: first the n OH end groups of the PEO precursors were derivatized into 2n2-bromoproprionate groups, the latter being subsequently used to initiate the ATRP of styrene (Scheme 10).

The well-defined character of all these branched architectures was checked after scission of the ester functions linking the PEO inner part to the PS moieties. The recovered PS arms were of low polydispersity index and their molar masses were found in good agreement with the values calculated by NMR on the corresponding star block and arborescent copolymers.

Preliminary investigations by ^{1}H NMR of these newly designed copolymers showed that they exhibited self-associating properties and responded differently to changes of the polarity of the medium in which they were placed; unimolecular micelles could be obtained in a specific solvent. Moreover, size exclusion chromatography with THF as eluent revealed different solution behavior when comparing arborescent-like PEO$_n$-b-PS$_{2n}$ and star block PEO$_n$-b-PS$_n$ homologues. This could be related to the fact that the hydrodynamic volume of the two macromolecular species might be different depending on the architecture considered. The dendritic PEO$_n$-b-PS$_{2n}$ species might exhibit a more globular shape in solution because of their branched structure resembling that of conventional dendritic polymers that contain only one monomer unit between two branching points, whereas the star block PEO$_n$-b-PS$_n$ copolymers likely exhibited a weaker segment density due to the presence of one single branching point (the central core) per molecule.

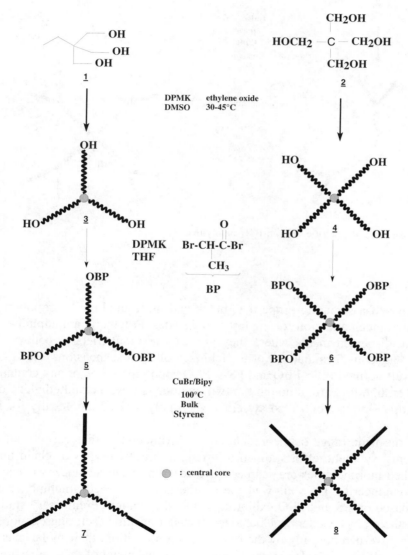

Scheme 9. Synthesis of star block PEO$_n$-b-PS$_n$ copolymers.

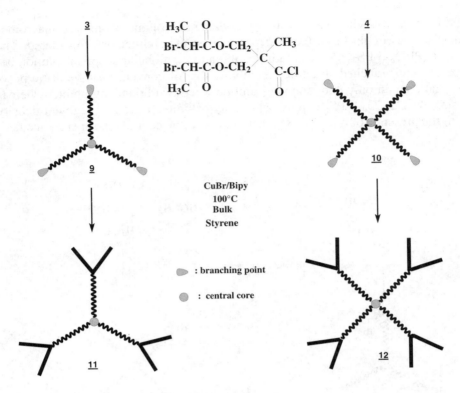

Scheme 10. Synthesis of dendritic PEO$_n$-b-PS$_{2n}$ copolymers.

5. CONCLUSION

Due to their self-associating properties, block and graft amphiphilic copolymers have attracted much attention for the last two decades. Polymeric amphiphiles are exploited in dispersed media where they are used as emulsifiers to control the rheology of aqueous formulations or to stabilize colloidal suspensions. However, access to well-defined PS-b-PEO and PS-g-PEO copolymers free of any contaminant is not straightforward although the two monomers (styrene and ethylene oxide) give truly 'living'/controlled systems, when polymerized anionically for instance.

Besides the wide range of theoretical and experimental studies carried out on these systems, synthetic efforts aimed at arranging the two types of chain units into branched morphologies are scarce. However, the recent advances of macromolecular engineering have allowed the synthesis of branched amphiphilic copolymers based on PS and PEO exhibiting original topology. Preliminary studies show that their physicochemical behavior (phase-transition, self-aggregation in solution) is governed not only by their chemical composition, their molar masses or their concentration, but also by the topological arrangement of their subchains.

Thus, tailoring the molecular architecture and distribution of chemical functionality in copolymers based on PS and PEO as well as adjusting the hydrophilic-lipophilic balance in these systems can lead to novel branched copolymers with unusual properties in solution and in the solid state.

The emergence of 'controlled' free radical polymerization techniques should make these syntheses easier to develop.

REFERENCES

1. (a) I. Piirma, "Polymeric Surfactants", Surfactant Science Series, M. Dekker Inc., New York, (1992). (b) R. Velichkova, D. Christova, *Prog. Polym. Sci.* **20**, 819 (1995). (c) A. Guyot, K. Tauer, *Adv. Polym. Sci.* **111**, 43 (1994). (d) B. Magny, I. Iliopoulos, R. Audebert, *Macromolecular Complexes in Chemistry and Biology*; Dubin et al., Eds. Springer Verlag: Berlin Heidelberg, p. 51 (1994). (e) Y. Y. Won, H. T. Davis, F. S. Bates, *Science*, **283**, 960 (1999).
2. (a) D. Zhao, Q. Huo, J. Feng, B. Chelmlka, G. Stucky, *J. Am. Chem. Soc.*, **120**, 6024 (1998). (b) A. Firouzi, D. Kumar, L. M. Bull, T. Besier, P. Sieger, Q. Huo, S. A. Walker, J. A. Zasadzinski, C. Glinka, J. Nicol, D. Margolese, G. D. Stucky, B. Chemlka, *Science*, **267**, 1138 (1995). (c) A. Firouzi; D. J. Schaefer, S. H. Tolbert, G. D. Stucky, B. Chmelka *J. Am. Chem. Soc.*, **119**, 9466 (1997).
3. G. R. Newkome, C. N. Moorefield, G. R. Baker, M. J. Saunders, S. H. Grossman, *Angew. Chem. Int. Ed. Engl.* **30**, 1178 (1991).
4. G. Riess, *Macromol. Chem., Suppl.*, **13**, 157 (1985).
5. (a) Z. Hruska, G. Hurtrez, S. Walter, G. Riess, *Polymer*, **33** (11), 2447 (1992). (b) Z. Hruska, M. A. Winnick, G. Hurtrez, G. Riess, *Polymer Communications*, **31**, 402 (1990).
6. R. P. Quirk, J. Kim, C. Kaush, M. Chun, *Polymer International*, **39**, 3 (1996).
7. (a) R. P. Quirk, J. Kim, K. Rodrigues, W. L. Mattice, *Makromol. Chem. Symp.*, **42/43**, 463 (1991). (b) T. N. Khan, R. H. Mobbs, C. Price, J. R. Quintana, R. B. Stubbetsfield, *Eur. Polym. J.*, **23**, 191 (1987). (c) F. Caldérara, Z. Hruska, G. Hurtrez, T. Nugay, G. Riess, *Makromol. Chem.*, **194**, 1411 (1993).
8. M. Niwa, N. Katsurada, T. Matsumoto, M. Okamoto, *J. Macromol. Sci., Chem.* **A25**, 445 (1988).
9. (a) M. K. Georges, R. P. N. Veregin, Kazmaier, P. M.; Hamer, G. K. *Trends Polym Sci.*, **2**, 66 (1994). (b) T. P. Davis, D. Kukulj, D. M. Haddleton, D. R. Maloney, *Trends Polym. Sci.*, **3**, 365 (1995). (c) C. J. Hawker, *Trends Polym. Sci.*, **4**, 183 (1996). (d) M. Sawamoto, M. Kamigaito, *Trends Polym. Sci.*, **4**, 371 (1996). (e) J. Chierafi, B. Y. K. Chong, F. Ercoles, J. Krstina, J. Jeffery, T. P. T. Le, R. T. A. Mayyadunne, G. F. Meijs, C. Moad, G. Moad, E. Rizzardo, S. H. Tang, *Macromolecules*, **31**, 5562 (1998).
10. (a) P. Pascal, G. Clouet, P. Corpart, D. Charmot, *J. Macromol. Sci., Chem.*, **A32**, 1341 (1995). (b) C. P. R. Nair, P. Chaumont, G. Clouet, *J. Macromol. Sci., Chem.*, **A27**, 791 (1990).
11. X. Chen, B. Gao, J. Kops, W. Batsberg, *Polymer*, **39**, 911 (1998).
12. K. Jankova, X. Chen, J. Kops, W. Batsberg, *Macromolecules*, **31**, 538 (1998).
13. B. Y. K. Chong, T. P. T. Le, G. Moad, E. Rizzardo, S. H. Tang, *Macromolecules*, **32**, 2071 (1999).
14. F. Candau, F. Afchtar-Taromi, P. Rempp, *Polymer*, **18**, 1253 (1977).
15. M. H. George, M. A. Majid, J. A. Barrie, I. Rezaian, *Polymer*, **28**, 1217 (1987).
16. I. Piirma, J. R. Lenzotti, *Br. Polym. J.*, **21**, 45 (1989).
17. Y. Wang, Q. Du, J. Huang, *Macromol. Rapid Commun.*, **19**, 250 (1998).
18. H. Derand, B. Wesslen, B. Wittgren, K. G. Walhund, *Macromolecules*, **26**, 8770 (1996).
19. A. R. Eckert, S. E. Webber, *Macromolecules*, **29**, 560 (1996).
20. (a) P. Jannasch, B. Wesslen, *J. Polym. Sci., Polym. Chem. Ed.*, **31**, 1519 (1993). (b) P. Jannasch, B. Wesslen, *J. Polym. Sci., Polym. Chem. Ed.*, **33**, 1465 (1995).

21. K. Se, K. Miyawaki, K. Hirahara, A. Takano, *J. Polym. Sci. Part A: Polym. Chem.* **36**(17) 3021 (1998).
22. Y. Gnanou, *Ind. J. Techn.* **31**, 317 (1993).
23. (a) P. Rempp, E. Franta, *Adv. Polym. Sci.* **58**, 1 (1984). (b) P. Rempp, P. Lutz, P. Masson, P. Chaumont, E. Franta, *Makromol. Chem.*, **13**, 47 (1985). (c) P. Rempp, P. Lutz, P. Masson, P. Chaumont, E. Franta, *Makromol. Chem.*, **8**, 3 (1984). (d) Y. Gnanou, P. Rempp, *Makromol. Chem.* **188**, 2111 (1987).
24. Y. Wang, J. Huang, *Macromolecules*, **31**, 4057 (1998).
25. (a) D. Chao, S. Itsuno, K. Ito, *Polymer Journal*, **23**, 9, 1045 (1991). (b) S. Kawaguchi, M. A. Winnick, K. Ito, *Macromolecules*, **29**, 4465 (1996). (c) K. Ito, K. Tanaka, H. Tanaka, G. Imai, S. Kawagushi, S. Itsuno, *Macromolecules*, **24**, 2348 (1991).
26. (a) G. S. Grest, L. J. Fetters, J. S. Huang, D. Richter, *Advances in Chemical Physics*; Wiley & Sons: New York, Vol. XCIV (1996). (b) N. Hadjichristidis, S. Pispas, M. Pitsikalis, H. Iatrou, C. Vlahos, *Advances in Polymer Science*, **142**, 71 (1998). (c) W. Burchard, *Advances in Polymer Science*, **143**, 113 (1998).
27. M. Pitsikalis, S. Pispas, J. W. Mays, N. Hadjichristidis, *Adv. Polym. Sci.* **135**, 1 (1998).
28. (a) B. A. David, D. J. Kinning, E. L. Thomas, L. J. Fetters, *Macromolecules*, **19**, 215 (1986). (b) N. Hadjichristidis, A. N. Guyot, L. J. Fetters, *Macromolecules*, **11**, 668 (1978). (c) M. Morton, T. E. Helminiak, S. D. Gadkary, F. J. Bueche, *Polym. Sci.*, **57**, 471 (1962). (d) J. Roovers, L. Zhou, P. M. Toporowski, M. Van Der Zwan, H. Iatrou, N. Hadjichristidis, *Macromolecules*, **26**, 4324 (1993).
29. (a) E. Cloutet, J. L. Fillaut, Y. Gnanou, D. Astruc, *Macromolecules*, **31,** 6748 (1998). (b) S. Jacob, I. Majoros, J. P. Kennedy, *Macromolecules*, **29**, 8631 (1996). (c) Schappacher, M.; Deffieux, A. *Macromolecules*, **25**, 6744 (1992). (d) S. Angot, K. S. Murthy, D. Taton, Y. Gnanou, *Macromolecules*, **31**, 7218 (1998). (e) J. L. Hedrick, M. Tröllsas, C. J. Hawker, B. Atthoff, H. Claesson, A. Heise, R. D. Miller, D. Mecerreyes, R. Jérome, Ph. Dubois, *Macromolecules*, **31**, 8691 (1998).
30. C. Tsitsilianis, D. Papanagopoulos, P. Lutz, *Polymer*, **36**, 19, 3745 (1995).
31. H. Xie, J. Xia, *Makromol. Chem.* **188**, 2543 (1987).
32. M. Gauthier, L. Tichagwa, J. Downey, S. Gao, *Macromolecules*, **29**, 519 (1996).
33. (a) Y. Tsukahara, S. Kohjiya, K. Tsutsumi, Y. Okamoto, *Macromolecules*, **27**, 1662 (1994).
34. (a) V. Héroguez, Y. Gnanou, M. Fontanille, *Macromolecules*, **29**, 4459 (1996). (b) V. Héroguez, Y. Gnanou, M. Fontanille, *Macromolecules*, **30**, 4791 (1997).
35. I. Berlinova, I. Dimitrov, I. Gitsov, *J. Polym. Sci., Polym. Chem. Ed.*, 673 (1997).
36. U. Bayer, R. Stadler, *Macromol. Chem. Phys.* **195**, 2709 (1994).
37. D. Taton, E. Cloutet, Y. Gnanou, *Macromol. Chem. Phys.*, **199**, 2501 (1998).
38. Y. Gnanou, S. Angot, D. Taton, V. Héroguez, M. Fontanille, *ACS, PMSE* **80**, 59 (1999).

Arborescent polymers: Designed macromolecules with a dendritic structure

MARIO GAUTHIER*

Institute for Polymer Research, Department of Chemistry, University of Waterloo, Waterloo, Ontario N2L 3G1, Canada

Abstract—Arborescent polymers are characterized by a tree-like or dendritic structure resulting from successive anionic grafting reactions. Synthetic routes to different types of well-defined arborescent homo- and copolymers with branching functionalities $f_w \approx 10$ to over 10 000 and weight-average molecular weights $M_w \approx 5 \times 10^4 - 10^8$ are discussed in this review. Styrene homopolymers with an arborescent structure are obtained in high yield by first coupling 1,1-diphenylethylene-capped polystyryl anions with a linear chloromethylated polystyrene substrate. The resulting comb-branched (generation G0) polymer is further chloromethylated and grafted with polystyryl anions to yield a G1 arborescent polystyrene, and so on. The synthetic method used for arborescent polystyrenes can be extended to the preparation of branched copolymers using both *grafting onto* and *grafting from* schemes. Polyisoprene, poly(2-vinylpyridine), and poly(*tert*-butyl methacrylate) copolymers can be prepared by direct coupling of the corresponding living anions with a suitably functionalized arborescent polystyrene grafting substrate (*grafting onto* method). It is also possible to synthesize copolymers containing poly(ethylene oxide), polydimethylsiloxane or polycaprolactone segments end-linked to the substrate chains using arborescent polystyrenes end-functionalized with hydroxyl groups as polyfunctional initiators (*grafting from* method). The physical properties of arborescent polymers in solution, in the molten and solid states more closely resemble those of hard spheres as the branching functionality increases, in particular for shorter side chains. Some specific examples of the unusual physical properties exhibited by arborescent polymers making them interesting for specialty applications are discussed.

Keywords: Arborescent polymers; anionic grafting; macromolecular engineering.

1. INTRODUCTION

Dendritic polymers form a class of macromolecules characterized by a cascade-branched structure, most often obtained in controlled polymerization reactions of polyfunctional monomers [1]. Two main types of dendritic polymers derived from small molecule building blocks can be distinguished. *Dendrimers* are synthesized according to a "generation" scheme consisting of protection, condensation, and deprotection reaction cycles. Because of the strict control achieved over the branching process, extremely narrow molecular weight distributions and well-

---

*E-mail: gauthier@uwaterloo.ca

defined structures can be obtained in these systems. Unfortunately, since the building blocks used are small molecules, many reaction cycles (generations) are necessary to generate high molecular weight materials. For dendrimers, the increase in molecular weight per generation is determined by the number of reactive groups carried by the monomeric units. For an AB_2 monomer, for example, the molecular weight should approximately triple for each generation. The complex reaction sequences involved in dendrimer syntheses are avoided with *hyperbranched* polymers, obtained in one-pot controlled self-condensation reactions. The products obtained under these conditions have a high molecular weight, but the molecular weight distribution typically corresponds to a Flory most probable distribution ($M_w/M_n \approx 2$).

A new approach was suggested independently by Tomalia *et al.* [2] and by Gauthier and Möller [3] in 1991 to alleviate some of the problems encountered in both dendrimer and hyperbranched polymer syntheses. The modified scheme relies on grafting reactions using *polymeric* building blocks rather than small molecules (Fig. 1). A linear polymer substrate suitably functionalized with grafting sites is first reacted with "living" polymeric chains to generate a comb-branched (generation G0) polymer. Further introduction of coupling sites on the comb polymer and grafting leads to the first generation (G1) arborescent polymer, and subsequently to higher generations. Because the grafting sites are randomly distributed on the substrate, the structure of the molecules is not as well-defined as for dendrimers, but the polydispersity achieved ($M_w/M_n = 1.1$-1.2 typically) compares favorably. The term "arborescent" is used to refer to the tree-like structure of the molecules [3], although the names "comb-burst" and "dendrigraft" have also been suggested for these types of materials [2].

The utilization of polymeric building blocks in the *aufbau* process leads to a very rapid increase in molecular weight per generation, since not only *polymeric*

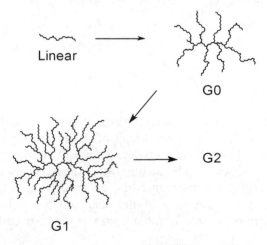

Figure 1. Schematic representation of the arborescent polymer synthesis.

units are grafted on the substrate, but also every side chain of the substrate contains many coupling sites. For example, for a series of reactions where the molecular weight of the branches (M_b) and the average number of side chains introduced per backbone chain (f) remains constant, the total molecular weight of an arborescent polymer of generation G can be expressed as:

$$M = \sum_{x=0}^{G+1} M_b f^x \qquad (1)$$

Based on equation (1), for a substrate containing 10 coupling sites per side chain ($f = 10$) grafted with branches of identical molecular weight, a 10-fold increase in molecular weight is expected for each generation.

Following is a review of our efforts aimed at the synthesis of polymers with an arborescent structure. The synthesis of macromolecules based on polystyrene side chains as well as copolymers incorporating a polystyrene core grafted with side chains of a different composition is discussed. Some of the physical properties making these materials potentially useful for specialty applications are also considered.

2. ARBORESCENT POLYSTYRENES

The synthetic scheme described in Fig. 1 can only be carried out over successive generations if the side chains introduced in each cycle can be further functionalized with coupling sites for the next grafting reaction. Polystyrene meets that criterion, since the aromatic rings can be easily derivatized to introduce a wide range of reactive substituents. Consequently, the synthesis of arborescent polymers was first demonstrated for styrene-based systems. Chloromethyl groups were selected as coupling sites, based on previous reports on the preparation of comb polymers [4–6]. However, two major hurdles needed to be overcome to apply the chloromethyl group coupling chemistry to the synthesis of arborescent polymers with well-defined features.

The chloromethylation of polystyrene with chloromethyl methyl ether is sensitive to cross-linking [7], since further reaction of the chloromethylated rings with other styrene units is difficult to avoid (Fig. 2a). Methylene bridging occurring under these conditions results in *intramolecular* and *intermolecular* cross-link formation. The latter can cause significant broadening of the molecular weight distribution, or lead to insoluble products in more extreme cases. Intermolecular cross-linking reactions are avoided in the arborescent polymer synthesis by selection of appropriate conditions for the chloromethylation process. Reactions in dilute solution (0.5-1% w/v polymer in CCl_4) with a large (10-fold) excess of chloromethyl methyl ether are essential to minimize cross-linking reactions. Moderately reactive Lewis acids such as $SnCl_4$ [3] or a catalyst complex of $AlCl_3$/1-nitropropane [8], in combination with the conditions described above,

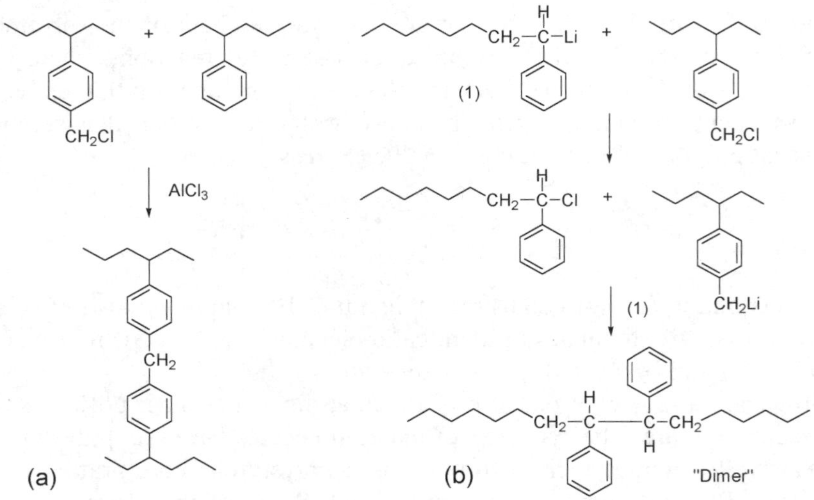

Figure 2. Side reactions in the synthesis of arborescent polystyrenes: (a) intermolecular cross-link formation during chloromethylation, (b) metal-halogen exchange reaction competing with coupling.

yield grafting substrates with chloromethylation levels of up to 40 mol% with no detectable molecular weight distribution broadening.

The other major problem encountered in the arborescent polystyrene synthesis is the occurrence of metal-halogen exchange reactions competing with coupling [9]. The effect of metal-halogen exchange on the coupling reaction of polystyryl anions and chloromethylated polystyrene is seen in Fig. 3a. The trimodal distribution obtained consists of the graft polymer (leftmost peak), deactivated side chains (rightmost peak) and "dimeric" chains (middle peak) with a molecular weight twice as large as the side chains. The reaction steps leading to the formation of the linear contaminants are depicted in Fig. 2b. Metal-halogen exchange initially produces relatively unreactive benzylic lithiated species and chlorine-terminated polystyrene chains. A portion of the halogen-capped chains may react further with "living" polystyryllithium species in the reaction mixture to produce the "dimeric" chains. The extent of the metal-halogen exchange reaction can be minimized relative to coupling under appropriate conditions. After capping of the polystyryl anions with a 1,1-diphenylethylene (DPE) unit, and grafting in the presence of tetrahydrofuran (THF) at a temperature of -30°C, a grafting efficiency of over 95% is achieved (Fig. 3b). This represents a remarkable improvement over the grafting efficiency of around 50% achieved under non-optimized conditions [4–6].

The synthesis of a comb polystyrene under the conditions described is summarized in Fig. 4. The stoichiometry of the coupling reaction is easily monitored by slow titration of the dark red-colored macroanions with a solution of the chloromethylated grafting substrate to the point where the color fades.

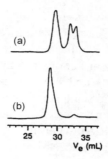

Figure 3. Size exclusion chromatograms obtained for crude grafting products from the reaction of (a) polystyryllithium and (b) DPE-capped polystyryllithium with linear chloromethylated polystyrene. Coupling efficiencies of 50% and 96% were determined for curves (a) and (b), respectively, by comparison of peak areas for the grafted and linear components (Adapted from [3], with permission from the American Chemical Society).

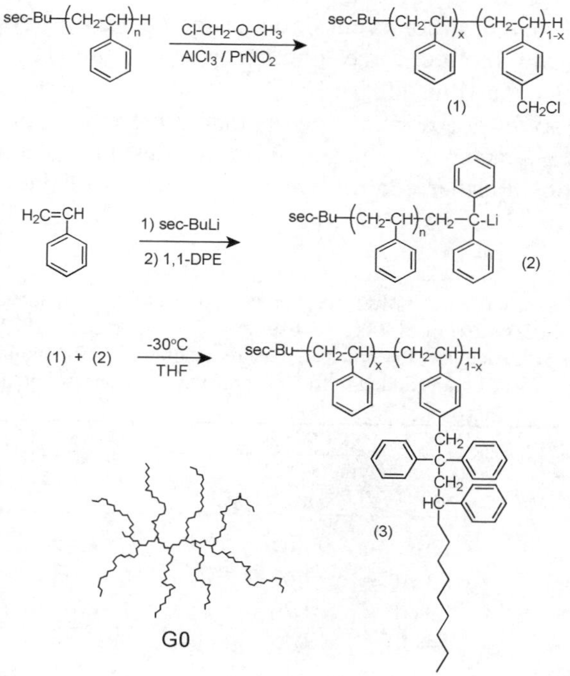

Figure 4. Reaction scheme for the synthesis of a comb-branched polystyrene in high yield.

The preparation of higher generations of arborescent polystyrenes involves removal of residual linear polymeric contaminant from the raw grafting product, chloromethylation and further grafting of macroanions. This "graft-on-graft" procedure leads to a very rapid increase in molecular weight and branching functionality over successive reactions. Characterization results [10] obtained for generations of arborescent polystyrenes synthesized using side chains with a molecular weight of either 5 000 or 30 000 are compared in Table 1. The absolute weight-average molecular weight (M_w^{LS}) and branching functionality (f_w) of the graft polymers, determined from light scattering measurements, increases in an approximately geometric fashion for successive grafting reactions. The increase is less pronounced for upper generation materials (in particular G3), presumably due to overcrowding effects limiting the accessibility of grafting sites in the reaction [3]. The apparent weight-average molecular weight of the graft polymers (M_w^{APP}), determined by size exclusion chromatography (SEC) analysis using a polystyrene standards calibration curve, is considerably lower than M_w^{LS}. This is expected, since arborescent polymers are much more compact than linear polymers of identical M_w, and are thus eluted later from the SEC column. The apparent polydispersity index (M_w^{APP}/M_n^{APP}) of the products remains relatively low over successive generations and is typically below 1.2, indicating a narrow molecular size distribution. The uniform size of arborescent polymer molecules was confirmed by direct observation using atomic force microscopy measurements of array-like packing in monomolecular films [11].

Arborescent polystyrenes are interesting as model branched polymers, because of their amorphous character and the possibility to design molecules with well-defined characteristics. Further control over the properties of these materials can

Table 1.
Characterization data for arborescent polystyrenes prepared using side chains with a molecular weight of either 5 000 (S05-series) or 30 000 (S30-series). The superscript APP refers to apparent molecular weights from SEC analysis; the other values are absolute. The branching functionality f_w is defined as the number of side chains added in the last grafting reaction. Adapted from [10]

	Branches		Graft polymers			
G	$M_w/10^3$	M_w/M_n	M_w^{APP}	M_w^{APP}/M_n^{APP}	M_w^{LS}	f_w
			S05 Series			
0	4.3	1.03	4.0×10^4	1.07	6.7×10^4	14
1	4.6	1.03	1.3×10^5	1.07	8.7×10^5	170
2	4.2	1.04	3.0×10^5	1.20	1.3×10^7	2900
3	4.4	1.05	4.4×10^5	1.15	9×10^7	17500
4	4.9	1.08	–	–	2×10^8	22000
	Branches		Graft polymers			
G	$M_w/10^4$	M_w/M_n	M_w^{APP}	M_w^{APP}/M_n^{APP}	M_w	f_w
			S30 Series			
0	2.8	1.15	2.1×10^5	1.12	5.1×10^5	18
1	2.7	1.09	5.9×10^5	1.22	9.0×10^6	310
2	2.7	1.09	–	–	1×10^8	3400
3	2.8	1.09	–	–	5×10^8	14300

be gained by chemical modification of the pendent phenyl rings, using a wide range of electrophilic substitution reactions [12]. Another avenue towards property control is the generation of copolymers incorporating monomer units other than styrene. This approach is illustrated with a few examples of arborescent graft copolymer syntheses.

3. COPOLYMERS OBTAINED BY THE *GRAFTING ONTO* TECHNIQUE

The generation-based grafting chemistry developed for arborescent polystyrenes relies on the presence of aromatic units enabling further functionalization with coupling sites. There is no requirement for the chains grafted in the last reaction to allow the introduction of grafting sites, however. This means that macroanions of a different type (e.g., polyisoprenyl anions) could be grafted in the last reaction, to yield a branched copolymer. This approach to the synthesis of graft copolymers is referred to as a *grafting onto* scheme. Different copolymer morphologies (and correspondingly distinct physical properties) may be attainable by this method, depending on the dimensions of the grafting substrate and the side chains used in the reaction (Fig. 5). When side chains with a molecular weight comparable to those in the core are used, a "layered" or "core-shell" copolymer morphology is expected. For long side chains, a structure analogous to star-branched polymers is obtained, since the core only serves as a small central linking unit accounting for a very small weight fraction of the sample.

The synthesis of copolymers based on a *grafting onto* scheme was first demonstrated for molecules incorporating an arborescent polystyrene core and polyisoprene side chains [13, 14]. Grafting polyisoprenyl anions onto a chloromethylated arborescent polystyrene substrate of generation G yields a copolymer of generation $G+1$. In analogy to the arborescent polystyrene synthesis, capping of the polyisoprenyl anions with DPE prior to coupling is necessary to minimize the occurrence of metal-halogen exchange. The addition of the capping agent also has the advantage of producing a deeper coloration than the uncapped polyisoprenyl anions, making it easier to monitor the stoichiometry of the grafting reaction. The synthesis of a G1 arborescent isoprene copolymer, by coupling DPE-capped polyisoprenyl anions with a chloromethylated comb (G0) polystyrene substrate, is illustrated in Fig. 6.

Side chains with a predominantly *cis*-1,4-microstructure or a mixed microstructure (1:1:1 ratio of 1,2-, 3,4- and 1,4-units) can be obtained, using either cyclohexane or THF as polymerization solvents, respectively. Depending on the molecular weight of the polyisoprene side chains and the number of grafting sites on the substrate, molecules containing a very high weight fraction of the elastomeric component may be obtained. For example, grafting polyisoprene side chains with a molecular weight of ca. 5 000 onto a polystyrene backbone with side chains of comparable molecular weight and a chloromethylation level of ca. 25 mol% yields a copolymer with a polyisoprene content of 84% w/w [14]. When

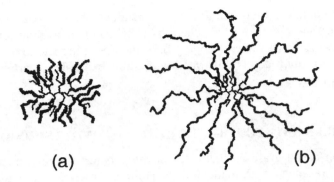

Figure 5. Arborescent copolymer morphology control: (a) "layered" or "core-shell" morphology, (b) star-like structure (Reprinted from [13], with permission from the American Chemical Society).

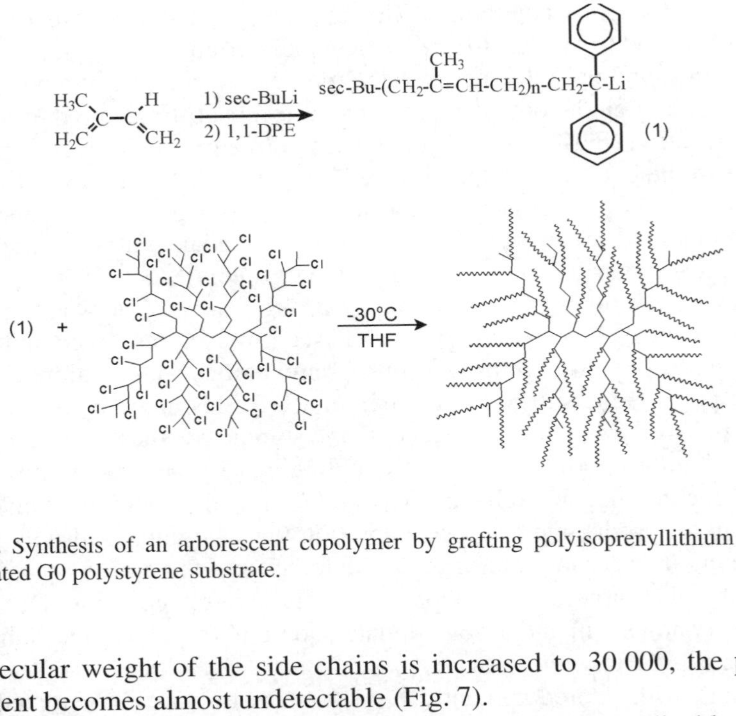

Figure 6. Synthesis of an arborescent copolymer by grafting polyisoprenyllithium onto a chloromethylated G0 polystyrene substrate.

the molecular weight of the side chains is increased to 30 000, the polystyrene component becomes almost undetectable (Fig. 7).

The method developed for isoprene copolymers is also applicable to the synthesis of copolymers carrying poly(2-vinylpyridine) side chains [15]. These molecules are particularly interesting because of their basic character, leading to arborescent *cationic* polyelectrolytes by simple quaternization of the pendent pyridine units (e.g., with HCl). These systems are synthesized by polymerization of 2-vinylpyridine in THF at -78°C in the presence of N,N,N',N'-tetramethylethylenediamine (TMEDA), and titration of the "living" anions with a solution of a chloromethylated polystyrene grafting substrate at -40°C. Due to the relatively

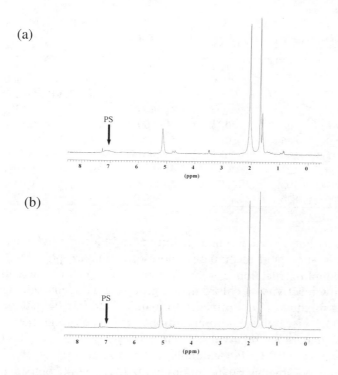

Figure 7. ^{1}H-NMR spectra for styrene-isoprene graft copolymers with (a) $M_w \approx 5\,000$ and (b) $M_w \approx 30\,000$ side chains (Adapted from [13], with permission from the American Chemical Society).

low reactivity of the poly(2-vinylpyridine) anions, no capping is necessary prior to grafting to avoid metal-halogen exchange. The red coloration of the anions also facilitates monitoring of the titration process. The addition of TMEDA to the reaction is beneficial in achieving "clean" polymerization and grafting conditions, by minimizing nucleophilic attack of the pyridine rings by the macroanions. This is confirmed by the residual yellow coloration, characteristic of ring addition, observed for reactions carried out in the absence of TMEDA. The products obtained with TMEDA, in contrast, are completely colorless.

Structure control was demonstrated in these systems by grafting poly(2-vinylpyridine) side chains with a molecular weight of either 5 000 or 30 000 onto chloromethylated polystyrene substrates with $M_w \approx 5\,000$ side chains of different generations (Table 2). Large increases in molecular weight are observed relative to the core polymers, and the copolymers are freely soluble in typical good solvents for poly(2-vinylpyridine) such as methanol and dilute aqueous HCl. The coupling efficiencies achieved in the 2-vinylpyridine copolymers are comparable to those observed for arborescent polystyrenes, and vary from 92% (grafting of side chains with $M_w = 5\,200$ onto a linear backbone) to 26% ($M_w = 34\,400$ side chains grafted onto a G1 polystyrene substrate).

Table 2.
Characterization results for styrene-2-vinylpyridine arborescent copolymers

Sample	Core M_w	f_w	Side Chains M_w^{APP}	Copolymer M_w	Composition /mol% P2VP	Coupling efficiency
PS-P2VP5	5 420	12	5 200	81 500	85.9	0.92
G0PS-P2VP5	66 700	108	5 820	718 000	86.6	0.89
G1PS-P2VP5	727 000	826	5 050	5 730 000	82.7	0.81
G2PS-P2VP5	5 030 000	2900	5 200	25 200 000	80.6	0.76
PS-P2VP30	5 420	12	27 200	415 000	94.8	0.80
G0PS-P2VP30	66 700	108	28 600	3 240 000	96.6	0.65
G1PS-P2VP30	727 000	826	34 400	26 800 000	91.8	0.26

Grafting of poly(*tert*-butyl methacrylate) anions onto chloromethylated polystyrene substrates was attempted to prepare arborescent *anionic* polyelectrolyte precursors, but the coupling reaction was found too inefficient to be useful. To compensate for the low reactivity of chloromethyl groups, a halogen exchange reaction with NaBr in a mixture of N,N-dimethylformamide and CH_2Br_2 can be carried out in nearly 100% yield under mild conditions [15]. The bromomethylated polystyrene substrate undergoes relatively efficient coupling with the poly(*tert*-butyl methacrylate) anions at 0°C. For example, a grafting efficiency of 77% is attained when grafting polymethacrylate anions with $M_w \approx 5\,000$ onto a linear polystyrene substrate, but the yield decreases for higher generation substrates and side chain molecular weights. Another problem complicating the synthesis of the *tert*-butyl methacrylate copolymers is the very weak coloration of the macroanions, making it impossible to rely on a titration procedure to monitor the coupling stoichiometry.

4. COPOLYMERS OBTAINED BY THE *GRAFTING FROM* TECHNIQUE

The examples discussed so far describe how arborescent copolymers can be generated by grafting a different type of macroanions *onto* a reactive substrate. Another route available to generate arborescent copolymers is the introduction on a substrate of reactive sites capable of initiating the polymerization of a new monomer (*grafting from* scheme). The initiating sites can be either distributed randomly or at specific loci in the molecule. The second approach, using reactive sites located at the chain ends of the grafting substrate, was preferred for the synthesis of arborescent copolymers.

An example of a reaction leading to amphiphilic copolymers with a polystyrene core and poly(ethylene oxide) segments is provided in Fig. 8 [8]. An arborescent polymer carrying hydroxyl groups at the chain ends is obtained using a bifunctional initiator, 6-lithiohexyl acetaldehyde acetal, to initiate the polymerization of styrene in the synthesis of the last core generation. After deprotection of the acetal

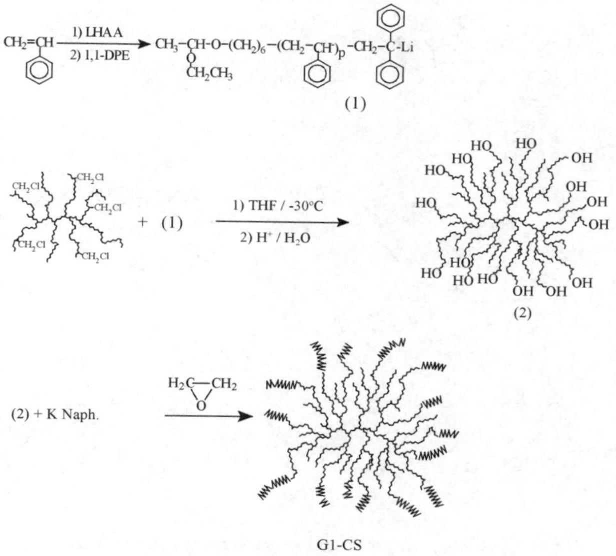

Figure 8. Synthesis of a G1 amphiphilic styrene-ethylene oxide arborescent copolymer by a *grafting from* scheme.

functionalities and titration with potassium naphthalide (as a strong, self-indicating base), the addition of purified ethylene oxide results in a chain extension reaction whereby a hydrophilic poly(ethylene oxide) block is grown directly from the polystyrene chain ends of the substrate. Initiating sites located only at the chain ends of the substrate result in a low concentration of ionic groups in the molecule, thereby maintaining relatively good solubility in polar solvents such as THF.

Initial shell growth attempts using the hydroxyl-functionalized arborescent cores were rather disappointing: For example, a G1 core with an apparent polydispersity $M_w^{APP}/M_n^{APP} = 1.08$, after titration with potassium naphthalide and addition of ethylene oxide, yielded a copolymer with $M_w^{APP}/M_n^{APP} = 3.90$. The source of the problem was determined to be a Wurtz-Fittig coupling reaction [8]. Residual chloromethyl sites in the core (due to incomplete grafting reactions) undergo rapid metal-halogen exchange with potassium naphthalide and intermolecular coupling with other core molecules (Fig. 9). The side reaction can be avoided by deactivating residual chloromethyl groups in the grafting substrate prior to the titration with potassium naphthalide. This is most conveniently achieved by a controlled metal-halogen exchange reaction with a large excess of *n*-butyllithium, followed by quenching of the benzylic lithiated species with water.

The structural features of the amphiphilic copolymer molecules are easily controlled by varying the conditions used for the reactions. For example, the structural rigidity of the polystyrene core can be increased by using shorter polystyrene

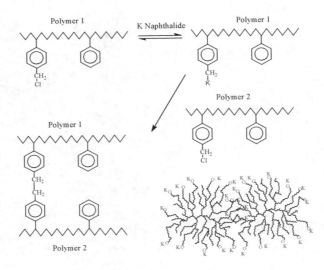

Figure 9. Metal-halogen exchange and Wurtz-Fittig coupling leading to dimerization in the titration of hydroxyl-functionalized cores with potassium naphthalide.

side chains in the synthesis. The thickness of the poly(ethylene oxide) layer increases with the amount of ethylene oxide added in the shell growth reaction. Characterization data are provided in Table 3 for two series of copolymers obtained by varying the amount of ethylene oxide used in the reaction [16, 17]. The copolymers with a higher weight fraction of ethylene oxide are characterized by a thicker hydrophilic shell, as expected. Comparison of the shell thickness [$\delta = D_h$(copolymer) - D_h(core), measured by dynamic light scattering] with the dimensions of linear poly(ethylene oxide) chains of comparable molecular weight in solution shows that the chains in the shell adopt an almost unperturbed (coil-

Table 3.
Comparison of the dimensions of amphiphilic styrene-ethylene oxide arborescent copolymers synthesized under different conditions. The weight fraction of PEO in the copolymers is provided in the sample name

Core polymers				
Sample	Branch M_w	M_w	f_w	D_h/nm
G1-LB	26800	2.1×10^6	62	49
G1-HB	29400	8.4×10^6	270	83
Copolymers				
Sample	M_{calc}	M_{PEO}	D_h/nm	δ/nm
G1-LB-0.065 PEO	2.3×10^6	2420	62	6.5
G1-LB-0.31 PEO	3.1×10^6	15500	113	32
G1-HB-0.08 PEO	9.1×10^6	2700	151	34
G1-HB-0.43 PEO	1.5×10^7	23400	213	65

like) conformation, with little extension of the backbone. This implies a relatively low packing density for the poly(ethylene oxide) chains at the surface of the molecules.

The *grafting from* scheme developed for the synthesis of the styrene-ethylene oxide copolymers is also applicable to the synthesis of other copolymers incorporating polydimethylsiloxane or poly(ε-caprolactone) segments [18]. The method used in both cases is analogous to that described in Fig. 8, substituting potassium naphthalide with lithium naphthalide, and ethylene oxide with either hexamethylcyclotrisiloxane (D_3) or ε-caprolactone as the chain extension monomer. To illustrate the synthesis of the siloxane copolymers, size exclusion chromatography (SEC) traces are compared in Fig. 10 for the hydroxyl-functionalized polystyrene core and for a polysiloxane copolymer. There is a significant shift in the position of the peak from an apparent (polystyrene equivalent) weight-average molecular weight $M_w = 127\ 000$ for the core to $M_w = 255\ 000$ for the copolymer. A small amount of linear polydimethylsiloxane is observed as a smaller peak at an elution volume $V_e \approx 24.5$ mL, due either to the addition of an excess of base or the presence of residual protic impurities in the sample.

Initial test reactions also give encouraging results when suitably purified ε-caprolactone is substituted as a monomer in the chain extension reaction (Fig. 11). In this case, however, the concurrent formation of a considerable amount of low molecular weight products is observed, as indicated by a broad peak in the SEC curve at $V_e \geq 24$ mL [18]. Interestingly, no low molecular weight side products are formed for the analogous polymerization reaction of ε-caprolactone initiated with lithium *n*-butoxide. Chain cleavage reactions by backbiting and intermolecular transesterification mechanisms have been previ-

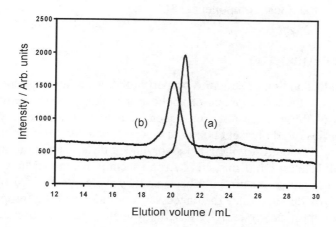

Figure 10. Comparison of SEC traces obtained for (a) a G1 hydroxyl-functionalized core polymer with $M_w \approx 5\ 000$ side chains, (b) polydimethylsiloxane graft copolymer obtained after titration of the core and addition of D_3.

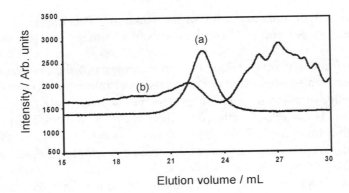

Figure 11. Comparison of SEC traces obtained for (a) a G1 hydroxyl-functionalized core polymer with $M_w \approx 5\,000$ side chains, (b) polyester graft copolymer obtained after titration of the core and addition of ε-caprolactone.

ously suggested to explain molecular weight distribution broadening in the polymerization of ε-caprolactone when lithium is the counterion [19]. It seems that *intramolecular* chain transfer reactions may be particularly favored in the synthesis of arborescent ε-caprolactone copolymers, because of the close proximity of the ionic sites in the substrate. Indeed, propagating centers remain mostly isolated from each other in a linear polymerization reaction, thus minimizing the chances of a transesterification reaction occurring between two chains. In contrast, for the arborescent substrates typically 100–1000 propagating centers are packed within a small molecular volume. The probability for interchain exchange reactions within the same molecule should be correspondingly higher, and explain at least in part the formation of the linear component [18].

5. PHYSICAL PROPERTIES

A thorough understanding of structure-property relations in arborescent polymers is important to allow the development of applications taking advantage of the special characteristics of these materials. Consequently, it is necessary to investigate the physical properties of arborescent homo- and copolymers by different techniques, to obtain a complete picture of their behavior under different conditions (solution, molten and solid states, thin films, etc.). Arborescent polystyrenes have been investigated in most detail so far not only because they were synthesized first, but also because they provide "baseline" data necessary to understand the properties of the more complex copolymer systems. Selected physical characterization investigations of arborescent polymers will be briefly reviewed, to illustrate important features of these molecules. These examples provide an incomplete picture of the physical properties of arborescent polymers, and the reader should refer to the individual papers for more information [10, 11, 16–18, 20–24].

A consistent finding of all arborescent polymer characterization studies is that the properties of the molecules are strongly linked to their structure. Depending on structure, properties typical of linear polymers, star polymers, or rigid spheres can be observed. Not surprisingly, properties resembling more closely those of rigid spherical particles are seen for polymers with higher branching functionalities (higher generations), in particular for short side chains.

The influence of side chain molecular weight on the solution properties of arborescent polystyrenes is depicted in Fig. 12 [10]. The hydrodynamic radii of two families of polymers, determined from intrinsic viscosity measurements, are compared for molecules containing side chains with a M_w of either 5 000 or 30 000. The dimensions are compared for each series of compounds in typical poor (cyclohexane) and good (toluene) polystyrene solvents. Expansion of the molecules with short side chains in toluene relative to cyclohexane is minimal. The polymers with longer side chains, in contrast, swell considerably. This shows that the solution behavior of arborescent polymer can vary considerably, depending on the structure of the molecules.

Another example, provided in Fig. 13, demonstrates the effect of branching functionality on film formation by arborescent polystyrene molecules [11]. Monomolecular films were obtained by spin casting of dilute polymer solutions on a freshly cleaved mica surface. Investigation of the films by atomic force microscopy (AFM) in the tapping mode reveals very different features for films prepared from G1 and G3 polymers with short ($M_w \approx 5\ 000$) side chains. The G1

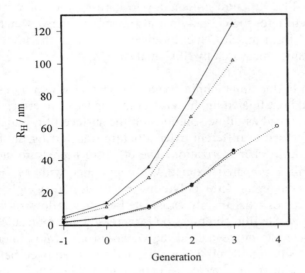

Figure 12. Effect of solvent quality on the hydrodynamic radius of arborescent polystyrene molecules of successive generations. The curves, from top to bottom, are for polymers with short side chains ($M_w \approx 5\ 000$) in toluene and in cyclohexane, and for polymers with long side chains ($M_w \approx 30\ 000$) in toluene and in cyclohexane (Reprinted from [10], with permission from Elsevier Science).

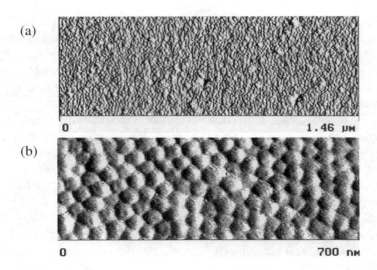

Figure 13. Comparison of film morphologies obtained from arborescent polystyrenes with $M_w \approx 5\,000$ side chains and a branching functionality (a) $f_w = 190$, (b) $f_w = 5\,650$.

polymer, with a branching functionality $f_w = 190$, only yields a film with a granular structure and no specific ordering, due to extensive interpenetration between adjacent molecules. The third-generation polymer ($f_w = 5\,650$), in contrast, packs in an array-like structure, in analogy to hard spheres of very uniform dimensions. The picture for the G3 polymer also confirms the very narrow size distributions inferred from the SEC measurements, based on the low values of apparent polydispersities measured by SEC analysis ($M_w^{APP}/M_n^{APP} = 1.13$ for the G3 sample).

Film formation by the amphiphilic styrene-ethylene oxide arborescent copolymers was likewise investigated, in this case at the air-water interface [16, 17]. The copolymers were spread as dilute solutions on the surface of a Langmuir balance trough, and compressed to different extents before transferring the film to a silicon substrate for AFM characterization. The effect of a pressure increase on the morphology of films prepared from a G1 copolymer with $M_w \approx 30\,000$ side chains, $f_w = 270$ containing 43% poly(ethylene oxide) by weight is shown in Fig. 14. The molecules exist mainly as isolated entities at low surface pressures ("gas-like" state). As the film compression is increased, the molecules associate to form pearl necklace-like aggregates at the water surface. Most interestingly, the aggregation process is completely reversible, and films removed before and after a compression-decompression cycle are indistinguishable. This unusual property clearly illustrates the high colloidal stability of amphiphilic arborescent copolymers. It is thought to be the direct consequence of the highly branched structure of the molecules, leading to efficient steric stabilization of the hydrophobic polystyrene core by the surrounding hydrophilic poly(ethylene oxide) shell.

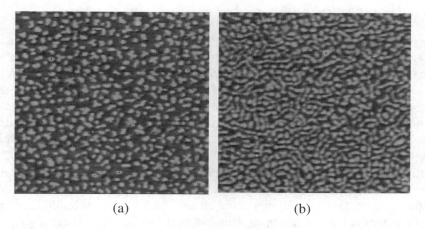

(a) (b)

Figure 14. Films of a styrene-ethylene oxide arborescent copolymer removed from a Langmuir trough after compression to a surface pressure (a) $\Pi = 2$ mN/m, (b) $\Pi = 9.5$ mN/m for 1 h.

6. POTENTIAL APPLICATIONS

The development of applications for arborescent polymers relies mainly on the exploitation of their unusual structural characteristics and properties. Based on the properties observed for arborescent polymers so far, it seems likely that these materials will be useful for a range of specialty uses. Arborescent molecules are characterized by a uniform size in the 10-300 nm range, and can be designed to achieve specific characteristics (e.g., controllable interpenetrability, structural rigidity, susceptibility to solvent quality). Furthermore, reactive chemical functionalities can be introduced on the side chains, or specifically at the chain ends of the molecules.

Preliminary investigations have already shown that arborescent isoprene copolymers have a good potential as polymer processing additives, by acting as lubricant on the walls of processing equipment [22]. Obvious applications for the water-soluble systems (ethylene oxide copolymers and polyelectrolytes derived from the 2-vinylpyridine and *tert*-butyl methacrylate copolymers) are in the area of microencapsulation, including controlled delivery, catalyst dispersion, microextraction and solubilization. The very high molecular weight, compact structure and ionizable nature of the arborescent polyelectrolytes should also be useful in the development of pH-sensitive reversible "smart" gels for biomedical and flow control applications. The copolymers with polydimethylsiloxane or polyisoprene segments are expected to have interesting elastomeric properties making them useful as network components, either on their own or when compounded with linear elastomers. One attractive feature of arborescent polymers for these applications is indeed the possibility to control the entanglement level of the molecules and thus tailor their relaxation characteristics [23].

7. CONCLUSIONS

Our work so far has demonstrated that the anionic polymerization and grafting techniques are powerful tools in the design of arborescent polymers with well-defined characteristics. Beyond the "simple" arborescent polystyrene molecules, a range of copolymers can be synthesized that provide highly branched materials with a wider spectrum of physical properties. The basic grafting technique used for arborescent polystyrenes is directly applicable to other monomers such as isoprene, 2-vinylpyridine, and *tert*-butyl methacrylate, and easily modified for the incorporation of reactive cyclic monomers like ethylene oxide, D_3, and ε-caprolactone. Detailed physical property investigations are necessary in determining the usefulness of the copolymers for different applications. This is one of the aspects of arborescent polymer research on which we are currently focusing.

While structure control has been demonstrated, we have so far not attempted to vary the shape of these polymers. For example, rod-like molecules could be generated by grafting a large number of short side chains onto a high molecular weight linear core. Subsequent grafting reactions should replicate the rod-like shape of the comb substrate on a larger scale.

One last aspect of interest is the development of new grafting methods. Chloromethyl groups are currently used as grafting sites in the reactions. This approach suffers from the fact that it requires chloromethyl methyl ether, a known carcinogen, for the functionalization of the substrate. We are currently developing alternate grafting methods that do not rely on chloromethyl methyl ether for the introduction of grafting sites. Concurrently, we are also interested in synthetic paths for arborescent polymers that do not rely on aromatic substitution chemistry (like the polystyrene systems), but that are rather applicable to other monomers such as dienes or (meth)acrylates.

REFERENCES

1. For a general introduction to dendritic polymers, see for example *Advances in Dendritic Molecules*, G. R. Newkome, Ed.; JAI: Greenwich, CT, 1994-6; Vols. 1-3.
2. D. A. Tomalia, D. M. Hedstrand and M. S. Ferrito, *Macromolecules*, **24**, 1435 (1991).
3. M. Gauthier and M. Möller, *Macromolecules*, **24**, 4548 (1991).
4. T. Altares, Jr., D. P. Wyman, V. R. Allen and K. Meyersen, *J. Polym. Sci., Part A*, **3**, 4131 (1965).
5. F. Cańdau and E. Franta, *Makromol. Chem.*, **149**, 41 (1971).
6. J. Pannell, *Polymer*, **12**, 558 (1971).
7. B. Green and L. R. Garson, *J. Chem. Soc., Sect. C*, 401 (1969).
8. M. Gauthier, L. Tichagwa, J. S. Downey and S. Gao, *Macromolecules*, **29**, 519 (1996).
9. M. Takaki, R. Asami and M. Ichikawa, *Polym. J. (Tokyo)*, **11**, 425 (1979).
10. M. Gauthier, W. Li and L. Tichagwa, *Polymer*, **38**, 6363 (1997).
11. S. S. Sheiko, M. Gauthier and M. Möller, *Macromolecules*, **30**, 2343 (1997).
12. For an overview of chemical modification reactions for polystyrene, see for example *Syntheses and Separations Using Functional Polymers*, D. C. Sherrington and P. Hodge, Eds., Wiley: Chichester, 1988.

13. R. A. Kee and M. Gauthier, *Polym. Mater. Sci. Eng.*, **77**, 176 (1997).
14. R. A. Kee and M. Gauthier, *Macromolecules*, **32**, 6478 (1999).
15. R. A. Kee and M. Gauthier, *Polym. Prepr.*, **40(2)**, 165 (1999).
16. L. Cao, M. Sc. Dissertation, University of Waterloo, Waterloo, 1997.
17. L. Cao, M. Gauthier, R. Rafailovich and J. Sokolov, *Polym. Prepr.*, **40(2)**, 114 (1999).
18. A. Khadir, Ph. D. Dissertation, University of Waterloo, Waterloo, 1999.
19. A. Duda, Z. Florjanczyk, A. Hofman, S. Slomkowski and S. Penczek, *Macromolecules*, **23**, 1640 (1990).
20. M. Gauthier, M. Möller and W. Burchard, *Macromol. Symp.*, **77**, 43 (1994).
21. R. S. Frank, G. Merkle and M. Gauthier, *Macromolecules*, **30**, 5397 (1997).
22. A. Khadir and M. Gauthier, *Polym. Mater. Sci. Eng.*, **77**, 174 (1997).
23. M. A. Hempenius, W. F. Zoetelief, M. Gauthier and M. Möller, *Macromolecules*, **31**, 2299 (1998).
24. M. Gauthier, J. Chung, L. Choi and T. T. Nguyen, *J. Phys. Chem. Part B.*, **102**, 3138 (1998).

Towards functional metallo-supramolecular assemblies and polymers

ULRICH S. SCHUBERT

Lehrstuhl für Makromolekulare Chemie, Technische Universität München, Lichtenbergstr. 4, 85747 Garching, Germany
Laboratory for Macromolecular and Organic Chemistry, Eindhoven Technical University, P.O. Box 513, 5600 MB Eindhoven, The Netherlands

Abstract—One of the main goals in today's chemistry and material science is the controlled fabrication of ordered architectures on nano- to mesoscopic length scales. Systems containing metal ions, which may potentially display new magnetic, photochemical or redox properties, are of special interest. A very promising approach comes from supramolecular chemistry utilizing the programmed self-assembly from mixtures of organic ligands and metal ions. In addition, such supramolecular building blocks can be used for the preparation of novel polymeric materials utilizing classical copolymerizations as well as controlled/living polymerization procedures.

1. INTRODUCTION

The design, synthesis and arrangement of highly ordered materials with nanometer dimensions containing special functional units is one main aim in today's polymer, supramolecular and material science [1–7]. Such architectures, e.g. on surfaces, in thin films or in bulk materials, could provide new thermal, mechanical, photochemical, electrochemical or magnetic properties and find potential applications in nanotechnology, e.g., considering molecular information storage devices. Systems containing metal ions are of particular interest. All of these attempts require a very precise control of the structures at very different length scales from molecular size to micrometers. At present, several approaches to such materials are used:

(1) In polymer chemistry phase separation of block copolymers or the formation of defined micelles are being used [8–15];
(2) Material scientists increase the precision for the mircrostructuring of surfaces or films [16–19];
(3) In supramolecular chemistry the self-assembly of suitable organic or inorganic molecules via non-covalent interactions is utilized.

The approach from supramolecular chemistry seems to open the most promising way towards the construction of nanometer size architectures. Recent work in this field has demonstrated that information stored in molecular components can be read out by non-covalent interactions to assemble the final architectures using hydrogen-bonding [20–23], metal-ligand [24–26] and cation [27] interactions (however, a large number of described systems contain infinite lattices or exist only in the solid state [28, 29]). Besides the design and synthesis of such systems, the ordered and stable arrangement of such architectures on surfaces or thin films is of special interest. Ordered layers of organized metal ions could provide, e.g., materials with functional units of a size smaller than quantum dots [30], with the additional advantage that they can be formed by spontaneous assembly instead of microfabrication. By addressing the metal ions photo- or electrochemically, it might be possible to inscribe patterns which could be read out non-destructively [1, 2, 25], offering a basis for the construction of information storage devices [31].

One interesting procedure in supramolecular chemistry uses molecular building blocks containing repeating subunits, with the capability to complex metal ions in a defined and predictable arrangement. In particular, multimetallic complexes of precise [m × n] nuclearity and two-dimensional geometry as models for information storage were developed. Their basic geometries may be termed *racks* [n]**R**, *ladders* [2n]**L**, and *grids* [m × n]**G**, where the nuclearity of the **R**, **L**, and **G** species is given by [n], [2n], and [m × n], in sequence of increasing complexity (see e.g. [1]). The ligand components are here oriented more or less orthogonal to each other [1, 32]. In these cases the metal ions serve as connecting centers for the structures and provide additional electrochemical, photochemical, and reactional properties. The choice of the ligands and metal ions represent the most important steps in the design of such coordination arrays. Transition metal ions with octahedral coordination geometry are expected to cover a wider range of usable elements and properties compared to tetrahedral metal ions used in ladder- [33] and grid-type [34] complexes before. As basic tridentate complexing unit, the well-known terpyridine, was chosen due to its property to form stable complexes with transition metal ions [35] and its interesting oxidative and reductive behavior [36–38]. To promote metal-metal interaction [39] and to enforce the alignment of the metal centers, a small metal-to-metal distance is required in the coordination arrays. These requirements are met by ligands consisting of fused terpyridine subunits containing bridging pyrimidine groups [1, 32].

Combining polymeric systems and modern living or controlled polymerization techniques, tailor-made materials with novel structure property relationships could become accessible. Besides linear structures, block copolymers and special architectures, as star-like structures or denditic systems, are of special interest. For this purpose, classical supramolecular building blocks, like bipyridine or terpyridine units, can be successfully applied. However, the necessary functionalization of the ligands as well as multi-gram syntheses are the main problems in this approach. In addition, the characterization of such new metallo-supramolecular assemblies and polymers leads to numerous problems caused by the charge, color, non-covalent interactions and relatively high molar masses of the compounds. Therefore, new analytical methods have to be applied or traditional methods have to be improved and adapted (see, e.g., approaches utilizing surface layer studies [40–42], electrospray (ESI) mass spectrometry [43–45] and NMR spectroscopy [46]).

In the present paper, a concept for the construction of metallo-supramolecular oligomeric and polymeric systems will be presented, together with some solutions concerning the "characterization" problems.

2. EXPERIMENTAL SECTION

General remarks: Materials were obtained from commercial suppliers and used without further purification. 2,2′:6′,2″-Terpyridine was obtained from Aldrich (No. 23,467-2) and recrystallized in hexane. ^{1}H and ^{13}C NMR spectra were recorded in CDCl$_3$ at 300 and 75 MHz, respectively, on a Bruker AC 300 spectrometer. Electronic absorption spectra were measured on a Varian Cary 3 spectrometer. Gel permeation chromatography (GPC) analysis was performed on a Waters Liquid Chromatograph system using Shodex GPC K-802S columns and a Waters Differential Refractometer 410 with chloroform as eluent. Calibration was conducted with polystyrene as standards. All MALDI-TOF-MS measurements were done using a Bruker BIFLEX III equipped with a 337 nm UV nitrogen laser producing 3-ns pulses. The MALDI-TOF-spectra were obtained in the reflective mode. Mass assignments were performed with unmanipulated spectra for an optimal correlation between observed and calculated mass.

Ligands: Experimental details of the synthetic procedures can be found elsewhere (see below). In general, commercially available amino-substituted pyridine precursors were brominated using well-described methods [47] in a special 4 l glass reactor (100 to 150 g scale, 85 to 90% yield). A different approach utilized the commercially available 6-chloronicotinic acid as already functionalized starting material [48]. Reaction with a mixture of PCl$_5$ and POCl$_3$ gave the corresponding acid chloride in high yields. Multiple treatment with PBr$_3$ at 160°C resulted in the brominated pyridine derivative. Reduction with NaBH$_4$ in water finally yielded the 2-bromo-5-hydroxymethylpyridine, which could be protected with a ^{t}butyldimethylsiloxy (TBDMS) protecting group [49, 50]. Alternatively a ^{t}butyloxy monoprotected butanediol spacer can be incorporated using a different method [51].

Complexation: The metal complexes were prepared by complexation of the corresponding free ligands using metal salts such as Cu(II) acetate or [Cu(I)(CH$_3$CN)$_4$](PF$_6$)) as described elsewhere (see in the main text). The complexes were in general isolated as PF$_6$ or AsF$_6$ salts and recrystallized from acetone/ether or acetonitrile/toluene solvent mixtures.

Polymerization: Experimental details for the general polymerization procedures can be found elsewhere. See, e.g., [52] for cationic polymerization, [53–55] for radical polymerization and [56] classical copolymerization.

3. RESULTS AND DISCUSSION

Grid-like metal coordination arrays: Recently a new class of coordination arrays presenting a two-dimensional [2 × 2] grid-type architecture was described. It is based on transition metal ions with octahedral coordination geometry and fused terpyridine ligands with alternating pyridines and pyrimidines (*bis*-tridentate) units (Figure 1) [57, 58]. The reaction of equimolar quantities of such ligands (e.g., 4,6-*bis*(6-(2,2′-bipyridyl))pyrimidine **L** or the methyl substituted analogs) [57] and cobalt(II) acetate in refluxing MeOH has been shown to lead exclusively to the formation of tetranuclear complexes $[M_4(L)_4]^{8+}$, as shown in Figure 1 [58]. Other transition metal ions, like Cd(II), Cu(II) or Zn(II), can also be used. These complexes were found to present interesting electronic, magnetic, and structural properties, such as electronic interactions between the metal centers and an antiferromagnetic transition at low temperatures [59–61]. The grid-type architecture of the complexes was confirmed by the determination of the crystal structure of $[Co_4(L)_4](SbF_6)_8$ (Figure 2) [58]. Furthermore, established techniques like MALDI-TOF mass spectrometry [62] and analytical ultracentrifugation [63–65] were adjusted or applied, for the first time, as universal characterization methods for metallo-supramolecular systems in solution and the solid state (Figures 3 and 4).

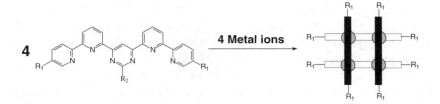

Figure 1. Schematic presentation of the complex formation (R_1 and R_2 = H or CH_3).

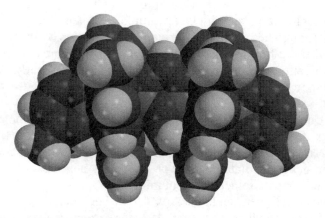

Figure 2. Molecular modeling (MAC Spartan Plus) of a cobalt(II) metal grid based on 4,6-*bis*(5″-methyl-(2′,2″-bipyrid-6′-yl))-2-methylpyrimidine ligands (PF_6 counter ions are omitted).

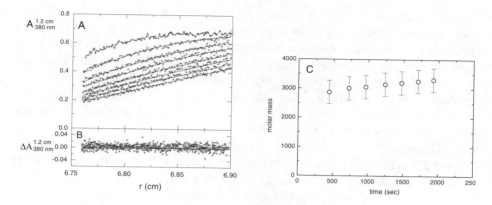

Figure 3. An approach to equilibrium experiment in the analytical ultracentrifuge: (A) Experimental data sets of the depletion at the meniscus, recorded at 380 nm and at intervals of 240 seconds (+), and curves fitted to the data. (B) Local differences between the experimental and the fitted data. (C) Time dependence of the apparent molar mass governing the depletion of the solute at the meniscus [64].

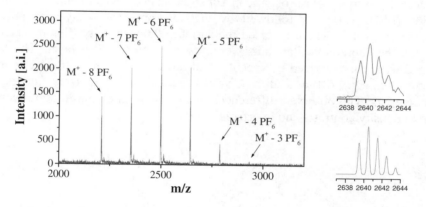

Figure 4. MALDI-TOF-MS measurements of the cobalt(II) metal grid based on 4,6-*bis*(5''-methyl-(2',2''-bipyrid-6'-yl))-2-methylpyrimidine ligands (matrix dithranol). Left top: Isotope pattern of the m/z 2640 peak; left bottom: Result of the simulation.

In order to further incorporate these ligands into polymers or larger assemblies, we developed a way to functionalize the R_1-positions of the final ligands using radical bromination reactions with NBS of the terminal methyl groups resulting in the *bis*functionalized ligand [57]. However, the *bis*brominated ligand could be isolated out of a mixture of different brominated products only at a yield of 27%, applying a multiple liquid chromatography methodology (which, in addition, cannot be used for the synthesis of larger quantities of the *bis*functionalized ligand). We therefore decided to develop an alternative synthetic approach using an already functionalized starting material. The commercially available 6-chloronicotinic acid was chosen as functionalized precursor. By applying a series of functional group interconversion reactions a couple of functionalized 2-chloro or 2-bromo-terminated pyridine building blocks became accessible. Following Stille-type coupling reactions, functionalized 4,6-*bis*(6-(2,2'-bipyridyl))-pyrimidine ligands and functionalized [n × n]-grids could be prepared (Figure 5). This opens new ways towards the synthesis of molecular assemblies and supramolecular polymers (Figure 6).

Several approaches were recently described towards the preparation of extended metallo-supramolecular architectures on surfaces. Besides the organization of hydroxy-terminated grids on a water trough using Langmuir-Blodgett (LB) technique and the following transfer onto substrates [66], unfunctionalized ligands were self-assembled at the air-water interface [67]. However, both methods were unable to provide ordered grid monolayers. In a different approach we used the self-assembly of the grids into thin monomolecular films by adsorption onto polyelectrolyte covered substrates. [1 × 1] and [2 × 2] metal coordination arrays were chosen as model systems for the film growth (Figure 7) [68–71] (cooperation with T. Salditt/University of München). First we prepared a thin cushion of polyelectrolytes on a glass or silicon substrate using poly(ethylenimine) hydrochloride (PEI). This was followed by the adsorption of a layer of poly(styrenesulfonate) (PSS) which reversed the surface charge density (Figure 7). In the next step the supramolecular units were adsorbed on the PSS layer. The layers obtained were characterized by specular and nonspecular x-ray reflectivity measurements (see e.g. [72]) as well as UV/VIS spectroscopy and x-ray fluorescence measurements. The observed differences in the oscillation amplitude reflects the different density contrasts and interfacial widths.

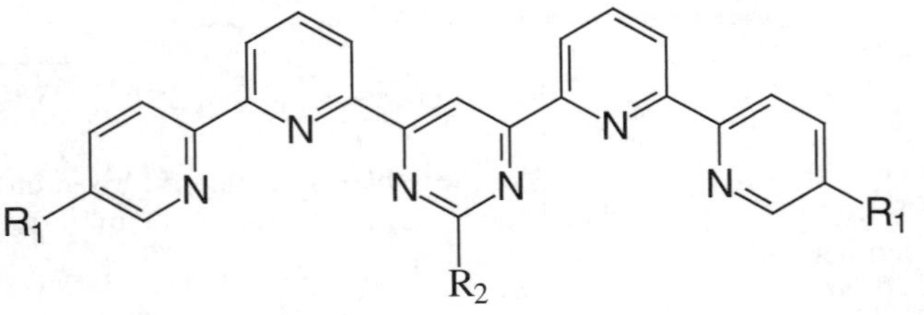

R_1 = CH$_2$OTBDMS, R_2 = H
R_1 = CH$_2$OH, R_2 = H
R_1 = CH$_3$, R_2 = Phenyl
R_1 = CH$_2$O(CH$_2$)$_4$tBu, R_2 = Phenyl

Figure 5. Schematic representation of the functionalized grid ligands (R' = TBDMS = ^{t}butyldimethylsilyl).

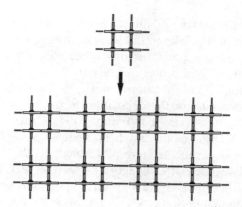

Figure 6. Schematic drawing of a potential arrangement of grids into two-dimensional layers using non-covalent or covalent polymerization.

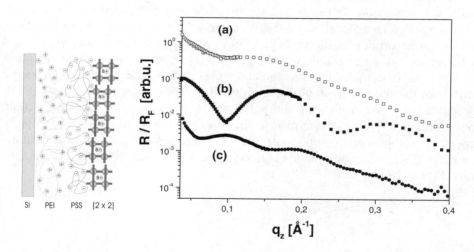

Figure 7. Left: Schematic drawing illustrating the build-up of the complex layers using [2 × 2] metal grids: Si, PEI and PSS refer to silicon substrate, poly(ethylenimine) and poly(styrene-4-sulfonate), respectively. Right: Typical x-ray reflectivity profiles $R(q_z)/R_F(q_z)$ of the self-assembled films after normalization. The three data sets are shifted by arbitrary factors for clarity and correspond to films composed of the polyelectrolyte layers PSS/PEI/Si (a) without any metal complexes, (b) with [1 × 1] Zn complexes, and (c) with [2 × 2] Cd complexes adsorbed on top. The curves reflect the laterally averaged density profiles of the film along the interface normal [68] (produced by Q. An and T. Saditt, University of München).

In addition, two other methods for the preparation of highly ordered monolayers of grids on surfaces were used, combined with scanning tunneling microscopy (STM) techniques to image and write in grid monolayers on solid surfaces with molecular resolution [73, 74] (cooperation with M. Möller/University of Ulm). For this purpose highly oriented pyrolitic graphite was treated with diluted solutions of the grids and the surface architecture was investigated by STM. Depending on the type of grid used, different molecular orientations on the surface could be observed. Again, the key point for this experiments was the preparation of monodisperse solutions of grid molecules, verification of the monodispersity being done by sedimentation equilibrium analysis in the analytical ultracentrifuge [73]. Two different methods for the deposition of the grids from these solutions on the graphite surface were applied (Figure 8): I) a drop of the solution was placed on a substrate, which had already been scanned using STM; after evaporation of the solvent the experiments were performed; II) the substrate was dipped into the solution and scanned after evaporation of the solvent. Using method I, monolayers of methyl-terminated grids could be prepared with a molecular pattern with nearly orthogonal periodicities of 1.4 nm × 2.6 nm (Figure 9). This pattern is consistent with the size of the metal grids oriented normally to the surface [58]. In addition, a periodicity of 0.6 nm could be observed correlating with the Co-Co distance within the grids (x-ray data: 0.64 nm [58]. In contrast to these findings, a deposition of grids with no terminal methyl groups led to a layer which showed a different molecular periodicity, namely 2.5 nm × 2.4 nm. These findings indicate an adsorption of the grids parallel to the surface. The factors which lead to differences in the orientation of the grids in those two cases are at present unknown. – In first molecular manipulation experiments the extraction of single grids from the monolayer could be performed by applying a short negative voltage pulse. However, following the pulse stable manipulated architectures could not be obtained, probably due to fast diffusion processes. Approaches aiming at generating stable metallo-supramolecular architectures on surfaces which allow reversible molecular manipulations are being pursued, utilizing photocrosslinkabe units (Figure 10).

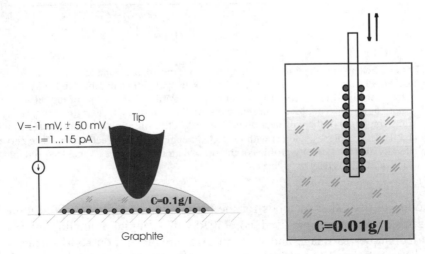

Figure 8. Direct adsorption of metal grids on surfaces. Schematic representation of the utilized methods.

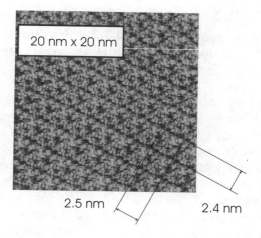

Figure 9. STM height image of a grid deposited on graphite [73] (produced by A. Semenov, J. P. Spatz and M. Möller, University of Ulm).

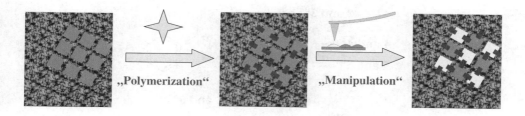

Figure 10. Schematic representation of a potential manipulation of crosslinked metal coordination arrays.

Metal containing block copolymers: An important aspect in the construction of metal containing block copolymers is the selection of suitable complexing units. N-Heterocyclic ligands have been known for nearly a century as very effective and stable complexation agents for transition metal ions [75, 76]. In particular, 2,2':6',2''-terpyridine ligands complexed with cobalt, copper, manganese, iron, chromium, nickel or ruthenium complexes [77–80] have been studied extensively, due to their electronic and redox properties [81–83]. However, only a few examples that use these units as building blocks for new block copolymers have been published in the last years (see e. g. [84–86]). This is mainly due to synthesis problems. We recently described a multi-gram synthesis of 5,5''-*bis*functionalized terpyridine ligands as building blocks for the preparation of 3-block and multi-block copolymers based on *bis*hydroxy-functionalized ligands [56]. From quantum mechanical calculation, the 5,5''-position seems to be the best suited for the introduction into polymers since it does not interfere with the complexation step and allow a linear incorporation of the terpyridines and the corresponding complexes into the polymer chain (Figure 11). To incorporate these building blocks into polymers, we first performed reactions of the *bis*hydroxy-terpyridine with a commercially available *bis*isocyanate-functionalized prepolymer, resulting in an AB multiblock copolymer (Figure 12). The incorporation of the terpyridine moieties was proven by GPC and MALDI-TOF-MS measurements. The result of the GPC investigation is shown in Figure 13 (GPC: M_n = 10500 g/mol, M_w = 16100 g/mol, M_w/M_n = 1.5, using $CHCl_3$ as eluent, RI-detection and polystyrene calibration). Treatment of a solution of the block copolymer with cobalt(II) acetate in chloroform immediately yielded a red-brown colored solution. UV/VIS spectroscopic investigations (Figure 14) revealed the formation of the corresponding terpyridine metal complexes and thus the formation of a supramolecular polymer/ion complex (for examples containing bipyridine metal complexes, see [87–89]). In addition, we recently could prepare *bis*isocyanate-functionalized terpyridine building blocks [90].

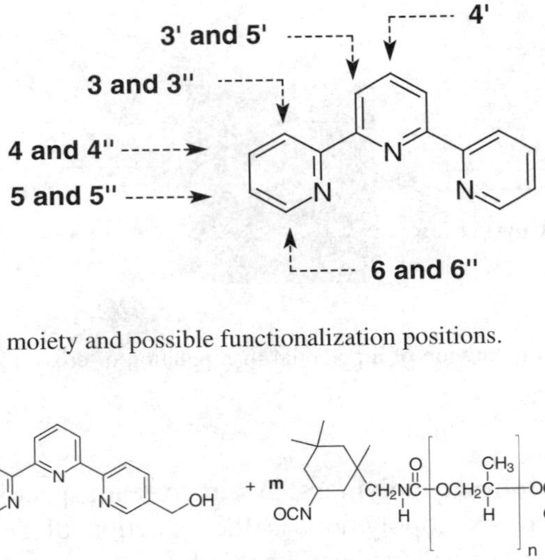

Figure 11. Terpyridine moiety and possible functionalization positions.

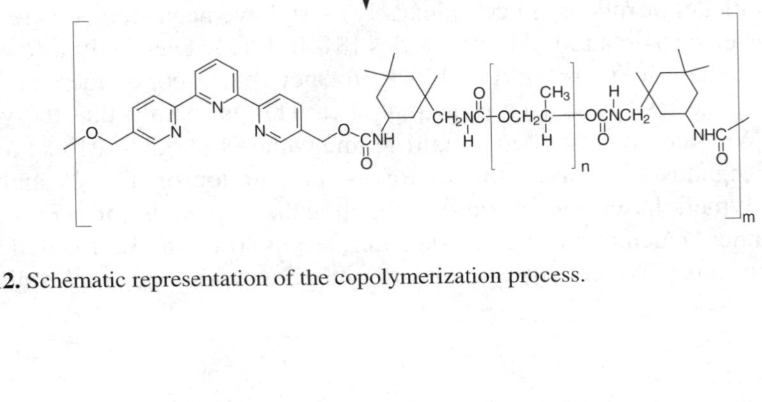

Figure 12. Schematic representation of the copolymerization process.

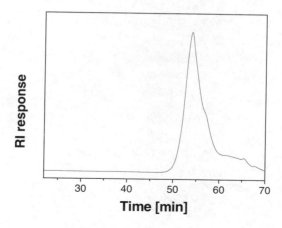

Figure 13. GPC curve of the terpyridine containing polymer (chloroform as eluent) [87].

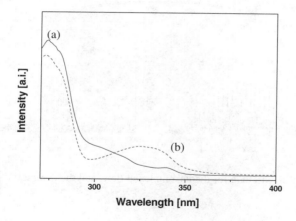

Figure 14. UV/VIS spectra of the terpyridine-containing polymer complexation in methanol: (a) polymer without metal ions; (b) after addition of cobalt(II) acetate tetrahydrate [56].

Metal binding telechelic copolymers: There is an alternative strategy towards metal complexing polymers based on terpyridine building blocks; it utilizes 4'-functionalized terpyridine units. Using Williamson ether condensation reactions we could prepare α,ω-functionalized metal complexing polyethylenoxide oligomers consisting of a terpyridine and a carboxyl endgroup (Figure 15) [91]. The addition of transition metal ions to these oligomers resulted in a dimerisation and therefore in a doubling of the molecular weight (Figure 16). The carboxyl end group opens further avenues for the attachment onto surfaces or other assemblies and polymers (Figure 17). Up to now, e.g., carboxy end groups were used for the attachment of the systems to surfaces, utilizing aminosilanes and amid coupling procedures.

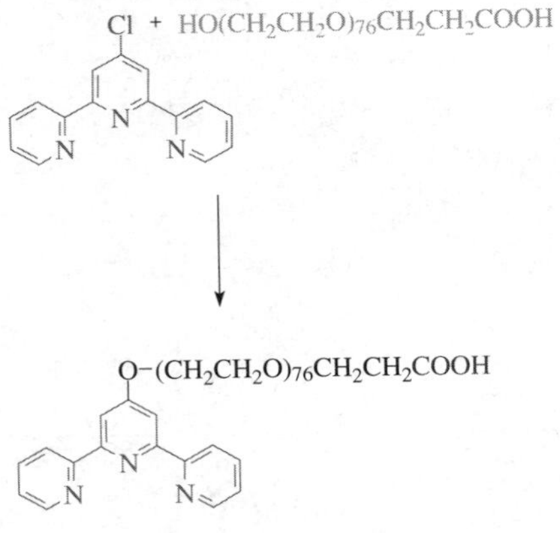

Figure 15. Schematic presentation of the telechelic preparation.

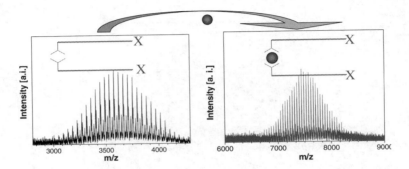

Figure 16. MALDI-TOF MS data for the uncomplexed (left) and the complexed (right) monofunctionalized polymers.

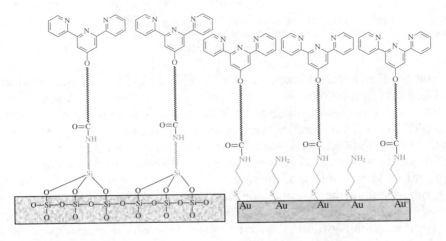

Figure 17. Potential attachment of α,ω-functionalized polymers on surfaces.

Metallo-supramolecular initiators for the living cationic polymerization: A very promising approach which combines metallo-supramolecular building blocks and polymer chemistry is the use of supramolecular systems as initiators for living or controlled polymerization reactions. In particular, the living polymerization of 2-oxazolines seems to be perfectly suited for this purpose, due to broad range of different polymer structures obtainable [92, 93]. We recently described a new approach towards bipyridine- and terpyridine-containing polymers utilizing metallo-supramolecular initiators for the living cationic polymerization of 2-oxazolines based on *bis*functionalized 6,6'-dimethyl-2,2'-bipyridine (Figure 18) [94], monofunctionalized 5,5'-dimethyl-2,2'-bipyridine [95, 96] or *bis*functionalized 5,5''-dimethyl-2,2':6',2''-terpyridine metal complexes [97] (see also results by Fraser et al. [98–100]). The living nature of the polymerization can be demonstrated by the linear relationship between the [monomer]/[initiator] ratio and the average molecular weight, and the narrow molecular weight distribution of the polymers (Figure 19). Besides homopolymers, block copolymers with amphiphilic properties and supramolecular segments based on different 2-oxazoline monomers with hydrophilic or hydrophobic characteristics

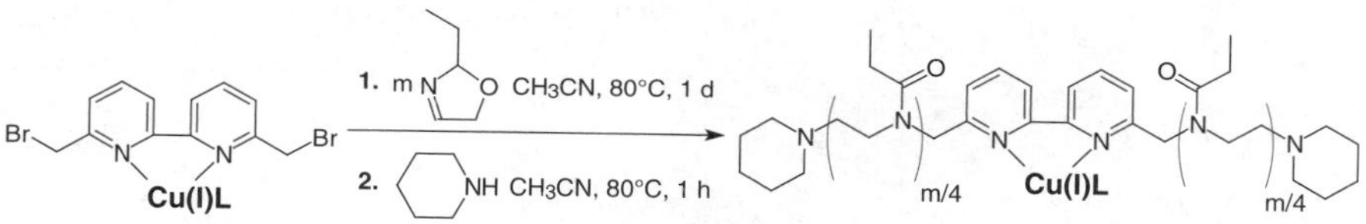

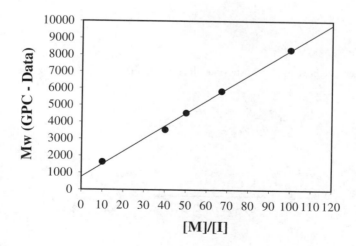

Figure 18. General polymerization procedure towards poly(oxazolines) containing 6,6'-*bis*functionalized 2,2'-bipyridine units.

Figure 19. Plot of the average molecular weight M_w versus [monomer]/[(Cu(I)(6,6'-(*bis*bromomethyl)-2,2'-bipyridine)$_2$)(PF$_6$)] ratio for the fragmented polymers ([I] refers to the initiating CH$_2$Br groups).

could be obtained (Figure 20). In general, this type of monomers can be polymerized by a living cationic mechanism using electrophiles such as bromomethyl endgroups, like in the present case the metallo-supramolecular complexes [94]. However, uncomplexed functionalized ligands alone are also able to initiate the polymerization. In this case the uncomplexed bipyridine unit can act as termination agent resulting in branched polymers as well as in some crosslinked polymers [94]. The function of the metal ion in the metallo-supramolecular complexes during the polymerization process is thus similar to a classical protecting group which prevents the growing polymer chains from termination processes (Figure 21).

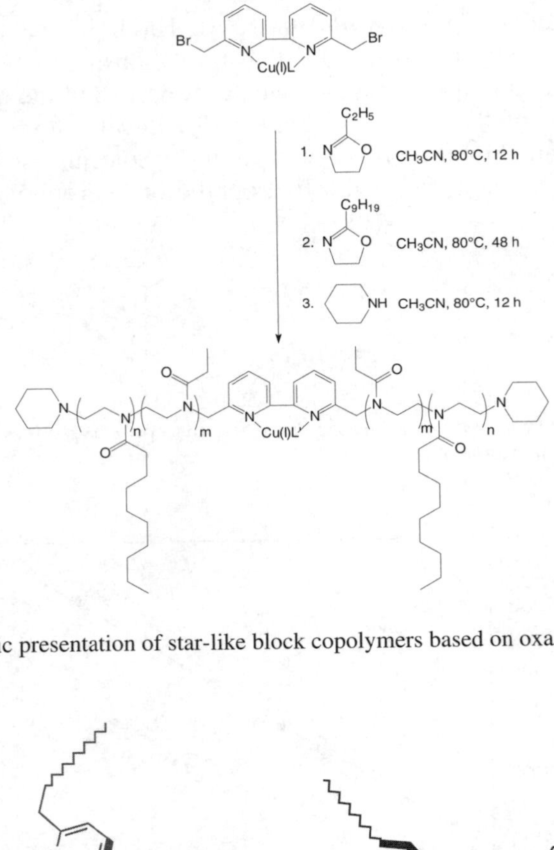

Figure 20. Schematic presentation of star-like block copolymers based on oxazolines.

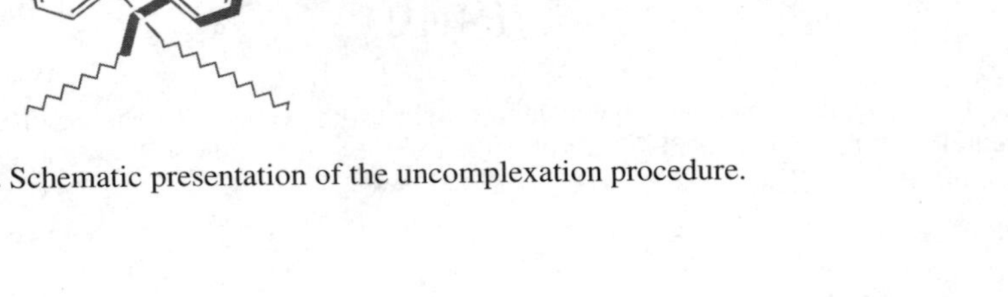

Figure 21. Schematic presentation of the uncomplexation procedure.

An interesting feature in this context is that polymers with higher molecular weight were found to be partially destroyed on the GPC column due to shear forces. As a consequence, two peaks were detected during GPC analysis (Figure 22). This situation was found for both homopolymers and block copolymers. Following extraction of the cation utilizing basic conditions, the higher molecular weight peak in the GPC traces disappeared, confirming the fragmentation of the star-like polymers on the GPC-columns. In addition, the termination (see e.g. [101]) of the living end groups during the polymerization with an amino-functionalized building blocks (e.g. terpyridines) opens the possibility for the construction of an A-Polymer-B type polymer with different metal binding sites (Figure 23). This method therefore offers a way to incorporate a single metal binding side into each polymer chain via the initiator unit and one or two metal binding sites at each end via the termination step, resulting in well-defined polymer architectures.

Metallo-supramolecular initiators for the living radical polymerization: Radical polymerization techniques are by far the most important procedures in industry and acedemic. However, in conventional radical polymerization processes control of molecular weight and especially molecular weight distribution is possi-

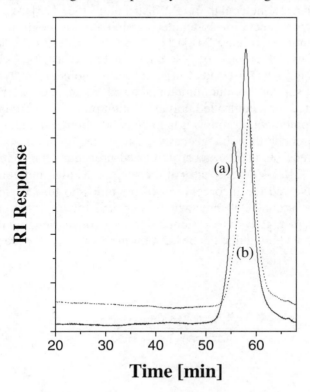

Figure 22. GPC curves of a typical oxazoline block copolymer: a) poly(ethyloxazoline) homopolymer, b) poly(ethyloxazoline)-*block*-poly(nonyloxazoline).

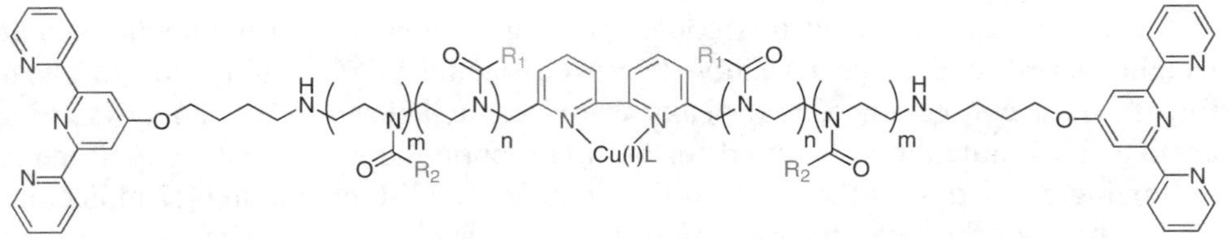

Figure 23. Schematic representation of a terpyridine terminated star-like poly(oxazoline).

ble in rare cases only [102]. In order to overcome the drawbacks of these procedures and to open new avenues for the design of polymers and block copolymers via radical methods, controlled ("living") free radical polymerization techniques have attracted much interest during the last years [103–122]. In particular, the so called **a**tom **t**ransfer **r**adical **p**olymerization (ATRP) technique allows the controlled polymerization of styrene and acrylate type monomers, the synthesis of block copolymer or other polymer architectures. We recently developed a new approach using already preformed metal complexes as catalysts, resulting in non-colored polymers with very low retaining copper content, predetermined molecular weight and narrow molecular weight distribution [53, 123]. We then tried to extend the concept of metallo-supramolecular initiators to controlled radical polymerization techniques (Figure 24) [124]. The GPC traces of the samples taken during the polymerization of styrene with [Cu(II)(4,4'-dimethyl-2,2'-bipyridine)$_3$](PF$_6$)$_2$ as catalyst and [Cu(I)(6,6'-*bis*bromomethyl-2,2'-bipyridine)$_2$]-(PF$_6$) as metallo-supramolecular initiator showed an increase of the molecular weight, as expected for a controlled polymerization procedure (Figure 25). Due to four initiating bromomethyl groups, it is to be expected that the molecular weight distribution is broader than in the case of a single initiation species like (1-bromoethyl)benzene (as used in most published procedure for controlled radical polymerization [53]). UV/VIS-spectra of the supramolecular unit containing polystyrene clearly showed the incorporation of the unit into the polymer. A π-π* -transition band at around 290 nm, which is typical for the initiator, is also detected in the polymer. Polystyrene samples from a conventional autopolymerization process do not show a distinct absorption in this range. In order to prove the

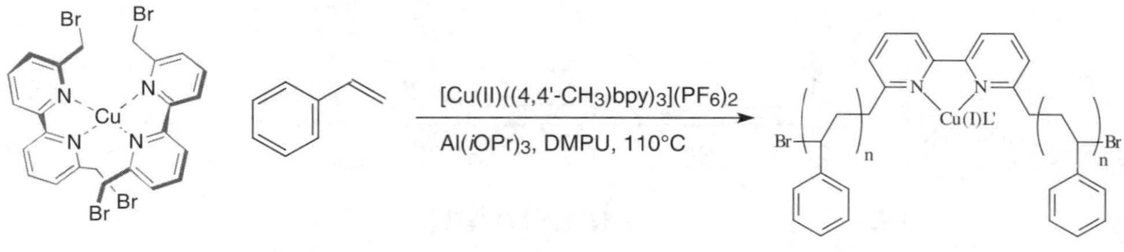

Figure 24. Schematic representation of the preparation of star-like poly(styrene) polymers (DMPU: 1,3-dimethyl-3,4,5,6-tetrahydro-2(1H)-pyrimidinon).

incorporation of the supramolecular unit into the polystyrene, we applied MALDI-TOF mass spectrometry. To overcome the problems resulting from higher molecular weight polymers in MALDI-TOF-MS, we choose samples which were taken after very small conversion rates with low molecular weights. A typical MALDI-TOF-MS for a polystyrene sample initiated with the metallo-supramolecular species is shown in Figure 26. The molecular weight found in MALDI-TOF-MS clearly revealed an incorporation of the bipyridine unit into the polymers (e.g. peak maximum at 5598 g/mol: 52 styrene units and one 6,6'-dimethyl-2,2'-bipyridine unit in the center of the polymer).

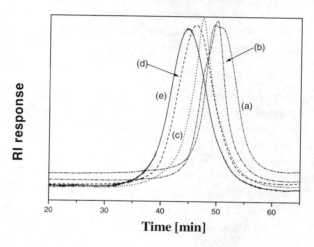

Figure 25. GPC curves (chloroform as eluent) for a few samples obtained during polymerization of styrene: (a) 10.5% conversion; (b) 15.5% conversion; (c) 20.6% conversion; (d) 45.8% conversion; (e) 47.6% conversion.

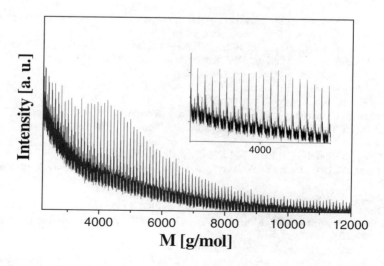

Figure 26. MALDI-TOF-MS of a typical poly(styrene) sample containing a central bipyridine unit.

4. CONCLUSIONS

The metallo-supramolecular systems described in this paper provide new avenues towards ordered molecular assemblies and supramolecular polymers with novel material properties. Key points in this research are the directed multi-gram synthetic routes towards suitable ligands, the directed self-assembly using transition metal ions and the application of special characterization methods. The combination of supramolecular building bocks and suitable polymers finally allows the construction of tailor-made materials. Besides the classical copolymerization or preparation of telechelics the combination with living or controlled polymerization techniques could be utilized. Further investigation will focus on the design and preparation of smart metallo-supramolecular materials, e.g. for potential applications in catalysis, coatings or reversible glues.

Acknowledgments

The research was supported partly by the *Bayerisches Staatsministerium für Unterricht, Kultus, Wissenschaft und Kunst*, the *Deutsche Forschungsgemeinschaft (DFG)* and the *Fonds der Chemischen Industrie*. I thank *Reilly Tar & Chem. Corp.* for contributing some of the ligands and precursors, and C. Eschbaumer, C. H. Weidl, G. Hochwimmer, O. Hien and M. Heller for most of the synthetic work and D. Schubert and D. Wasserberg for help in the preparation of the manuscript. The results were obtained in cooperation with the groups of M. Möller (University of Ulm), T. Salditt (University of München) and D. Schubert (University of Frankfurt). I thank Prof. Dr. J.-M. Lehn and Prof. Dr.-Ing. O. Nuyken for there support.

REFERENCES

1. J.-M. Lehn, *Supramolecular Chemistry - Concepts and Perspectives*, VCH, Weinheim, 1995.
2. J.-M. Lehn, *Macromol. Chem., Macromol. Symp.* 1993, **69**, 1.
3. J.-M. Lehn, *Angew. Chem.* 1990, **102**, 1347; *Angew. Chem., Int. Ed. Eng.* 1990, **29**, 1304.
4. G. M. Whitesides, J. P. Mathias, C. T. Seto, *Science* 1991, **254**, 1312.
5. D. S. Lawrence, T. Jiang, M. Levett, *Chem. Rev.* 1995, 2229.
6. G. M. Whitesides, E. E. Simanec, J. P. Mathias, C. T. Seto, D. N. Chin, M. Mammen, D. M. Gordon, *Acc. Chem. Res.* 1996, **28**, 37.
7. C. Piguet, G. Bernardinelli, G. Hopfgarten, *Chem. Rev.* 1997, **97**, 2005.
8. J. H. Golden, F. J. DeSalvo, J. M. J. Fréchet, J. Silcox, M. Thomas, J. Elman, *Science* 1996, **273**, 782.
9. S. I. Stupp, S. Son, H. C. Lin, L. S. Li, *Science* 1993, **259**, 59.
10. S. I. Stupp, V. LeBonheur, K. Walker, L. S. Li, K. E. Huggins, M. Keser, A. Amstutz, *Science* 1997, **276**, 384.
11. J. P. Spatz, A. Roescher, M. Möller, *Adv. Mater.* 1996, **8**, 337.
12. J. P. Spatz, S. Mößmer, M. Möller, *Chem. Eur. J.* 1996, **2**, 1552.
13. J. P. Spatz, S. Mößmer, M. Möller, *Angew. Chem.* 1996, **108**, 1673; *Angew. Chem., Int. Ed. Engl.* 1996, **35**, 1510.
14. M. Antonietti, S. Henke, A. Thünemann, *Adv. Mater.* 1996, **8**, 41.
15. M. Antonietti, C. Göltner, *Angew. Chem.* 1997, **109**, 944; *Angew. Chem., Int. Ed. Engl.* 1997, **36**, 910.

16. Y. Xia, J. J. McClelland, R. Gupta, D. Qin, X.-M. Yhao, L. L. Sohn, R. J. Celotta, G. M. Whitesides, *Adv. Mater.* 1997, **9**, 147.
17. Y. Xia, E. Kim, M. Mrksich, G. M. Whitesides, *Chem. Mater.* 1996, **8**, 601.
18. D. Qin, Y. Xia, G. M. Whitesides, *Adv. Mater.* 1997, **9**, 407.
19. M. L. Renak, G. C. Bazan, D. Roitman, *Adv. Mater.* 1997, **9**, 392.
20. D. S. Lawrence, T. Jiang, M. Levett, *Chem. Rev.* 1995, 2229.
21. S. C. Zimmerman, F. Zeng, D. E. C. Reichert, S. V. Kolotuchin, *Science* 1996, **271**, 1095.
22. M. Kotera, J.-M. Lehn, J.-P. Vigneron, *J. Chem. Soc., Chem. Comm.* 1994, 197.
23. R. M. Grotzfeld, N. Branda, J. J. Rebek, *Science* 1996, **271**, 487.
24. J.-M. Lehn, A. Rigault, J. Siegel, J. Harrowfield, B. Chevrier, D. Moras, *Proc. Natl. Acad. Sci. USA* 1987, **84**, 2565.
25. P. N. W. Baxter, J.-M. Lehn, A. DeCian, J. Fischer, *Angew. Chem.* 1993, **105**, 92; *Angew. Chem., Int. Ed. Engl.* 1993, **32**, 69.
26. E. C. Constable, *Prog. Inorg. Chem.* 1994, **42**, 67.
27. D. A. Dougherty, *Science* 1996, **271**, 163.
28. B. F. Abrahams, M. J. Hardie, B. F. Hoskins, R. Robson, G. A. Williams, *J. Am. Chem. Soc.* 1992, **114**, 10641.
29. E. C. Constable, A. J. Edwards, D. Phillips, P. R. Raithby, *Supramol. Chem.* 1995, **5**, 93.
30. R. C. Ashoori, *Nature* 1996, **379**, 413.
31. J. R. Friedman, M. P. Sarachik, J. Tejada, R. Ziolo, *Phys. Rev. Lett.* 1996, **76**, 3830.
32. G. S. Hanan, C. R. Arana, J.-M. Lehn, D. Fenske, *Angew. Chem.* 1995, **107**, 1191; *Angew. Chem., Int. Ed. Engl.* 1995, **34**, 1122.
33. P. N. W. Baxter, G. S. Hanan, J.-M. Lehn, *J. Chem. Soc., Chem. Comm.* 1996, 2019.
34. P. N. W. Baxter, J.-M. Lehn, J. Fischer, M.-T. Youinou, *Angew. Chem.* 1994, **106**, 2432; *Angew. Chem., Int. Ed. Engl.* 1994, **33**, 2284.
35. G. Morgan, F. H. Burstall, *J. Chem. Soc.* 1937, 1649.
36. W. R. McWhinnie, J. D. Miller, *Adv. Inorg. Chem. Radiochem.* 1969, **12**, 135.
37. E. C. Constable, *Adv. Inorg. Chem. Radiochem.* 1986, **30**, 69.
38. J. R. Kirchhoff, D. R. McMillin, P. A. Marnot, J.-P. Sauvage, *J. Am. Chem. Soc.* 1985, **107**, 1138.
39. P. J. Steel, *Coord. Chem. Rev.* 1990, **106**, 227.
40. S. R. Cohen, I. Weissbuch, R. Popovitz-Biro, J. Majewski, H. P. Mauder, R. Lavi, L. Leiserowitz, M. Lahav, *Israel J. Chem.* 1996, **36**, 97.
41. F. Stevens, D. J. Dyer, D. M. Walba, *Angew. Chem.* 1996, **108**, 955; *Angew. Chem., Int. Ed. Engl.* 1996, **35**, 900.
42. D. M. Cyr, B. Venkataraman, G. W. Flynn, *Chem. Mater.* 1996, **8**, 1600.
43. E. Leize, A. Van Dorsselaer, R. Krämer, J.-M. Lehn, *J. Chem. Soc., Chem. Commun.* 1993, 990.
44. F. Marquis-Rigault, A. Dupont-Gervais, A. Van Dorsselaer, J.-M. Lehn, *Chem. Eur. J.* 1996, **2**, 1395.
45. M. Przybylsky, M. O. Glocker, *Angew. Chem.* 1996, **108**, 878; *Angew. Chem., Int. Ed. Engl.* 1996, **35**, 806.
46. G. S. Hanan, C. R. Arana, J.-M. Lehn, G. Baum, D. Fenske, *Chem. Eur. J.* 1996, **2**, 1292.
47. P.-M. Windscheif, F. Vögtle, *Synthesis* 1994, 87.
48. U. S. Schubert, C. H. Weidl, J.-M. Lehn, *Design. Monomer. Polym.* 1999, **2**, 1.
49. E. J. Corey, *J. Am. Chem. Soc.* 1972, **94**, 6190.
50. P. J. Kocienski, *Protecting Groups*, Thieme, Stuttgart, 1994.
51. U. S. Schubert, C. Eschbaumer, C. H. Weidl, *Synlett* 1999, 342.
52. G. Hochwimmer, O. Nuyken, U. S. Schubert, *Macromol. Rapid Commun.* 1998, **19**, 309.
53. U. S. Schubert, G. Hochwimmer, C. E. Spindler, O. Nuyken, *Macromol. Rapid Commun.* 1999, **20**, 251.
54. U. S. Schubert, G. Hochwimmer, C. E. Spindler, O. Nuyken, *Polym. Bull.* 1999, **43**, 319.
55. U. S. Schubert, C. E. Spindler, C. Eschbaumer, O. Nuyken, *Polym. Preprint* 1999, **40(2)**, 416.

56. U. S. Schubert, C. Eschbaumer, C. H. Weidl, *Design. Monom. Polym.* 1999, **2**, 185.
57. G. S. Hanan, U. S. Schubert, D. Volkmer, E. Riviere, J.-M. Lehn, N. Kyritsakas, J. Fischer, *Can. J. Chem.* 1997, **75**, 169.
58. G. S. Hanan, D. Volkmer, U. S. Schubert, J.-M. Lehn, G. Baum, D. Fenske, *Angew. Chem.* 1997, **109**, 1929; *Angew. Chem., Int. Ed. Engl.* 1997, **36**, 1842.
59. O. Waldmann, J. Hassmann, P. Müller, G. S. Hanan, D. Volkmer, U. S. Schubert, J.-M. Lehn, *Phys. Rev. Lett.* 1997, **78**, 3390.
60. O. Waldmann, J. Hassmann, R. Koch, P. Müller, G. S. Hanan, D. Volkmer, U. S. Schubert, J.-M. Lehn, *Mat. Res. Soc. Symp. Proc.* 1998, **448**, 841.
61. O. Waldmann, J. Hassmann, P. Müller, D. Volkmer, U. S. Schubert, J.-M. Lehn, *Phys. Rev. B* 1998, **58**, 3277.
62. U. S. Schubert, C. Eschbaumer, *J. Incl. Phenom.* 1999, **35**, 101.
63. D. Schubert, J. A. van den Broek, B. Sell, H. Durchschlag, W. Mächtle, U. S. Schubert, J.-M. Lehn, *Prog. Colloid Polym. Sci.* 1997, **107**, 166.
64. D. Schubert, C. Tziatzios, P. Schuck, U. S. Schubert, *Chem. Eur. J.* 1999, **5**, 1377.
65. C. Tziatzios, H. Durchschlag, B. Sell, J. A. van den Broek, W. Mächtle, W. Haase, J.-M. Lehn, C. H. Weidl, C. Eschbaumer, D. Schubert, U. S. Schubert, *Prog. Colloid Polym. Sci.* 1999, **109**, 114.
66. U. S. Schubert, J.-M. Lehn, J. Hassmann, C. Y. Hahn, N. Hallschmidt, P. Müller, in *Functional Polymers* (Eds.: A. O. Patil, D. N. Schulz, B. M. Novak), *ACS Symp. Ser.* 1998, **704**, 248.
67. I. Weissbuch, P. N. W. Baxter, S. Cohen, H. Cohen, K. Kjaer, P. B. Howes, J. Als-Nielsen, G. S. Hanan, U. S. Schubert, J.-M. Lehn, L. Leiserowitz, M. Lahav, *J. Am. Chem. Soc.* 1998, **120**, 4850.
68. T. Salditt, Q. An, A. Plech, C. Eschbaumer, U. S. Schubert, *Chem. Commun.* 1998, 2731.
69. U. S. Schubert, C. Eschbaumer, Q. An, T. Salditt, *Polym. Preprints* 1999, **40(1)**, 414.
70. U. S. Schubert, C. Eschbaumer, Q. An, T. Salditt, *J. Incl. Phenom.* 1999, **35**, 35.
71. T. Salditt, Q. An, A. Plech, J. Peisl, C. Eschbaumer, C. H. Weidl, U. S. Schubert, *Thin Solid Films* 1999, **354**, 208.
72. T. P. Russel, *Mat. Sci. Reports* 1990, **5**, 171.
73. A. Semenov, J. P. Spatz, M. Möller, J.-M. Lehn, B. Sell, D. Schubert, C. H. Weidl, U. S. Schubert, *Angew. Chem.* 1999, **111**, 2701; *Angew. Chem. Int. Ed.* 1999, **38**, 2547.
74. A. Semenov, J. P. Spatz, J.-M. Lehn, C. H. Weidl, U. S. Schubert, M. Möller, *Appl. Surface Sci.* 1999, **144/145**, 456.
75. E. C. Constable, *Metals and Ligand Reactivity*, VCH, Weinheim, 1996.
76. G. Wilkinson, R. D. Gillard, J. A. McCleverty (Eds.), *Comprehensive Coordination Chemistry*, Pergamon Press, Oxford, 1987, **2**.
77. B. N. Figgis, E. S. Kucharski, A. H. White, *Aust. J. Chem.* 1983, **36**, 1527.
78. R. Bhula, D. C. Weatherburn, *Aust. J. Chem.* 1991, **44**, 303.
79. B. N. Figgis, E. S. Kucharski, A. H. White, *Aust. J. Chem.* 1983, **36**, 1563.
80. A. T. Baker, H. A. Goodwin, *Aust. J. Chem.* 1985, **38**, 207.
81. W. R. McWhinnie, J. D. Miller, *Adv. Inorg. Chem. Radiochem.* 1969, **12**, 135.
82. E. C. Constable, *Adv. Inorg. Chem. Radiochem.* 1986, **30**, 69.
83. E. C. Constable, A. M.W. C. Thompson, *New. J. Chem.* 1992, **16**, 855.
84. K. Hanabusa, K. Nakano, T. Koyama, H. Shirai, N. Hojo, A. Kurose, *Makromol. Chem.* 1990, **191**, 391.
85. K. Hanabusa, A. Nakamura, T. Koyama, H. Shirai, *Makromol. Chem.* 1992, **193**, 1309.
86. K. T. Potts, D. A. Usifer, *Macromolecules* 1988, **21**, 1985.
87. C. D. Eisenbach, U. S. Schubert, *Macromolecules* 1993, **26**, 7372.
88. C. D. Eisenbach, W. Degelmann, A. Göldel, M. Heinlein, M. Terskan-Reinold, U. S. Schubert, *Macromol. Symp.* 1995, **98**, 565.
89. C. D. Eisenbach, A. Göldel, M. Terskan-Reinold, U. S. Schubert, *Makromol. Chem. Phys.* 1995, **196**, 1077.

90. U. S. Schubert, C. Eschbaumer, C. H. Weidl, *Polymeric Materials Science and Engineering (ACS), Proceedings* 1999, **80**, 191.
91. U. S. Schubert, C. Eschbaumer, *Polym. Preprints* 1999, **40(2)**, 1070.
92. S. Kobayashi, *Progr. Polym. Sci.* 1990, **15**, 751.
93. Y. Chujo, T. Saegusa; in *"Ring Opening Polymerization"*, Ed. D. I. Brunelle; Hanser, München, 1993.
94. G. Hochwimmer, O. Nuyken, U. S. Schubert, *Macromol. Rapid Commun.* 1998, **19**, 309.
95. U. S. Schubert, O. Nuyken, G. Hochwimmer, *Division of Polymeric Materials: Science and Engineering, Preprints* 1999, **80**, 193.
96. U. S. Schubert, C. Eschbaumer, G. Hochwimmer, *Tetrahedron Lett.* 1998, **39**, 8643.
97. U. S. Schubert, C. Eschbaumer, O. Nuyken, G. Hochwimmer, *J. Incl. Phenom.* 1999, **35**, 23.
98. J. J. S. Lamba, C. L. Fraser, *J. Am. Chem. Soc.* 1997, **119**, 1801.
99. J. J. S. Lamba, J. E. McAlvin, B. P. Peters, C. L. Fraser, *Div. Polym. Chem., Polym. Preprints* 1997, **38(1)**, 193.
100. J. E. Collins, C. L. Fraser, *Macromolecules* 1998, **31**, 6715.
101. O. Nuyken, G. Maier, A. Groß, H. Fischer, *Macromol. Chem. Phys.* 1996, **197**, 83.
102. G. Moad, D. H. Solomon, *The Chemistry of Free-Radical Polymerization*, Pergamon, Oxford, 1995.
103. M. K. Georges, R. P. N. Veregin, P. N. Kazmaier, G. K. Hamer, *Polym. Mater. Sci. Eng.* 1993, **68**, 6.
104. M. K. Georges, R. P. N. Veregin, P. N. Kazmaier, G. K. Hamer, M. Saban, *Macromolecules* 1994, **27**, 7228.
105. C. J. Hawker, *J. Am. Chem. Soc.* 1994, **116**, 11185.
106. H. Uegaki, Y. Kotani, M. Kamigaito, M. Sawamoto, *Macromolecules* 1997, **30**, 2249.
107. J. Ueda, M. Matsuyama, M. Kamigaito, M. Sawamoto, *Macromolecules* 1998, **31**, 557.
108. T. Ando, M. Kato, M. Kamigaito, M. Sawamoto, *Macromolecules* 1996, **29**, 1070.
109. T. Nishikawa, M. Kamigaito, M. Sawamoto, *Macromolecules* 1999, **32**, 2204.
110. C. Granel, P. Dubois, R. Jerome, P. Teyssie, *Macromolecules* 1996, **29**, 8576.
111. M. Husseman, E. E. Malmstrom, M. McNamara, M. Mate, D. Mecerreyes, D. G. Benoit, J. L. Hedrick, P. Mansky, E. Huang, T. P. Russell, C. J. Hawker, *Macromolecules* 1999, **32**, 1424.
112. D. Colombani, M. Steenbock, M. Klapper, K. Müllen, *Macromol. Rapid Commun.* 1997, **18**, 243.
113. M. Steenbock, M. Klapper, K. Müllen, C. Bauer, M. Hubrich, *Macromolecules* 1998, **31**, 5223.
114. J.-S. Wang, K. Matyjaszewski, *Macromolecules* 1995, **28**, 7901.
115. T. Patten, J. Xia, T. Abernathy, K. Matyjaszewski, *Science* 1996, **272**, 866.
116. K. Matyjaszeski, T. Patten, J. Xia, *J. Am. Chem. Soc.* 1997, **119**, 674.
117. K. Matyjaszewski, D. A. Shipp, J.-S. Wang, T. Grimaud, T. E. Patten, *Macromolecules* 1998, **31**, 6836.
118. V. Percec, K. Barboiu, A. Neumann, J. C. Ronda, M. Zhao, *Macromolecules* 1996, **29**, 3665.
119. G. Kickelbick, H. J. Paik, K. Matyjaszewski, *Macromolecules* 1999, **32**, 2941.
120. J. H. Xia, K. Matyjaszewski, *Macromolecules* 1999, **32**, 2434.
121. G. M. DiRenzo, M. Messerschmidt, R. Mülhaupt, *Macromol. Rapid Commun.* 1998, **19**, 381.
122. M. Destarac, J. M. Bessière, B. Boutevin, *Macromol. Rapid Commun.* 1997, **18**, 967.
123. U. S. Schubert, G. Hochwimmer, C. E. Spindler, O. Nuyken, *Deutsches Patent* DE 199 09 624.4, patent pending.
124. U. S. Schubert, G. Hochwimmer, *Polym. Preprints* 1999, **40(2)**, 340.

Syntheses and functions of glycopeptide surface-modified dendrimers

MASAHIKO OKADA,* KEIGO AOI and KANAME TSUTSUMIUCHI

Department of Applied Molecular Biosciences, Graduate School of Bioagricultural Sciences, Nagoya University, Chikusa-ku, Nagoya 464-8601, Japan

Abstract—This article briefly reviews our recent investigation on the design, syntheses and functions of glycopeptide surface-modified dendrimers. First, a general methodology is described for the synthesis of glycopeptide-modified dendrimers, "Sugar Balls", by the reaction of D-glucose-, D-galactose-, or *N*-acetyl-D-glucosamine-carrying L-serine *N*-carboxyanhydrides (glycoNCA's) with primary amino groups on the surface of poly(amido amine) dendrimer or by the ring-opening polymerization of glycoNCA's using primary amine-terminated dendrimer as a macroinitiator. The second section is concerned with synthesis and self assembly of amphiphilic surface block dendrimers, in particular, half sugar balls having a hydrophilic hemisphere surface covered with sugar moieties and a hydrophobic hemisphere surface covered with *n*-hexyl groups. The final section deals with some functions of glycopeptide surface-modified dendrimers, such as encapsulation of low molecular weight compounds, complexation of DNA, and molecular recognition as evaluated by the hemagglutination inhibition assay.

Keywords: Dendrimer; glycopolymer; molecular recognition; self-assembly; α-amino acid *n*-carboxyanhydride; ring-opening polymerization; macroinitiator.

1. INTRODUCTION

Naturally occurring glycoconjugates such as glycoproteins and glycolipids play important and essential roles in diverse biological functions such as intercellular communication, transportation of proteins from one part to another of a cell, and rendering proteins resistant to enzymatic degradation but allowing them to be recognized by certain protein receptor [1–3]. With the advent of precise synthetic methodologies in glycopeptide chemistry in recent years [4–7], a number of artificial glycoconjugates have been designed and synthesized [8–10]. There are two major purposes of synthesizing such artificial glycoconjugates. One is to provide structurally well-defined models necessary for biochemical and immunochemical studies on the molecular level. The other is to develop new polymeric materials displaying biological functions similar, or even superior, to those of naturally occurring glycoconjugates.

---

* To whom correspondence should be addressed. E-mail: okada@agr.nagoya-u.ac.jp

Of the artificial glycoconjugates based on synthetic polymers, vinyl polymers with pendant sugar moieties have been most widely investigated because of their facile preparation. In view of the fact that naturally occurring multiantennary oligosaccharides show a much higher affinity to multisubunit receptors than simple monosaccharide-receptor binding, a proper geometric arrangement of terminal sugar residues of natural multiantennary oligosaccharides appears to be a decisive factor for their highly specific molecular recognition ability [11]. Taking this into consideration, we have designed intersugar space-regulated globular glycoconjugates using a dendrimer skeleton [12]. At the outset, we synthesized maltose or lactose surface-modified dendrimers by the reactions of the corresponding sugar lactones with the primary amino groups at the periphery of poly(amido amine) dendrimers. We named these sugar-persubstituted dendrimers "Sugar Balls", because the surface of the globular poly(amido amine) dendrimers is covered with sugar residues covalently bonded to the dendrimer skeleton. Recently, different types of sugar-containing dendrimers have been reported by Roy et al. [13–17] and Stoddart et al. [18–21].

In this article, we describe briefly our recent investigation on the design, syntheses and functions of glycopeptide-modified dendrimers, based on the ring-opening reaction or oligomerization of D-glucose-, D-galactose-, or N-acetyl-D-glucosamine-carrying L-serine N-carboxyanhydrides (hereafter referred to as glycoNCA's).

2. SYNTHESIS OF GLYCOPEPTIDE-MODIFIED DENDRIMERS

Ring-opening polymerization of NCA's is the most frequently used methodology to synthesize homopolypeptides of α-amino acids. Chemistry of ring-opening polymerization of NCA's has been extensively studied [22, 23] and is well established at present. In contrast to the general tendency that anionic polymerization of NCA's carrying a bulky substituent is often accompanied by side reactions such as hydantoic acid formation [24], we have found that glycoNCA's **1** undergo n-hexyl amine-initiated polymerization without side reactions, irrespective of the bulky sugar moiety, to give the corresponding polypeptides **2** of a narrow molecular weight distribution [25–31]. Deacetylation of **2** with hydrazine under mild conditions gives poly(L-serine)s with controlled chain lengths having the corresponding sugar unit in each repeating unit. In addition, our kinetic analysis disclosed that the rate of initiation was a few thousand times as high as that of propagation in the polymerization of glycoNCA **1a** initiated with n-hexyl amine.

As described above, the primary amine-initiated ring-opening polymerization of glycoNCA's has two important features: First, the polymerization proceeds without side reactions to yield polypeptides of a controlled chain length. Secondly, the initiation is much faster than the propagation. Judging from these features, it is very likely that star-shaped glycopeptide-dendrimer conjugates **3** are formed, when we use primary amine-terminated poly(amido amine) dendrimer as a multifunctional macroinitiator for the polymerization of glycoNCA's.

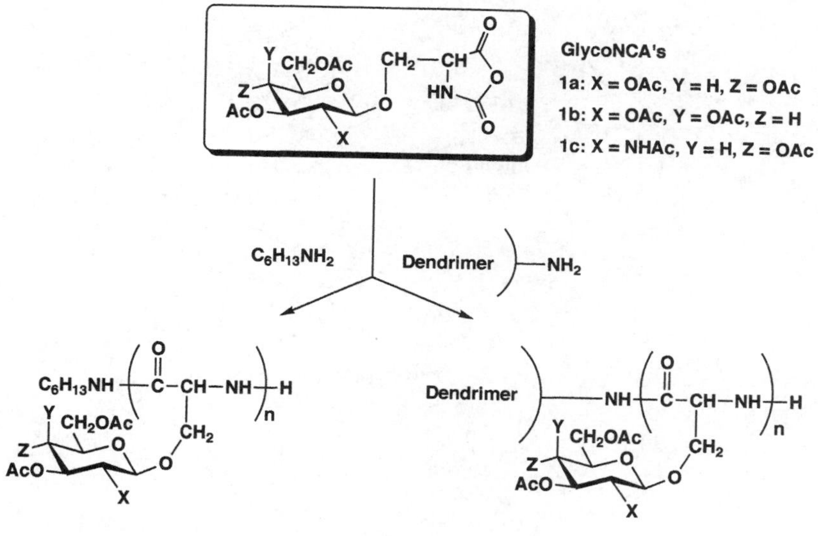

Scheme 1. Ring-opening polymerization of glycoNCA's initiated with primary amines.

Reaction of glycoNCA **1a** with primary amine-terminated poly(amido amine) dendrimer (**2**, G=3.0) was carried out at -30°C for 15 min using a slightly excess amount of glycoNCA to the primary amino group (Scheme 2). The M_w/M_n value of the product was close to that of the dendritic macroinitiator (M_w/M_n=1.02) [32]. The observed molecular weight agreed well with the theoretical value calculated on the assumption that one molecule of glycoNCA reacted with each primary amino group of the dendrimer skeleton. These results together with spectroscopic data indicate that each primary amino group on the dendrimer surface reacted with only one molecule of glycoNCA **1** under the reaction conditions to give a mono(glycopeptide)-type sugar ball precursor (**4a**, G=3.0), and that the initiation of the ring-opening polymerization of the glycoNCA **1** is much faster than the propagation. Mono(glycopeptide) surface-modified dendrimer was readily obtained by deprotecting the acetyl groups by the hydrazine treatment in methanol.

Oligomerization of glycoNCA **1a** with poly(amido amine) dendrimer in chloroform at 27°C gave acetyl-protected oligo(glycopeptide)-modified dendrimer **6a** in quantitative yields (Scheme 2) [33]. The M_w/M_n values determined by size exclusion chromatography (SEC) were 1.0_3-1.1_1, reflecting a rapid initiation/slow propagation system. Removal of the acetyl protecting groups yielded oligo(glycopeptide) surface-modified dendrimer **7a**. Scheme 2 shows a representative example using glycoNCA **1a** and poly(amido amine) dendrimer of the third generation (G=3.0). The above reaction occurred in a similar manner for any combinations of other NCA's including **1b** and **1c** and poly(amido amine) dendrimer or poly(trimethyleneimine) dendrimer of different generations, giving rise

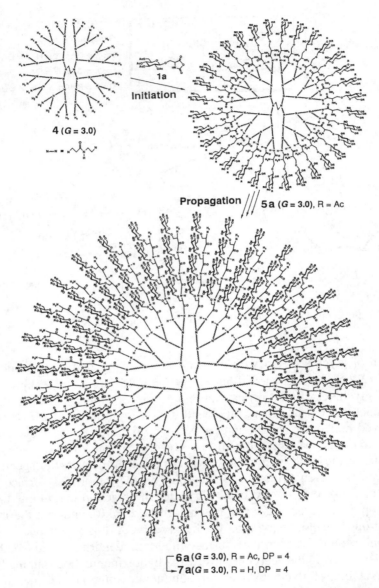

Scheme 2. Synthesis of mono(glycopeptide)-modified dendrimer and oligo(glycopeptide)-modified dendrimer from poly(amido amine) dendrimer and glycoNCA **1a**.

to the corresponding mono(glycopeptide) or oligo(glycopeptide) surface-modified dendrimers. For example, we synthesized poly(trimethyleneimine) dendrimer-*block*-(polysarcosine)$_{64}$ by ring-opening polymerization of sarcosine NCA with primary amine-terminated poly(trimethylene-imine) dendrimer of the fourth generation as a macroinitiator [34].

However, a question may arise as to whether the chain length of the polypeptide growing from a dendrimer surface can be controlled or not. To clarify this point, we have designed and synthesized a glycopeptide-type sugar ball possessing an ester linkage between each branch end of a poly(amido amine) dendrimer skeleton and the glycopeptide chain therefrom. All the ester linkages were cleaved by the treatment of the glycopeptide-type dendrimer with hydrazine in methanol at ambient temperature and the glycopeptide chains were isolated. Acetyl protecting groups were also removed by this treatment, and the isolated glycopeptide chains became insoluble in water when the chain lengths were relatively high, whereas glycopeptide of shorter chains were water-soluble. In the former case, they were acetylated again and their molecular weight distributions were evaluated by SEC in DMSO. For comparison, a similar experiment was carried out, under otherwise the identical conditions, using n-hexyl amine as an initiator which had been confirmed to induce living polymerization of glycoNCA's **1**. The observed molecular weight distributions ($M_w/M_n=1.1_4$-1.1_8) of the glycopeptides obtained in the dendrimer initiator system are nearly equal to those of the glycopeptides formed in the n-hexylamine initiator system. This means that the glycopeptide chains growing from the dendrimer surface are of a controlled chain length (Scheme 3).

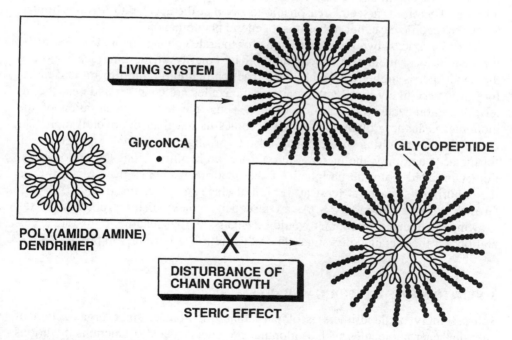

Scheme 3. Formation of oligo(glycopeptide)-modified dendrimer of a controlled chain length by living polymerization of glycoNCA with dendrimer as a macroinitiator.

Kinetic analysis has revealed an unusual acceleration phenomenon in the oligomerization of glycoNCA initiated with a dendrimer initiator in chloroform. Thus, the conversion of glycoNCA **1a** initiated with a poly(amido amine) dendrimer (G=5.0) was found to be appreciably higher than that initiated with *n*-hexylamine under similar conditions. The propagation rate constant at each degree of polymerization was estimated on the assumption that the rate of initiation was much higher than that of propagation. The analysis showed that the shorter the oligopeptide chain and the higher the generation of the dendrimer initiator, the greater was the acceleration. For example, the propagation rate constants within the interface region of 10Å from the poly(amido amine) dendrimer surface of the fifth and sixth generations were respectively about 30 times and 80 times higher than that in the *n*-hexylamine initiated system.

A similar acceleration phenomenon was found in the oligomerization of sarcosine NCA using a poly(amido amine) dendrimer with primary amino groups at the periphery as an initiator [34]. Since sarcosine NCA does not possess a sugar moiety, the observation means that the acceleration observed in the dendrimer initiator system is not ascribable to any specific interaction between the sugar moiety of glycoNCA and the dendrimer surface. In addition, a similar acceleration phenomenon was observed also in the oligomerization of GlycoNCA using poly(trimethyleneimine) dendrimer instead of poly(amido amine) dendrimer as an initiator. Therefore, it seems reasonable to say that the acceleration is not limited to a very specific case, but rather it is a general phenomenon.

There are at least two possible reasons for the rate enhancement: The first is the increase in local monomer concentration. The polarity of the interface region of the dendrimer is higher than that of the reaction medium (chloroform) and therefore, it is very likely that NCA monomer molecules are concentrated near the interface region where the reaction occurs. Thus, the local concentration of the monomer near the growing chain end becomes higher than the overall monomer concentration, resulting in the enhanced rate of oligomerization. Another reason for the acceleration is the increase in collision probability. Since the oligomerization is initiated from the periphery of the dendrimer, chains grow radially from the dendrimer surface, at least at the initial stage of the chain growth ("Radial-Growth Polymerization" [33]). As a consequence, the collision probability of the growing chain end to monomer would be enhanced, thus leading to the rate acceleration. Further detailed investigation on this subject is currently in progress.

3. SYNTHESIS OF HALF-SUGAR BALLS

Self-assembly of dendrimers is of particular importance, since organization of globular macromolecules such as globular proteins is a key to functions in various life systems. Artificial assemblies of dendrimers have been demonstrated by Fréchet *et al.* using a surface-block dendrimer [35] and by Zimmerman *et al.* using a dendrimer with a key moiety acting as a recognition and assembly site [36]. In addition, a variety of amphiphilic dendritic-linear block copolymers have been synthesized to investigate their self assemblies in solutions [37–42].

Synthesis of amphiphilic surface block dendrimers by the convergent method was reported by Fréchet and his colleagues [43]. We have recently developed two synthetic methodologies, i.e., a divergent/convergent joint method [44] and a divergent/divergent method [45], for sugar-containing amphiphilic surface block poly(amido amine) dendrimers (Scheme 4).

Starting from mono(benzyloxycarbonyl)-protected ethylenediamine, amphiphilic dendrimers of up to the fourth generation were synthesized by a divergent/divergent method. First, a hydrophobic hemisphere block was synthesized by the poly(amido amine) dendrimer construction and at a given generation, the terminal ester groups were allowed to react with n-hexylamine to synthesize a hydrophobic hemisphere block. Then, after deprotection of the core, the second hydrophilic hemisphere block was synthesized by the dendrimer construction followed by the reaction of glycoNCA **1c** with the primary amino groups at the periphery of the hemisphere block at a given generation. Amphiphilic surface block poly(amido amine) dendrimers **6** (G=2.0, 3.0 and 4.0) containing N-acetyl-D-glucosamine residues on the hemisphere surface were successfully synthesized by this procedure (Scheme 5).

Surface tension of aqueous solutions of these amphiphilic dendrimers was measured by the drop-weight method [46] to investigate their self assemblies. All these amphiphilic dendrimers showed two critical monomer concentrations (CMC). This is strongly indicative of two-step formation of micelles, i.e., small assemblies consisting of a few molecules are formed at the first step, and they are converted to larger micelles at higher concentrations.

Amphiphilic dendrimers containing hydroxyethyl groups instead of sugar groups as hydrophilic parts were synthesized and the surface tension of their aqueous solutions was measured by the same method. Also in these cases, two CMC's were observed. The CMC values of the hydroxyethyl-terminated dendrimers were lower than those of the glycopeptide surface-modified dendrimers of the same generation. The higher hydrophilicity and hence higher water-solubility of the glycopeptide surface-modified dendrimers compared with those

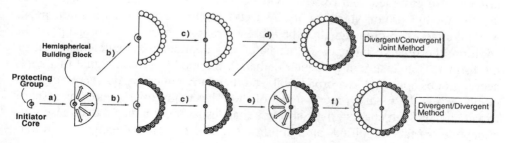

Scheme 4. Synthetic routes of amphiphilic surface block dendrimers.

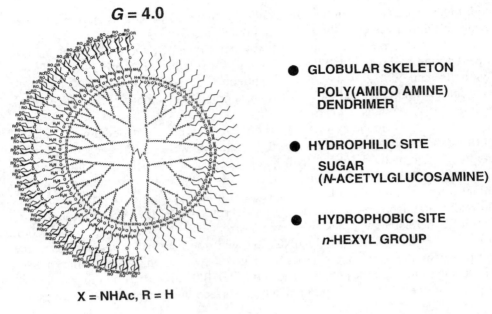

Scheme 5. Structure of amphiphilic surface block glycopeptide-modified dendrimer (G=4.0).

of the hydroxyethyl-terminated dendrimers are responsible for the higher CMC's for the former dendrimers. Judging from the findings that the dendrimers of the different structures formed micelles in two steps, properly designed amphiphilic dendrimers would form organized assemblies of a higher order. Unless the generation is very high, dendrimers are not stiff, but they are flexible enough to change their shapes according to environment. It would appear that the flexibility is an essential factor for the formation of organized assemblies from dendrimer molecules. In contrast to the symmetrical structures of conventional dendrimers, the aforementioned surface-block amphiphilic dendrimers are anisotropic, and hence they are expected to be promising elemental structures.

^{1}H NMR measurement of N-acetyl-D-glucosamine-modified surface block dendrimers also provided confirmative evidence for micelle formation. In heavy water, the methyl proton signal of the N-acetyl groups appeared as a sharp peak, while the methyl proton signal of the n-hexyl groups appeared as a broad peak. In addition, the relative intensity of the latter was appreciably smaller than that of the former. Conversely in perdeuterioethanol, the methyl proton signal of the N-acetyl groups appeared as a broad peak and its relative intensity decreased considerably, while the methyl proton signal of the n-hexyl groups appeared as a sharp triplet signal. Such characteristic spectral changes strongly indicate that these glycopeptide surface-modified amphiphilic dendrimers form micelles in water, whereas they form reverse micelle-like assemblies in ethanol.

4. SOME FUNCTIONS OF GLYCOPEPTIDE SURFACE-MODIFIED DENDRIMERS

One of the chief objects to design glycopeptide surface-modified dendrimers is to develop target-directing materials applicable for biomedical purposes on the basis of sugar residues at the periphery as a recognition site. For this purposes, we have done fundamental investigations on encapsulation of low molecular weight compounds and complexation of DNA.

Meijer *et al.* reported a "dendritic box", that is, poly(trimethyleneimine) dendrimer encapsulating low molecular weight compounds such as Bengal Rose inside its cavity by surface modification with a *tert*-butoxycarbonyl protected amino acid [47]. The glycopeptide surface-modified poly(amido amine) dendrimers described in the previous sections are characterized not only by their relatively large inner cavity but also by the sugar residues at the periphery acting as recognition sites. It is therefore highly expected that they are potential candidates for target-directing drug carriers, if low molecular weight compounds are effectively encapsulated into the inner cavity. As a preliminary approach to this object, encapsulation of sodium 8-anilinonaphthalene-1-sulfonate (ANS), a fluorescence probe, was examined using glycopeptide surface-modified dendrimers [48]. Thus, ANS and poly(amido amine) dendrimers (G=6.0–8.0) were mixed in methanol, and after evaporation of the solvent, the residue was allowed to react with peracetylated *N*-acetyl-D-glucosamine-carrying L-serine NCA in chloroform at -30°C. ANS molecules adhered onto the glycopeptide surface-modified dendrimer surfaces were thoroughly removed by repeated reprecipitation, and finally the protecting groups were removed by the treatment with hydrazine in methanol at ambient temperature (Scheme 6).

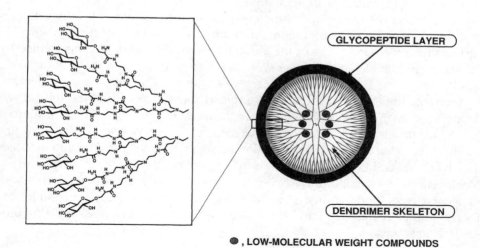

Scheme 6. Encapsulation of ANS by surface modification of amine-terminated poly(amido amine) dendrimer with glycoNCA.

The resulting glycopeptide surface-modified dendrimer showed fluorescence peak at 495nm, suggesting the entrapment of ANS in the cavity. This was further confirmed by distracting the dendrimer skeleton by alkaline hydrolysis and quantitatively determining the ANS molecules liberated from the dendrimer by means of UV spectroscopy. These analyses showed that the higher the generation, the greater was the number of ANS molecules encapsulated inside a dendrimer molecule. The UV and fluorescence analyses together with gel filtration chromatography of the capsule indicated the presence of ANS molecules in the internal cavity of the mono(glycopeptide)-modified dendrimer capsules.

Poly(amido amine) dendrimers contain a defined number of amino groups on the surface of the polymer that are positively charged at physiologic pH and therefore these molecules could interact with biologically relevant polyanions including nucleic acids. Gene transfection was investigated using poly(amido amine) dendrimer [49–52]. Highly efficient transfection of a broad range of eukaryotic cells and sell lines was achieved with minimal cytotoxicity using the DNA/dendrimer complexes. The capability of a dendrimer to transfect cells appeared to depend on the size, shape, and number of primary amino groups on the surface of the dendrimer [51].

As a first step toward application as a gene carrier, we have investigated complexation of plasmid DNA with glycopeptide surface-modified dendrimers. The complex formation between them was ascertained by electrophoresis, CD spectroscopy, and fluorescence measurement. In addition, thermal stability and nuclease resistance of DNA were enhanced by the complexation with the glycopeptide-modified dendrimers.

To evaluate molecular recognition ability of the aforementioned glycopeptide surface-modified dendrimers, interactions of N-acetyl-D-glucosamine-carrying dendrimers with wheat germ agglutinin (WGA) lectin which specifically recognizes an N-acetyl-D-glucosamine moiety were investigated by a hemmagglutination inhibition assay. The minimum inhibition concentrations of the glycopeptide surface-modified dendrimers of the third and fourth generations as represented by the concentration of sugar residues were found to be lower than those of the same dendrimers of the fifth and sixth generations with a higher surface sugar density. Furthermore, this trend was found more remarkable for a series of the star-shaped dendrimers having oligopeptide chains of the average degree of polymerization of 3.5-3.7 on their surfaces. Therefore, the recognition ability seems to be strengthened by the favorable spatial orientation of sugar residues for recognition rather than by the so-called cluster effect. The aforementioned glycopeptide-type surface block dendrimers and also the ANS encapsulated glycodendrimers retained a high molecular recognition ability. It is therefore expected that the glycopeptide surface-modified dendrimers may be a promising candidate for target-directing biomedical materials.

5. CONCLUDING REMARKS

We have described the outline of the synthesis and functions of glycopeptide surface-modified dendrimers. These are typical examples of functionalization of dendrimers. Optimum design of functional dendrimers of this type would lead to new materials applicable for biomedical purposes, such as nanocapsules, target-directing gene-vectors, and drug delivery systems.

Nearly fifteen years have passed since the first paper on poly(amido amine) dendrimers by Tomalia *et al.* [53] was published. Currently, importance of dendrimers has been increasingly well-recognized not only in fundamental sciences but also in practical applications. In particular, much interest is now being paid to creation of new functional materials by effectively utilizing the characteristic architectures and properties of dendrimers that differ from those of conventional linear polymers, i.e., a globular shape that varies dynamically depending on the environment, functional groups at the periphery oriented in a geometrically controlled fashion, a relatively large inner cavity accommodating small molecules, etc. Above all, dendrimers themselves could be regarded as unimolecular micelles organized by covalent bonds. Supramolecular organization of such dendrimer molecules as building blocks would lead to potential applications in different areas of materials science such as molecular electronics, information processing, biomolecular engineering and so on.

Acknowledgments

The authors sincerely thank Professor T. Imae of Nagoya University for her helpful discussion and our students, K. Itoh, H. Noda, A. Yamamoto, T. Kobayashi, and M. Ohno for their continuous collaboration. The present work was financially supported in part by Grant-in-Aid (Nos. 09450349 and 11121216) from the Ministry of Education, Science, Sports, and Culture of Japan.

REFERENCES

1. Y. C. Lee and R. T. Lee in: *Glycoconjugates*, H. J. Allen, and E. C. Kisailus (Eds), p. 121. Marcel Dekker, New York (1992).
2. A. Varki, *Glycobiology*, **3**, 97, (1993).
3. R. Dwek, *Chem. Rev.*, **96**, 683 (1996).
4. H. G. Garg and R. W. Jeanloz, *Adv. Carbohydr. Chem. Biochem.*, **43**, 135 (1985).
5. H. G. Garg, K. von dem Bruch, and H. Kunz, *Adv. Carbohydr. Chem. Biochem.*, **50**, 277 (1994).
6. S. Cohen-Ansfield and P. T. Lansbury, *J. Am. Chem. Soc.*, **115**, 10531 (1993).
7. C.-H. Wong, R. H. Halcomb, Y. Ichikawa, and T. Kajimoto, *Angew. Chem. Int. Ed. Engl.*, **34**, 412 and 521 (1995).
8. R. T. Lee and Y. C. Lee, in: *Neoglycoconjugates: Preparation and Applications* Y. C. Lee and R. T. Lee (Eds), p. 23. Academic Press, San Diego (1994).
9. M. Okada, in: *Polymeric Materials Encyclopedia*, J. C. Salamone (Ed.), p. 2834. CRC Press, Boca Raton (1996).
10. R. Roy, *Top. Curr. Chem.*, **187**, 241 (1997).
11. Y. C. Lee and R. T. Lee, *Acc. Chem. Soc.*, **28**, 321 (1995).
12. K. Aoi, K. Itoh, and M. Okada, *Macromolecules*, **28**, 5391 (1995).

13. D. Zanini, W. K. C. Park, and R. Roy, *Tetrahedron Lett.*, **36**, 7383 (1995).
14. D. Zanini and R. Roy, *J. Am. Chem. Soc.*, **119**, 2088 (1996).
15. M. Llinares and R. Roy, *Chem. Commun.*, 2119 (1997).
16. D. Pagel and R. Roy, *Bioconjugate Chem.*, **8**, 714 (1997).
17. R. Roy and J. M. Kim, *Angew. Chem. Int. Ed.*, **38**, 369 (1999).
18. P. R. Ashton, S. E. Boyd, C. L. N. Brown, N. Jayaraman, S. A. Nepogodiev, and J. F. Stoddart, *Chem. Eur. J.*, **2**, 1115 (1996).
19. P. R. Ashton, S. E. Boyd, C. L. Brown, S. A. Nepogodiev, E. W. Meijer, H. W. I. Peerlings, and J. F. Stoddart, *Chem. Eur. J.*, **3**, 974 (1997).
20. N. Jayaraman, S. A. Neogodiev, and J. F. Stoddart, *Chem. Eur. J.*, **3**, 1193 (1997).
21. P. R. Ashton, E. F. Hounsell, N. Jayaraman, T. M. Nilsen, N. Spencer, and M. Young, *J. Org. Chem.*, **63**, 3429 (1998).
22. Y. Imanishi, in: *Ring-Opening Polymerization*, K. J. Ivin and T. Saegusa (Eds), p. 523. Elsevier, London (1984).
23. H. R. Kricheldorf, *α-Amino Acid N-Carboxyanhydrides and Related Heterocycles*, p. 59. Springer, Berlin (1987).
24. H. R. Kricheldorf, in: *Comprehensive Polymer Science*, G. C. Eastmond, A. Ledwith, S. Russo, and P. Sigwalt (Eds), Vol. 3, p. 531, Pergamon, Oxford (1989).
25. K. Aoi, K. Tsutsumiuchi, and M. Okada, *Macromolecules*, **27**, 875 (1994).
26. M. Okada and K. Aoi, *Macromolecular Report*, **A32** (Suppls 5&6), 907 (1995).
27. K. Tsutsumiuchi, K. Aoi, and M. Okada, *Macromol. Rapid Commun.*, **16**, 749 (1995).
28. K. Aoi, K. Tsutsumiuchi, E. Aoki, and M. Okada, *Macromolecules*, **29**, 4456 (1996).
29. K. Tsutsumiuchi, K. Aoi, and M. Okada, *Macromolecules*, **30**, 4013 (1997).
30. M. Okada, K. Aoi, and K. Tsutsumiuchi, *Proc. Japan Acad.*, **73**, Ser. B, 205 (1997).
31. K. Aoi, K. Tsutsumiuchi, and M. Okada, *Kobunshi Ronbunshu*, **54**, 649 (1997).
32. K Aoi, K. Tsutsumiuchi, A. Yamamoto, and M. Okada, *Macromol. Rapid Commun.*, **19**, 5 (1998).
33. K. Aoi, K. Tsutsumiuchi, A. Yamamoto, and M. Okada, *Tetrahedron*, **53**, 15415 (1997).
34. K. Aoi, T. Hatanaka, K. Tsutsumiuchi, M. Okada, and T. Imae, *Macromol. Rapid. Commun.*, **20**, 378 (1999).
35. J. M. J. Fréchet, *Science*, **263**, 1710 (1994).
36. S. C. Zimmerman, F. Feng, D. E. C. Reichert, and S. V. Kolotuchin, *Science*, **271**, 1095 (1996).
37. T. M. Chapman, G. L. Hillyer, E. J. Mahan, and K. A. Shaffer, *J. Am. Chem. Soc.*, **116**, 1119 (1994).
38. J. M. J. Fréchet and I. Gitsov, *Macromol. Symp.*, **98**, 441 (1995) and the references cited therein.
39. M. R. Leduc, C. J. Hawker, J. Dao, and J. M. J. Fréchet, *J. Am. Chem. Soc.*, **118**, 11111 (1996).
40. J. C. M. van Hest, D. A. P. Delnoye, M. W. P. L. Baars, C. Elissen-Roman, M. H. P. van Gendeen, and E. W. Meijer, *Chem. Eur. J.*, **2**, 1616 (1996) and the references cited therein.
41. K. Aoi, A. Motoda, M. Okada, *Macromol. Rapid Commun.*, **18**, 945 (1997).
42. J. Iyer, K. Fleming, and P. Hammond, *Macromolecules*, **31**, 8757 (1998).
43. C. J. Hawker, K. L. Wooley, J. M. J. Fréchet, *J. Chem. Soc., Perkin Trans.* I., 1287 (1993).
44. K. Aoi, K. Itoh, and M. Okada, *Macromolecules*, **30**, 8072 (1997).
45. K. Aoi, A. Motoda, M. Ohno, K. Tsutsumiuchi, M. Okada, and T. Imae, *Polym. J.*, **31**, 1021 (1999).
46. H. Okuda, T. Imae, and S. Ikeda, *Colloid Surfaces*, **27**, 187 (1987).
47. J. F. G. A. Jansen, E. M. M. de Brabander-van den Berg, and E. M. Meijer, *Science*, **266**, 1226 (1994).
48. K. Tsutsumiuchi, K. Aoi, and M. Okada, *Polym. J.*, **31**, 935 (1999).
49. J. Haensler and F. C. Szoka, Jr., *Bioconjugate Chem.*, **4**, 372, (1992).
50. M. X. Tang, G. T. Redemann, and F. C. Szoka, Jr., *Bioconjugate Chem.*, **7**, 703 (1996).
51. J. Kukovska-Lattalo, A. U. Bielinska, J. Johnson, R. Spindler, D. A. Tomalia, and J. R. Baker, Jr., *Proc. Matl. Acad. Sci., U. S. A.*, **93**, 4897 (1996).
52. A. U. Bielinska, J. Kukovska-Lattalo, J. Johnson, D. A. Tomalia, and J. R. Baker, Jr., *Nucleic Acid Res.*, **24**, 2176 (1996).
53. D. A. Tomalia, H. Baker, J. R. Dewald, M. Hall, G. Kallos, S. Martin, J. Roeck, J. Ryder, and P. Smith, *Polym. J.*, **17**, 117 (1985).

"Perfectly branched" hyperbranched polymers

GERHARD MAIER,[1,*] CHRISTINA ZECH,[1] BRIGITTE VOIT[2] and HARTMUT KOMBER[2]

[1]Technische Universität München, D-85747 Garching, Germany
[2]Institute of Polymer Research IPF Dresden e.V., D-01069 Dresden, Germany

Abstract—A strategy for the synthesis of hyperbranched polymers with a degree of branching DB = 1 ("perfectly branched" hyperbranched polymers) based on AB_2 monomers is proposed. It requires a monomer with 2 B-groups with coupled reactivity. The "criss-cross"-cycloaddition reaction between azines and unsaturated compounds was selected as suitable reaction, and consequently the synthesis of AB_2 monomers based on aromatic aldazines with an unsaturated function as A-group was targeted. The preparation of 4-methoxy-4'-(3"-N-maleimidopropoxy)-benzaldazine was completed successfully. Molar mass determination of the resulting polymer indicated values for $\overline{M}_n = 5700$ g · mol^{-1} by GPC with polystyrene calibration and $\overline{M}_n = 16700$ g · mol^{-1} by GPC with light scattering detector. NMR spectroscopy did not show any defect structures (linear repeating units). MALDI-TOF mass spectroscopy was unsuccessful. No suitable matrix for laser desorption could be found up to now. Characterization of the physical properties is in progress.

Keywords: Hyperbranched polymers; degree of branching; perfectly branched; criss-cross-cycloaddition; AB_2-monomers; azine.

1. INTRODUCTION

Hyperbranched polymers are often compared to dendrimers because of certain similarities, such as the highly branched structure, large number of functional groups, good solubility, and low viscosity [1–4]. However, there are also important differences [2–4]. The most prominent ones are the irregular architecture of the hyperbranched polymers and the resulting statistical distribution of the functional groups all over the molecules in contrast to the perfectly controlled structure of the dendrimers. By definition [2–4], the degree of branching of dendrimers is DB = 1, since the sequential synthesis in consecutive steps requires complete conversion in each step and hence complete branching at each repeating unit. In contrast, the degree of branching of hyperbranched polymers prepared from a monomer of the AB_n-type is determined purely by statistics, when certain assumptions are made. These are: equal reactivity and

---

* To whom correspondence should be addressed. Technische Universität München, Lichtenbergstraße 4, D-85747 Garching, Germany.

equal accessibility of all end groups (terminal or linear; near the focal group or at the "outside"), exclusion of loop formation, reactions exclusively between A- and B-groups (no reaction between two A- or two B-groups). For an AB_2-monomer, only half of the B-groups can be converted. Based on these assumptions a degree of branching of DB = 0.5 is predicted by calculations for hyperbranched polymers prepared in batch reactions [1, 5–8].

Attempts to influence the degree of branching of hyperbranched Polymers as well as the theoretical background of such strategies have received considerable interest recently. Continuous monomer addition [6–8], addition of polyfunctional core molecules [5, 8–13], introduction of preformed branch points [14, 15], and neighbor group effects (activation of the second B-group by reaction of the first one) [16, 17] have been studied. However, while all these methods do indeed allow to increase DB to a certain extent, none of them absolutely excludes the formation of linear repeating units.

At this point it is important to note that a degree of branching of DB = 1 is not necessarily indicative of a controlled, symmetric architecture. As Kim [18] pointed out, for a dendritic structure based on an AB_2 monomer, at a degree of polymerization corresponding to generation 4 of a perfect dendrimer, there are already more than 500 isomeric structures with DB = 1. At generation 5, there are already more than 500 million isomers. Feast [16] recently drew attention to the fact that the geometry of some of these is far from the spherical shape of dendrimers (Scheme 1, "quasi-linear" structure).

Despite this situation it is of interest to study the possibility to design a one-step batch reaction, which ensures complete branching (i.e. DB = 1) of a hyperbranched polymer.

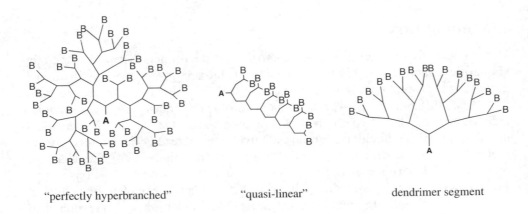

"perfectly hyperbranched"   "quasi-linear"   dendrimer segment

Scheme 1. Some isomeric structures [16] of an AB_2-based polymer with DB = 1.

2. REQUIREMENTS FOR A SUITABLE TYPE OF REACTION

From the viewpoint of synthesis, it is clear that complete branching can only be achieved when the reactions of the B-groups in an AB_n-monomer are coupled to each other. For an AB_2 monomer, the following set of requirements should be sufficient:

(1) The reaction of the first B-group of a repeating unit with an A-group must be reversible.
(2) The intermediate formed in this reaction must be unstable.
(3) On reaction of the second B-group of the repeating unit, the intermediate must be transformed into a stable final product.
(4) This second step must be irreversible under the reaction conditions employed.
(5) Side reactions must be avoided.

In a scenario as described by the above set of requirements, the instability of the intermediate formed by the reaction of the first B-group results in the absolute necessity of a consecutive reaction step. If side reactions are absent, there are only two possibilities: either the first step is reversed, and the starting materials reform, or the second B-group reacts with another A-group in order to form a stable end product. Thus, either both B-groups of a repeating unit react, and a branched unit is formed, or no B-group reacts at all, and a terminal unit is formed. Consequently, the resulting polymer consists exclusively of branched and terminal units (besides the focal group). According to all definitions, this corresponds to $DB = 1$.

The only reaction fulfilling these requirements we have found so far is the "criss-cross"-cycloaddition reaction [19–23] between azines and unsaturated compounds. It is shown in Scheme 2 using a benzaldehyde azine derivative and an isocyanate as example.

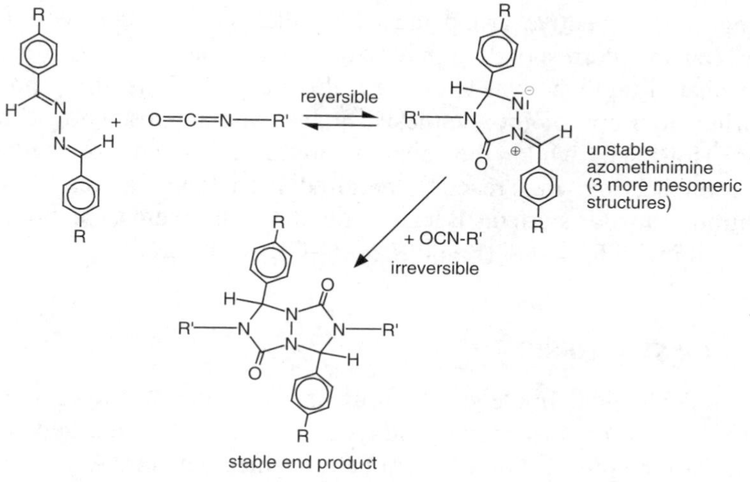

Scheme 2. "Criss-cross"-cycloaddition reaction [19–23].

3. CHOICE OF THE AB$_2$ MONOMER

The azine is difunctional in the "criss-cross"-cycloaddition reaction, and therefore it can serve as the two B-groups in an AB$_2$ monomer. The A-group can be basically any unsaturated function. Based on our earlier work on linear polymers [22, 23] prepared by this type of reaction we chose the following monomers (Scheme 3):

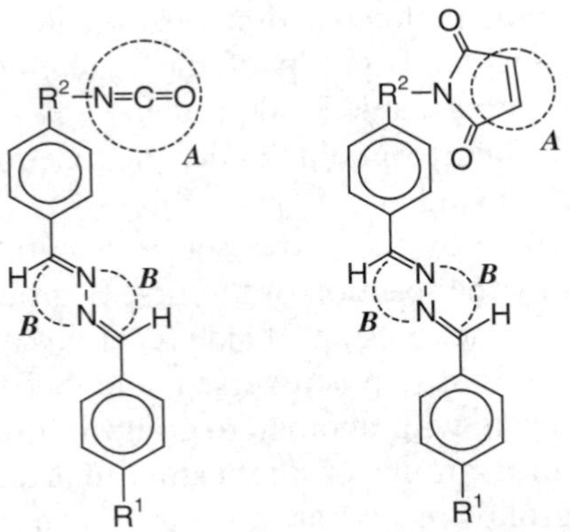

Scheme 3. AB$_2$ monomers for hyperbranched polymers by "criss-cross"-cycloaddition.

The synthesis of monomer **1** requires the use of a protective group for the isocyanate, since the isocyanate would react with the hydrazine, which is necessary for the formation of the azine. This synthesis proved to be quite challenging. The protective group must be rather stable, since the nucleophilic hydrazine (or the corresponding hydrazone) can react with labile protective groups. A phenyl urethane was chosen for this purpose. This group proved to be stable during all steps of the synthesis, and it was supposed to be removable thermally. However, while the phenyl urethane started to regenerate the isocyanate on heating, this reaction required high temperatures, at which the cycloaddition already started. Because of these problems, monomer **2** was targeted finally with R^1 = -OCH$_3$ and R^2 = -O-CH$_2$-CH$_2$-CH$_2$-.

4. MONOMER SYNTHESIS

The use of the maleic imide as A-group in monomer **2** allows the synthesis without the need for a protective group strategy. However, the general problem associated with the preparation of asymmetric azines remains. Asymmetric azines tend to equilibrate in the presence of water under the conditions of azine formation. This results in the inevitable formation of both symmetric azines besides the desired asymmetric one. Because of the symmetry, one of the symmetric azines usually crystallizes easily, and hence all workup procedures result in the isolation of this symmetric azine due to a shift of the equilibrium. Rigid exclusion of water is required to avoid this. In addition, a synthesis method has to be employed, which does not involve the formation of water as byproduct of the condensation of the aldehydes with hydrazine. After investigation of a number of potentially suitable strategies, the one shown in Scheme 4 proved to be successful [24].

Alkoxy substituents at the phenyl rings of the azine increase the electron density and hence the rate of the "criss-cross"-cycloaddition reaction. Therefore, 4-methoxy benzaldehyde is used for one part of the azine, and the A-group is introduced via a propyloxy function in the other part of the azine. The 4-(3'-hydroxy)propoxybenzaldehyde is converted into the Schiff-base with aniline, and 4-methoxy benzaldehyde is transformed into the hydrazone.

The condensation of the hydrazone with the Schiff-base (imine) instead of the aldehyde itself avoids the formation of water during the reaction. Instead, aniline is the condensation product. Finally, it proved to be useful to introduce the maleic imide in the last step via a Mitsunobu-type reaction. The use of polymer bound triphenyl phosphine allows easy separation of the phosphine oxide during workup. Also, the use of di-*tert.*-butyl azodicarboxylate instead of the more common diethyl derivative allows removal of the corresponding hydrazide by sublimation.

NMR spectroscopy and elemental analysis are in accordance with the expected structure of monomer **2**. Mass spectroscopy (EI) was used to check the absence of the symmetric azines. Scheme 5 shows the mass spectrum and the proposed fragmentation pattern. The molar mass of monomer **2** is 391 g · mol^{-1}.

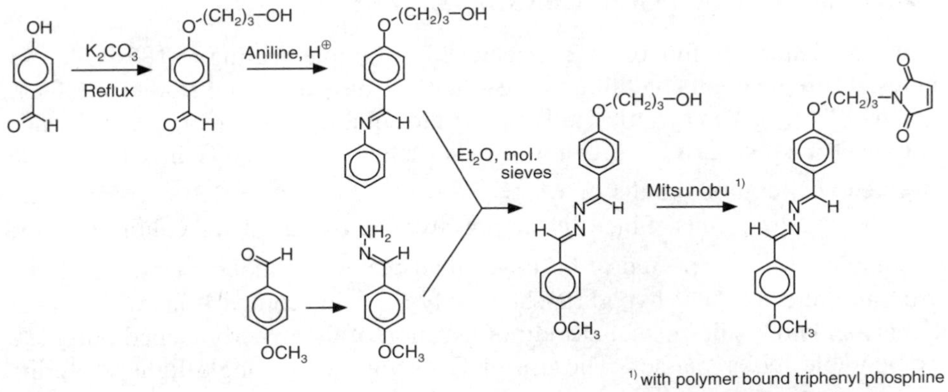

Scheme 4. Synthesis of monomer **2**.

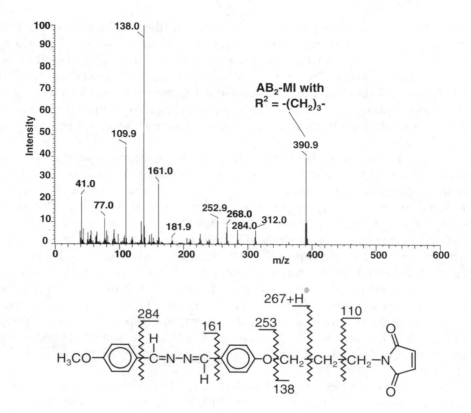

Scheme 5. EI-mass spectrum and proposed fragmentation pattern of monomer **2**.

5. POLYMER SYNTHESIS AND CHARACTERIZATION

The polymerization of monomer **2** is achieved by simple heating to 185 °C for 15 minutes. These reaction conditions were adapted from our earlier studies on linear polymers [22, 23]. The resulting polymers **3** are soluble in many organic solvents, including THF and CHCl$_3$. Molar mass determination by GPC in CHCl$_3$ with polystyrene calibration results in $\overline{M}_n$ = 5700 and $\overline{M}_w/\overline{M}_n$ = 2.2, corresponding to 14–15 repeating units. Since linear polystyrene was used for calibration, the $\overline{M}_n$ value could be expected to be lower than the actual molar mass. The more dense structure of the hyperbranched polymers in comparison with linear polymers should result in smaller radii of gyration of the hyperbranched polymers at comparable molar masses. The use of GPC with a low angle light scattering detector gave a value of $\overline{M}_n$ = 16700, corresponding to 43 repeating units. The use of samples of known concentration allows in principle the online determination of the refractive index increment and hence, in combination with the distribution obtained by GPC, the determination of absolute molar masses.

However, at these low masses detection by light scattering may be underestimating the lower mass fractions, and the real mass may be lower than indicated by this method. Determination of molar masses by absolute methods is in progress.

The number of repeating units can be estimated from the ^{1}H NMR spectrum. The ratio of terminal units and branched units is 2:1 for a trimer and approaches 1:1 for $\overline{P}_n \to \infty$. Thus, from the proton NMR spectrum in DMSO-d_6 14 repeating units are calculated, which may be indeed close to the real value.

The determination of defect structures is important, since any defects will introduce linear units. The most revealing method to find defects in the present case would be MALDI-TOF mass spectroscopy. If the polymerization proceeds as intended, in each addition step two monomer units must be added. The presence of linear structures would require the addition of only one monomer unit. Thus, in the MALDI-TOF mass spectrum only one series of lines, separated by the mass of two monomer units should be seen. Unfortunately, we were not able to obtain an MALDI-TOF mass spectrum so far. All matrixes used so far did not result in the detection of any signal, even at the highest laser intensity.

Scheme 6 shows possible defect structures formed by side reactions of the unstable azomethinimine intermediate.

The formation of **4** by sigmatropic 1,4-H-shift is unlikely because of the transoid conformation fixed due to the exocyclic structure of the azomethinimine. The cyclic transition state required for a pericyclic sigmatropic shift is not

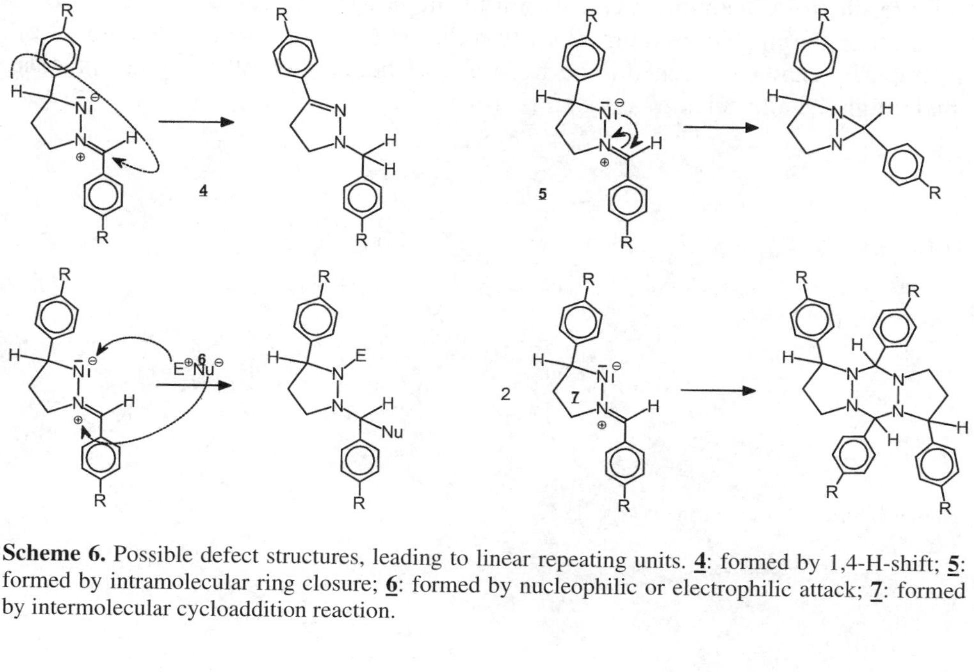

Scheme 6. Possible defect structures, leading to linear repeating units. **4**: formed by 1,4-H-shift; **5**: formed by intramolecular ring closure; **6**: formed by nucleophilic or electrophilic attack; **7**: formed by intermolecular cycloaddition reaction.

possible. Intramolecular ringclosure to the diaziridine **5** should be reversible at the high reaction temperature employed. The addition of nucleophiles and electrophiles (structure **6**) can be excluded by working with dry solvents and pure starting materials under inert atmosphere. The [4+4]-cycloaddition to the tricyclic structure **7** is thermally not allowed. Thus, these defect structures should not be present to a large extent.

In principle, one should be able to identify the defect structures by ^{13}C NMR spectroscopy. The situation is complicated here by the presence of a large number of isomeric structures of the repeating unit. Scheme 7 shows the trimer as a model for the repeating units in the polymer.

The central tetracyclic system formed by the cycloaddition reaction contains 6 chiral centers. In addition, the central NN-bond can exist in *cis* or in *trans* configuration. The maleic imide rings must be fused to the central bicyclic ring system in *cis* configuration, but with respect to each other, they can be in *syn* or *anti* position. When the central NN-bond is *cis* configured, there is also *endo* and *exo* isomerism of the two maleic imide rings. Thus, a large number of isomers is possible. However, due to steric hindrance only a few can exist in reality. Based on the ^{13}C NMR spectrum of the polymer and model compounds, 4 isomers have been identified so far. A publication describing the exact assignment procedure is in preparation. Scheme 8 shows the ^{13}C NMR spectrum, and Scheme 9 the structure of the hyperbranched polymer **3**.

Besides the signals of the repeating unit (various isomers), there are some minor signals, marked **A, a, b, c, d,** and **e**. **A** results form the *tert.*-butyl groups of residual di-*tert.*-butyl azodicarboxylate from the monomer synthesis. The signals **a** to **e** result from incomplete conversion of the last Mitsunobu coupling reaction to introduce the maleic imide group. From the ^{1}H NMR spectrum, the amount of this impurity can be estimated to be 1.3 mol-%. There are no other signals present which might be attributed to any defect structures.

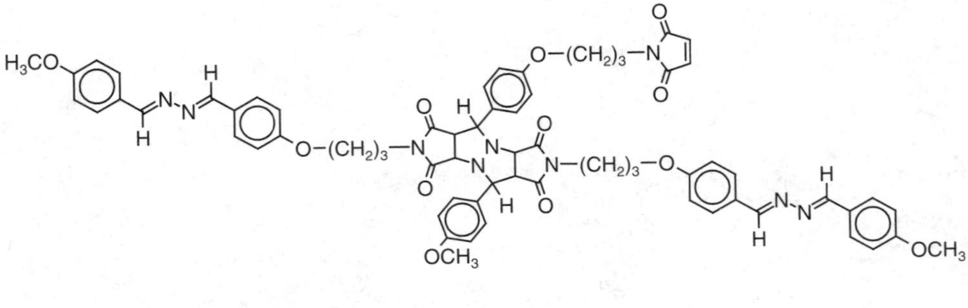

Scheme 7. Structure of the trimer.

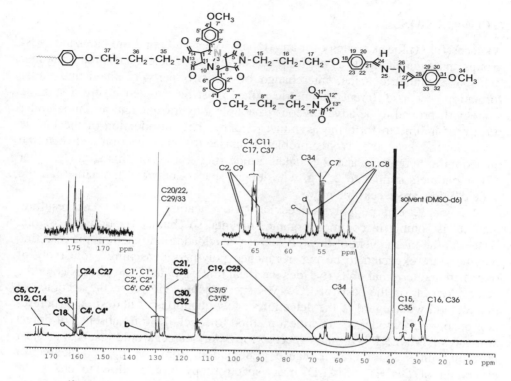

Scheme 8. ^{13}C NMR spectrum of the hyperbranched polymer **3**.

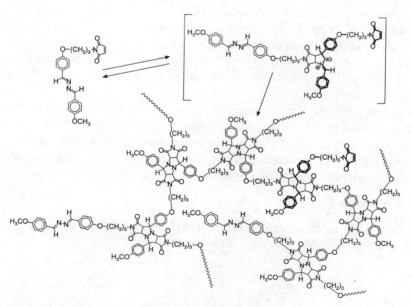

Scheme 9. Structure of the hyperbranched polymer **3**.

6. CONCLUSIONS

A successful synthesis strategy for an AB_2 monomer with an aromatic azine as B-groups and a maleic imide as A-group was developed. Since the desired monomer **2** is an asymmetric azine, interchange of the two party of the azine under formation of two different symmetric azines must be avoided, using a suitable strategy. It proved to be advantageous to couple a hydrazone and an imine under release of aniline to form the asymmetric azine. The introduction of the maleic imide in the last step avoids problems caused by the reaction between the nucleophilic hydrazone and the maleic imide ring. Based on this strategy, other AB_2 monomers similar to **2**, e.g. with longer spacers between the azine and the maleic imide, are accessible now.

Monomer **2** can be polymerized easily by heating in bulk. The resulting material is soluble in common organic solvents, so characterization is possible. Based on the mechanism of the "criss-cross"-cycloaddition reaction, the resulting polymers are expected to be hyperbranched polymers consisting exclusively of branched and terminal units (besides the focal unit). No defect structures could be detected so far on the basis of ^{13}C NMR spectroscopy. However, the accuracy of this method with regard to the detection of low quantities of undesired structures is probably only moderate. The ideal method to check the total absence of defect structures is MALDI-TOF mass spectroscopy. Any defect would result in a linear repeating unit, and therefore cause a deviation from the desired degree of branching of DB = 1. MALDI-mass spectroscopy would allow to check the presence of an additional series of oligomers besides the expected one. If the reaction proceeds as expected, the oligomer signals should be separated by the mass of two monomer units. A defect (linear repeating unit) would cause an additional series with peaks separated by only on monomer mass unit. Unfortunately, so far it was not possible to obtain a signal in MALDI-TOF mass spectroscopy. Even at the highest laser intensity, the polymer chains were not desorbed and ionized. The search for a suitable matrix material and ionization conditions is continued.

At present, we have no indication that the material obtained from the polymerization of monomer **2** contains linear defect structures. However, we are still searching for final proof. A question we are studying at present is the solution viscosity behavior of the material. It remains to be seen if the solution viscosity is more close to a dendrimer, or to a conventional hyperbranched polymer.

Acknowledgements

The authors wish to thank the Deutsche Forschungsgemeinschaft and the Fonds der Chemischen Industrie for financial support.

EXPERIMENTAL

*Synthesis of the AB$_2$-monomer **2***

Synthesis of 4-(3'-hydroxy)propoxybenzaldehyde. A solution of 35.00 g (0.25 mol) of 3-bromo-1-propanol and 27.48 g (0.225 mol) of 4-hydroxybenzaldehyde in 600 ml of dry THF was heated to reflux for 20 h in the presence of 103.64 g (0.75 mol) of K$_2$CO$_3$, which was suspended in the solution by vigorous stirring. After cooling to room temperature the solid components were filtered off and the solvent was evaporated under reduced pressure. Excess 3-bromo-1-propanol was removed at 80°C in vacuum (5 · 10^2 Pa). The colored viscous product was obtained in 98% yield and was converted into the Schiff-base without further cleaning.

^{1}H NMR (DMSO-d$_6$): δ = 1.89 (-C$\underline{H}$$_2$-, 2H, m), 3.57 (-C$\underline{H}$$_2$-OH, 2H, t), 4.16 (-OC$\underline{H}$$_2$-, 2H, t), 4.60 (-O$\underline{H}$, 1H, s), 7.10 ($\underline{H}$3, 2H, d), 7.85 ($\underline{H}$2, 2H, d), 9.86 (-C$\underline{H}$O, 1H, s).

^{13}C NMR (DMSO-d$_6$): δ = 32.2 (-$\underline{C}$H$_2$-), 57.3 (-$\underline{C}$H$_2$-OH), 65.4 (-O$\underline{C}$H$_2$-), 115.1 ($\underline{C}$3), 129.7 ($\underline{C}$1-CHO), 132.0 ($\underline{C}$2), 163.9 ($\underline{C}$4-OCH$_2$-), 191.4 (-$\underline{C}$HO).

IR (film): 3421 (OH), 3100-2882 (CH), 2830/2743 (C$\underline{H}$=O), 1684 (C=O), 1601/1576/1509 (C=C), 1261/1161 (C-O-C), 1157 (C-O), 833 (CH) cm^{-1}.

Synthesis of the Schiff-base. 39.73 g (0.22 mol) of 4-(3'-hydroxy)-propoxybenzaldehyde and catalytic amounts of para toluene sulfonic acid mono hydrate were dissolved in 500 ml dry diethyl ether. To the stirred solution 27.24 g (0.29 mol) of freshly distilled aniline were added dropwise over a period of 30 min. The mixture was refluxed for 18 h and after cooling to room temperature stirred for another 4 h. The precipitated amorphous white powder was isolated by filtration and washed with dry diethyl ether for several times. The combined solutions were concentrated in vacuum and the precipitate was isolated and washed with dry diethyl ether. The white product had a melting point of 86.9°C (yield: 70 %).

^{1}H NMR (CDCl$_3$): δ = 1.79 (-O$\underline{H}$, 1H, s broad), 2.10 (-C$\underline{H}$$_2$-, 2H, q), 3.88 (-C$\underline{H}$$_2$-OH, 2H, t), 4.19 (OC$\underline{H}$$_2$-, 2H, t), 6.99 ($\underline{H}$3, 2H, d), 7.19-7.24 ($\underline{H}$3'/$\underline{H}$4', 3H, m), 7.38 ($\underline{H}$2', 2H, m), 7.86 ($\underline{H}$2, 2H, d), 8.38 (-N=C$\underline{H}$-, 1H, s).

^{13}C NMR (CDCl$_3$): δ = 31.9 (-$\underline{C}$H$_2$-), 60.0 (-$\underline{C}$H$_2$-OH), 65.6 (-O$\underline{C}$H$_2$-), 114.7 ($\underline{C}$3), 120.8 ($\underline{C}$2'), 125.6 ($\underline{C}$4'), 129.2 ($\underline{C}$3'), 129.2 ($\underline{C}$1-CH=N-), 130.6 ($\underline{C}$2), 130.6 ($\underline{C}$1'-N=CH-), 159.7 (-$\underline{C}$H=N-), 161.6 ($\underline{C}$4-OCH$_2$-).

IR (KBr): 3192 (OH), 3100-2850 (CH), 1625 (C=N), 1604/1590/1514 (C=C), 1263/1171 (C-O-C), 1070 (C-O), 829 (CH), 753/691 (CH) cm^{-1}.

Synthesis of 4-methoxyphenylhydrazone. To a stirred solution of 33.44 g (0.67 mol) of hydrazine hydrate in 200 ml of water a solution of 15.17 g (0.11 mol) of 4-methoxybenzaldehyde in 400 ml diethyl ether was added dropwise. After stirring for 2h at room temperature the mixture was heated to reflux temperature for 12 h. The mixture was allowed to cool down again to room temperature and the organic layer was separated and dried over MgSO$_4$. The solvent was slowly removed under reduced pressure (2 · 10^4 Pa) to yield 72 % of a yellow oil.

^{1}H NMR (CDCl$_3$): δ = 3.79 (-OCH$_3$, 3H, s), 5.44 (-NH$_2$, 2H, s), 6.87 (H3', 2H, d), 7.47 (H2', 2H, d), 7.71 (-N=CH-, 1H, s).

^{13}C NMR (CDCl$_3$): δ = 55.3 (-OCH$_3$), 114.1 (C3'), 127.8 (C2'), 128.0 (C1'-CH=N-), 143.2 (-N=CH-), 160.1 (C4'-OMe).

IR (film): 3375/3202 (NH), 3100-2850 (CH), 1625 (C=N), 1606 (NH), 1606/1511/1463 (C=C), 1248/1169 (C-O-C), 832 (CH) cm^{-1}.

Synthesis of the asymmetric 4-(3'-hydroxy)propoxy-4'-methoxy-benzaldazine. A suspension of 5.00 g (19.50 mmol) of powdered Schiff-base (as prepared above) and 3.00 g of molecular sieves (4 Å) in 350 ml of dry diethyl ether was stirred vigorously under nitrogen, when 3.52 g (23.45 mmol) of the hydrazone (as prepared above) were added. The suspension was stirred moderately in order not to destroy the molecular sieves for another 12 h. The product suspension was decanted off the molecular sieves and afterwards the solid product was isolated by filtration, washed with dry diethyl ether and dried in vacuum. The filtrate was combined with the washing solutions and the resulting solution was concentrated under reduced pressure. The precipitate was filtered off for several times, washed with dry diethyl ether and dried in vacuum. The product had a melting point of 185.1°C (yield: 74 %).

^{1}H NMR (DMSO-d$_6$): δ = 1.88 (-CH$_2$-, 2H, m), 3.58 (-CH$_2$-OH, 2H, m), 3.82 (-OCH$_3$, 3H, s), 4.11 (OCH$_2$-, 2H, t), 4.56 (-OH, 1H, t), 7.04 (H3/H3', 4H, m), 7.81 (H2/H2', 4H, m), 8.62 (-N=CH-, 2H, s).

^{13}C NMR (DMSO-d$_6$): δ = 32.2 (-CH$_2$-), 55.5 (-OCH$_3$), 57.4 (-CH$_2$-OH), 65.0 (-OCH$_2$-), 114.5 (C3'), 115.0 (C3), 127.3/127.4 (C1-CH=N-/C1'-CH=N-), 130.1 (C2/C2'), 160.6 (CH=N-), 161.3 (C4-OCH$_3$), 161.8 C4-OCH$_2$-).

IR (KBr): 3276 (OH), 3100-2850 (CH), 1624 (C=N), 1605/1510 (C=C), 1253/1172 (C-O-C), 1060 (C-O), 837 (CH) cm^{-1}.

Synthesis of the AB$_2$ monomer 4-(3'-maleimido)propoxy-4'-methoxy-benzaldazine **2** *via Mitsunobu reaction.* 1.87 g (6.00 mmol) of the asymmetric hydroxy functionalized azine as prepared above was dissolved in 100 ml of dry THF together with 0.58 g (6.60 mmol) of maleic imide. To the solution 2.20 g of polymer bound triphenyl phosphine (6.60 mmol PPh$_3$) was added and the resulting suspension was stirred for 1h in order to swell the cross-linked polystyrene carrier polymer of the triphenyl phosphine. To the mixture a solution of 1.52 g (6.60 mmol) di-*tert.*-butylazo dicarboxylate in 50 ml of dry THF was added dropwise at -5°C. Afterwards the mixture was allowed to warm up to room temperature and was stirred for 60 h.

The suspended polystyrene particles were removed by filtration under nitrogen and the solution was concentrated by evaporation of approx. 90 % of the solvent. The concentrated solution was added dropwise to 400 ml of dry diethyl ether in order to precipitate soluble traces of the polystyrene carrier. The precipitate was removed by filtration and the solvents were removed by distillation. From the resulting bright yellow mixture the product was isolated by sublimation of di-*tert.*-butyl dicarboxylate hydrazine. As polymerization started at about 140°C – below the visible melting of the solid yellow product at about 165°C – no melting point of the monomer could be determined. The yellow monomer was obtained in a yield of 87 %.

^{1}H NMR (DMSO-d$_6$): δ = 1.98 (-C$\underline{H}_2$-, 2H, m), 3.60 (-C$\underline{H}_2$-MI, 2H, t), 3.83 (-OC$\underline{H}_3$, 3H, s), 4.05 (-OC$\underline{H}_2$-, 2H, t), 6.95-7.10 (H3/H3', 4H, m), 7.02 (=C$\underline{H}$-, 2H, s), 7.77 (H2/H2', 4H, d), 8.62 (-N=C$\underline{H}$-, 2H, s).

^{13}C NMR (DMSO-d$_6$): δ = 27.7 (-$\underline{C}$H$_2$-), 34.8 (-CH$_2$-MI), 55.5 (-O$\underline{C}$H$_3$), 65.8 (-O$\underline{C}$H$_2$-), 114.6 ($\underline{C}$3'), 115.0 ($\underline{C}$3), 126.8 ($\underline{C}$1-CH=N-/$\underline{C}$1'-CH=N-), 130.1 ($\underline{C}$2/$\underline{C}$2'), 134.7 (=CH-), 160.6 (-N=$\underline{C}$H-), 161.0 ($\underline{C}$4'-OCH$_3$), 161.8 ($\underline{C}$4-OCH$_2$), 171.2 (C=O).

IR (KBr): 3100-2910 (CH), 1696 (C=O), 1624 (C=N), 1600/1500/1450 (C=C), 1252/1166 (C-O-C), 820 (CH), 695 (-HC=CH-) cm^{-1}.

MS: m/z = 391 (M$^\oplus$), 364 (M$^\oplus$-C$_2$H$_2$), 253 (M$^\oplus$-CH$_2$-CH$_2$-CH$_2$-MI), 138 (CH$_2$-CH$_2$-CH$_2$-MI), 110 (138-C$_2$H$_4$).

Synthesis of the hyperbranched polymer **3**. 0.10 g (0.26 mmol) of the AB$_2$ monomer **2** were evacuated for several times in order to remove traces of water or oxygen. The polymerization was carried out in a reaction vessel with a nitrogen inlet. After filling the reaction vessel with nitrogen gas the polymerization was started by immersing the reaction vessel into an oil bath with a constant temperature of 185°C. After 5 min viscosity increased so much that stirring was no longer possible. After a over all reaction time of 15 min the polymerization was stopped by cooling down to room temperature. The mixture was dissolved in dry chloroform and precipitated in dry hexane. The precipitate was isolated by filtration and dried in vacuum. The yield was quantitative.

GPC: $\overline{M}_n$ = 5700; $\overline{M}_w$ = 12500 (CHCl$_3$, polystyrene calibration). $\overline{M}_n$ = 16700 (CHCl$_3$, light scattering detector).

DSC: T_g = 136°C.

IR (KBr): 3100-2850 (CH), 1781/1708 (C=O), 1624 (C=N), 1605/1511 (C=C), 1252/1168 (C-O-C), 832 (CH) cm^{-1}.

^{1}H NMR (DMSO-d$_6$): δ = 1.5-2.1 (H8", H16, m), 3.1-4.8 (H1, H2, H4, H8, H9, H11, H15, H17, H7", H9", m), 3.8 (H7', H34, m), 6.6-7.5 (H2', H3', H5', H6', H2", H3", H5", H6", m), 7.0 (H19, H23, H30, H32, H12', H13', m), 7.8 (H20, H22, H29, H33, m), 8.3 (H24, H27, m).

REFERENCES

1. P. J. Flory, *J. Am. Chem. Soc.* **74**, 2718 (1952)
2. J. M. J. Fréchet, C. J. Hawker, "Synthesis and Properties of Dendrimers and Hyperbranched Polymers", in: *Comprehensive Polymer Chemistry* (S. L. Agarwal, S. Russo, Eds.), Pergamon Press, Oxford 1996, Second Supplement, p. 71.
3. B. I. Voit, *Acta Polymer.* **46**, 87 (1995).
4. G. R. Newkome, C. N. Moorefield, F. Vögtle, *Dendritic Molecules*, VCH, Weinheim, 1996.
5. D. Hölter, A. Burgath, H. Frey, *Acta Polymer.* **48**, 30 (1997).
6. R. Hanselmann, D. Hölter, H. Frey, *Macromolecules* **31**, 3790 (1998).
7. D. Hölter, H. Frey, *Acta Polymer.* **48**, 298 (1997).
8. W. Radke, G. Litvinenko, A. H. E. Müller, *Macromolecules* **31**, 239 (1998).
9. E. Malmström, F. Liu, R. H. Boyd, A. Hult, *Polym. Bull. (Berlin)* **32**, 679 (1994).
10. H. R. Kricheldorf, G. Löhden, *Macromol. Chem. Phys.* **196**, 1839 (1995).
11. H. R. Kricheldorf, O. Stöber, D. Lübbers, *Macromolecules* **28**, 2118 (1995).
12. E. Malmström, M. Johansson, A. Hult, *Macromolecules* **28**, 1698 (1995).
13. W. J. Feast, N. M. Stainton, *J. Mater. Chem.* **5**, 405 (1995).
14. E. Malmström, M. Trollsås, C. J. Hawker, *Polym. Mat. Sci. Eng.* **77**, 151 (1997).
15. C. J. Hawker, F. Chu, *Macromolecules* **29**, 4370 (1996).
16. L. J. Hobson, W. J. Feast, *Chem. Commun.* **1997**, 2067.
17. L. J. Hobson, W. J. Feast, A. M. Kenwright, *Polym. Mat. Sci. Eng.* **77**, 220 (1997).
18. Y. H. Kim, *Macromol. Symp.* **77**, 21 (1994).
19. J. R. Bailey, N. H. Moore, *J. Am. Chem. Soc.* **39**, 279 (1917).
20. J. R. Bailey, A. T. McPherson, *J. Am. Chem. Soc.* **39**, 1322 (1917).
21. K. Burger et al., *Angew. Chem.* **86**, 481–484 (1974).
22. G. Maier, *Macromol. Chem. Phys.* **197**, 3067 (1996).
23. G. Maier, A. Fenchl, G. Sigl, *Macromol. Chem. Phys.* **198**, 137 (1996).
24. G. Maier, C. Zech, B. Voit, H. Komber, *Macromol. Chem. Phys.* **199**, 2655 (1998).

Preparation and properties of heteroarm polymers and graft copolymers of poly(ethylene oxide) and polystyrene by macromonomer method

YASUHISA TSUKAHARA,* MASAHIRO TAKATSUKA, KAZUHIKO HASHIMOTO and KYOJI KAERIYAMA

Department of Chemistry and Materials Technology, Kyoto Institute of Technology, Matsugasaki, Kyoto 606-8585, Japan

Abstract—Heteroarm polymers as well as graft copolymers of poly(ethylene oxide), PEO, and polystyrene, PSt, were prepared by the macromonomer method to investigate the influence of polymer architecture on polymer properties. Macromonomers used were ω-methacryloyl poly(ethylene oxide), MA-PEOs, and methacryloyloxyethyl polystyrene, MA-PSts, which were prepared by the living anionic polymerization technique. The MA-PEOs were copolymerized with MA-PSt or styrene monomer using AIBN in benzene to obtain the heteroarm polymers and graft copolymers composed of the same polymer constituents. Polarization microscopic observation and differential scanning calorimetry measurement for solvent cast film specimens revealed that the melting point, Tm, and the crystallinity, Xc, of PEO component were influenced by the existence of PSt chains within the molecule. Tm and Xc decreased with increase in the PSt content in the copolymers, where decrease in Xc was more prominent than Tm and was influenced by the polymer architecture. Hysteresis of contact angles on the film specimens toward water were also influenced by the polymer architecture.

Keywords: Macromonomer; copolymacromonomer; heteroarm polymer; branched polymer; star block polymer; poly(ethylene oxide); polystyrene.

1. INTRODUCTION

In accordance with recent development in polymer chemistry, it is possible now to prepare various kinds of well-defined branched polymers composed of different polymer chains, such as A_nB_n or ABC star block copolymers [1–7]. These heteroarm polymers show unique phase separated morphology and bulk properties different from the conventional linear block copolymers [8, 9]. AB-type heteroarm polymers have been mostly prepared by the two-step coupling reaction of living polymer of each constituent with multi-functional chlorosilane compounds. On the other hand, copolymerizations of different macromonomers also produce heteroarm polymers of various branching architecture. Gnanou *et al.* prepared

---

* To whom correspondence should be addressed. E-mail: tsukah@ipc.kit.ac.jp

heteroarm polymers of poly(ethylene oxide), PEO, and polystyrene, PSt, by ring-opening metathesis copolymerization of norbornene-terminated PEO and PSt macromonomers [10]. Ishizu et al. recently reported radical chain copolymerization behavior of maleate-terminated PEO and vinylbenzyl-terminated PSt macromonomers [11]. We have been also investigated preparation an properties of the heteroarm polymers by radical chain copolymerizations of methacrylate-terminated macromonomers of PEO, PSt, and polydimethylsiloxane [12]. We report here some results on the preparation and properties of the heteroarm polymers of PEO and PSt. The properties of the heteroarm polymers were compared with those of corresponding graft copolymers composed of the same constituents to investigate the effect of polymer architecture.

2. EXPERIMENTAL

ω-Methacryloyl poly(ethylene oxide) macromonomers, MA-PEOs, and methacryloyl-oxyethyl polystyrene macromonomers, MA-PSts, were synthesized according to Scheme 1. That is, MA-PEOs were synthesized by living anionic ring opening polymerization of ethylene oxide, EO, with potassium t-butoxide in

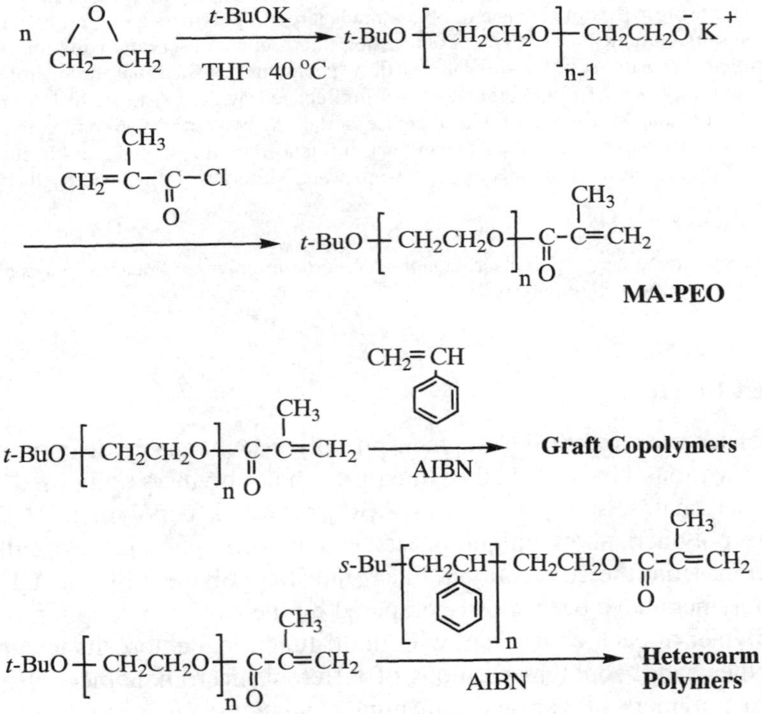

Scheme 1. Preparations of graft copolymers and heteroarm polymers of poly(ethylene oxide) and polystyrene.

THF followed by termination with methacryloyl chloride [13], and MA-PSts were synthesized by living anionic polymerization of styrene with s-BuLi in toluene/THF (9/1) mixed solvent followed by addition with ethylene oxide and termination with methacryloyl chloride using the break seal method [14–17]. Monomers and solvents used were purified and dried by the usual method. Obtained macromonomers were purified by precipitation and characterized by gel permeation chromatograph (GPC) (Tosoh HLC-802A) and ^{1}H-NMR (GE QE-300). End group functionality of the macromonomers were confirmed by the maximum conversion of the macromonomer in radical copolymerization with MMA in benzene using AIBN. Preparation conditions and characteristics of the macromonomers are shown in Table 1.

MA-PEO macromonomers were copolymerized with MA-PSt macromonomer using azobisisobutyronitrile (AIBN) initiator in benzene. Copolymerization products were purified by precipitation-extraction procedures with benzene-diethylether ether and THF-water remove unreacted macromonomer and comonomer. The purification were repeated several times until the sharp peak in the GPC curve, corresponding to the unreacted macromonomer, disappeared. MA-PEOs were also copolymerized with styrene, St, monomer with AIBN in benzene to obtain conventional graft copolymers poly[St-co-(MA-PEO)]s for comparison. These copolymerization conditions are shown in Table 2.

Copolymer compositions of the heteroarm polymers as well as the graft copolymers were determined by ^{1}H-NMR. The molecular weights were determined by GPC and a membrane osmometer (Jupiter model 230). The membrane osmometer was operated with toluene at 35°C. Thermal behavior was measured by a DSC (Perkin Elmer) at heating rate 10°C/min and crystalline morphology was observed by a polarization microscope (Nikon OPTIPHOTO POLwith Microflex AFX-II) at room temperature (20°C). Contact angle toward water on the cast film specimen of the copolymer was measured by a contact angle meter (Kyowa-Kaimen Kagaku CA-A) at room temperature.

Table 1.
Preparation conditions and characteristics of Macromonomers

Run	Code	Monomer (mmol)	[I] (mmol)	[MACl][a] (mmol)	Time (h)	Mn Calcd	Mn GPC	Mw/Mn	F[b]
1[c]	MA-PEO4900	1.0	10.0	40.0	12	4400	4900	1.14	0.97
2[c]	MA-PEO14500	1.0	3.7	20.5	15	11890	14500	1.08	0.87
3[d]	MA-PSt6100	0.48	9.0	30.0	2	5340	6100	1.03	0.92
4[d]	MA-PSt14000	0.48	4.0	15.0	2	1230	14000	1.05	0.97

[a] MACl: Methacryloyl chloride.
[b] End group functionality determined by ^{1}H-NMR and the maximum conversion in polymerization of the macromonomer with MMA in benzene at 60°C for 24hrs.
[c] Polymerizations were carried out in THF at 40°C.
[d] Polymerizations were carried out in Toluene/ THF (9/1) at –78°C.

Table 2.
Preparation Conditions of Heteroarm polymers and Graft polymers [a)]

Run	MA-PEO (mmol)	MA-PSt (mmol)	Styrene (mmol)	AIBN (mmol/L)	Benzene (ml)
1	0.327	0.209	—	20	1.2
2	0.204	0.071	—	20	1.2
3	0.082	0.114	—	20	1.2
4	0.069	0.071	—	20	1.2
5	0.124	0.164	—	20	1.2
6	0.306	—	1.92	20	3.4
7	0.306	—	4.81	20	4.3
8	0.306	—	6.24	20	4.3
9	0.306	—	9.62	20	5.0
10	0.306	—	12.5	20	5.6

[a)] Copolymerizations were carried out at 60°C for 24 hours.

3. RESULTS AND DISCUSSION

Figure 1 shows the schematic representation of the heteroarm polymers and the graft copolymers in this study. The copolymerization reaction mixtures of MA-PEO with styrene monomer in benzene were transparent. However, the reaction mixtures of MA-PEO and MA-PSt macromonomers were opaque from the beginning of copolymerization due to the phase separation between the unlike macromonomers, even the presence of benzene as a mutual good solvent. Therefore, magnetic stirring was necessary during the copolymerization. Figure 2 shows an example of the isolation procedure of the heteroarm polymers prepared from MA-PEO4900/MA-PSt14000. GPC curves for the crude copolymerization product, the

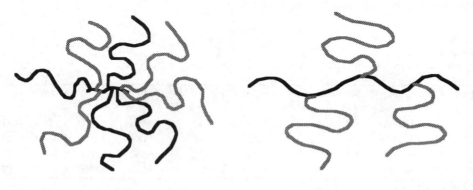

Figure 1. Schematic representation of heteroarm polymers and graft copolymers of poly(ethylene oxide) and polystyrene.

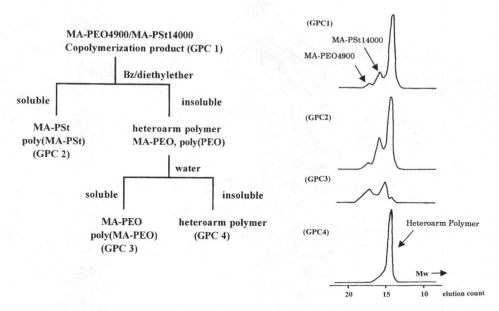

Figure 2. An example of the purification procedure to isolate the heteroarm polymers prepared by copolymerization of MA-PEO4900 and MA-PSt14000.

soluble part in diethylether, that in water and the heteroarm polymers after the purification procedure are also shown in Figure 2. The peaks corresponding to the unreacted macromonomers in the crude sample almost disappear in the final product. In the case of graft copolymers, it was also possible to remove the unreacted macromonomer completely. ^{1}H-NMR spectra of the purified copolymer samples showed the signal corresponding to the methoxy protons of PEO component at 3.7 ppm and that corresponding to the phenyl protons of PSt component at around 6.6–7.2 ppm, from the relative signal intensity of which, the average compositions of the copolymers were determined. The characteristics of these polymers are shown in Table 3. The PEO composition in the heteroarm polymers ranges from 22.3 to 86.3 wt% and that in the graft copolymers ranges from 21.5 to 55.3 wt%.

Figures 3 and 4 show polarization microphotographs for the solvent cast films of the heteroarm polymers and those of the graft polymers. The photograph of polymerization product of MA-PEO alone, i.e., poly(macromonomer), is also shown in Figure 3 for comparison. For the poly(MA-PEO), the large spherulite of diameter about 53 mm is seen and each spherulite is tightly contacting to the neighboring spherulites. On the other hand, the spherulites in the heteroarm polymers as well as the graft copolymers were much smaller as seen in Figures 3 and 4. This might indicate the influence of the PSt chains existing within the molecule, where the PSt chain prevents easily the crystallization of the PEO chains. It was rather difficult to observe the spherulites for the sample of the PEO

Table 3.
Characterization of Heteroarm Polymers and Graft Copolymers

Sample Code[a]	Average Composition[b]				$Mw^{c)} \times 10^3$	$Mn \times 10^3$		$Mw/Mn^{c)}$
	mol%		wt%			$(A)^{c)}$	$(B)^{d)}$	
	PEO	PSt	PEO	PSt				
MA-PSt14000/MA-PEO4900	93.7	6.3	86.3	13.7	240.1	155.1	—	1.55
MA-PSt14000/MA-PEO4900	72.5	27.5	52.7	47.3	102.1	66.4	—	1.54
MA-PSt14000/MA-PEO4900	54.9	45.1	34.0	66.6	143.0	91.9	193.7	1.56
MA-PSt14000/MA-PEO14500	58.5	42.5	37.3	62.6	82.0	60.8	—	1.35
MA-PSt6100/MA-PEO8060	40.4	54.6	22.3	77.7	788.3	204.6	—	3.85
MA-PEO4900/St	89.6	10.4	78.5	21.5	67.7	31.3	—	2.16
MA-PEO4900/St	84.2	15.8	69.3	30.7	81.3	53.3	73.3	1.53
MA-PEO4900/St	79.5	20.5	63.5	36.5	65.1	39.8	—	1.64
MA-PEO4900/St	75.6	24.3	56.8	43.2	—	—	—	—
MA-PEO4900/St	65.7	34.3	44.7	55.3	31.9	22.2	—	1.51

[a] MA-PSt14000/MA-PEO4900 means heteroarm polymers of MA-PEO4900 and MA-PSt14000, MA-PEO4900/St means graft copolymers of MA-PEO4900 and Styrene.
[b] Determined by ^{1}H-NMR.
[c] Determined by GPC using a standard polystyrene calibration curve.
[d] Determined by a membrane osmometer in toluene.

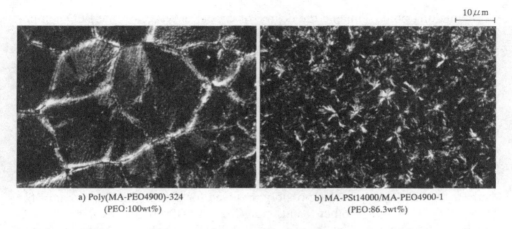

a) Poly(MA-PEO4900)-324
(PEO:100wt%)

b) MA-PSt14000/MA-PEO4900-1
(PEO:86.3wt%)

Figure 3. Polarization microphotographs for cast film specimens of (a) a poly(macromonomer) of poly(MA-PEO4900) and (b) a heteroarm polymer of poly[(MA-PSt14000) -co- (MA-PEO4900)]. Under cross nicol.

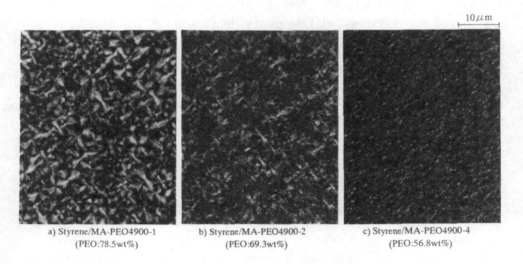

a) Styrene/MA-PEO4900-1
(PEO:78.5wt%)

b) Styrene/MA-PEO4900-2
(PEO:69.3wt%)

c) Styrene/MA-PEO4900-4
(PEO:56.8wt%)

Figure 4. Polarization microphotographs for cast film specimens of graft copolymers of poly[St-co-(MA-PEO4900)] with different PEO compositions. The weight fraction of the PEO component is (a) 78.5wt%, (b) 69.3wt%, and (c) 56.8wt%. Under cross nicol.

composition less than around 50wt%. Comparing of the spherulites in the photographs for the heteroarm polymers of 86.3 wt% PEO composition and graft copolymers of 78.5wt% PEO composition shows the crystallization of the PEO chains in the heteroarm polymers is likely to be more prevented and the spherulite is coarser to some extent. This might be the result of the difference in the polymer architecture.

The melting point, Tm, and the degree of crystallinity, Xc, of the PEO component in the polymers determined from DSC measurements are summarized in Table 4. Variations of Tm and Xc with the PEO composition are shown in Figure 5. The value measured for a linear PEO (Mn=1.6 x 10^5, Mw/Mn=1.07) is also included for comparison. Tm of high molecular weight linear PEOs is reported as 60–76°C [18], which depends on the existence of defects in the crystal lattice such as the folding of the polymer chains and also on the molecular weight. Tm of the perfectly crystalline linear PEO with fully linear unfolded chain are reported as 75 ± 3 °C [19]. The linear PEO in Table 4 shows Tm=61°C, while the macromonomer MA-PEO4900 shows Tm=41.0 °C which is much lower than the value of the linear PEO due to the low molecular weight. However, influence of the methacryloyl end group is also possibly involved for the low melting point of the MA-PEO4900. On the other hand, poly(MA-PEO4900) shows Tm=45.2 °C which is somewhat higher the Tm of MA-PSt4900. The PEO chains in the poly(MA-PEO) are connected to the central backbone chain at one chain end and this might influence the crystallization and melting behavior of the poly(macromonomer). This suggests that the intramolecular crystallization might be involved in the crystallization of the poly(macromonomer).

It is seen from Figure 5 that Tm values of heteroarm polymers as well as graft copolymers are lower than those of MA-PEO macromonomer and the poly(MA-PEO). Tm gradually decrease as the content of the polystyrene component increases. Sanchez and Eby investigated the crystallization of random copolymers composed of crystalline and noncrystalline components [20]. Nishi and Wang reported the reduction of Tm with increase in the noncrystalline PMMA in the compatible blends of crystalline poly(vinylidene fluoride) and noncrystalline

Table 4.
Results of DSC Measurements

Sample Code [a]	EO-unit [b] (wt%)	Tm (°C)	ΔH_f (J/g)	Crystallinity (%) [c]	
				(A)	(B)
MA-PEO4900	97.0	41.0	108	57.3	50.0
poly(MA-PEO4900)	97.0	45.2	121	64.4	56.0
MA-PEO4900/MA-PSt14000	83.7	41.0	88.5	47.0	41.0
MA-PEO4900/MA-PSt14000	51.1	38.8	51.5	27.3	23.9
MA-PEO4900/MA-PSt14000	32.9	36.1	25.2	13.3	11.7
MA-PEO4900/St	76.0	38.9	106	56.3	49.1
MA-PEO4900/St	66.3	37.9	80.8	42.9	37.4
MA-PEO4900/St	43.4	37.0	60.0	31.8	27.8

[a] MA-PEO4900/MA-PSt14000 means heteroarm polymers of MA-PEO4900 and MA-PSt14000, MA-PEO4900/St means graft copolymers of MA-PEO4900 and St.
[b] Based on the ethylene oxide unit.
[c] Crystallinity was calculated from ΔH_f using (A) the heat of fusion for PEO 8.29 kJ/mol [22] and (B) 9.2 kJ/mol [19] on the base of the ethylene oxide unit.

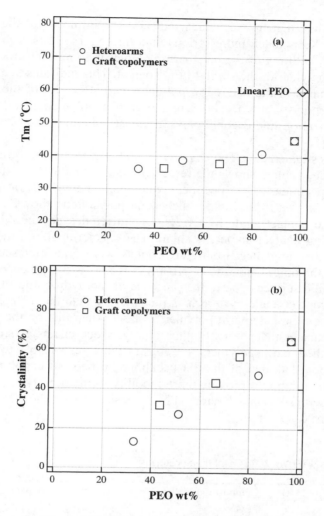

Figure 5. Variation of (a) Tm and (b) Xc of the heteroarm polymers and the graft copolymers with the PEO composition.

PMMA [21]. In this study, the similar tendency was observed in both the heteroarm polymers and graft copolymers qualitatively. These indicate that the molecular mixing due to the chain connectivity between crystalline and noncrystalline chains influence much more the melting behavior than the corresponding simple immiscible blends.

The degree of crystallinity was calculated by the endothermic peak area at the melting point using the heat of fusion for PEO 8.29 kJ/mol [22] and 9.2 kJ/mol [19] on the basemole of the ethylene oxide unit. In Figure 5, Xc of the heteroarm polymers and graft copolymers also decreases with increase in the content of the noncrystalline polystyrene component. However, the degree of lowering in Xc is much steeper than that in Tm, which indicates the influence of the polystyrene component on Xc is much more serious than that on Tm. It is seen that the degrees of the crystallinity of the heteroarm polymer is somewhat lower than those of the graft copolymer of the similar PEO content. This indicates that the polystyrene branch chains in the heteroarm polymers can affect Xc of the PEO chains more than the polystyrene backbone of the graft copolymers. This is consistent with the polarization microscopic observation of spherulite morphology in Figures 3 and 4.

We also investigated the effect of polymer architecture on the wettability on the polymer surface. Table 5 shows the results of contact angle of a water droplet on the samples specimens. The specimens were prepared by solvent casting from chloroform. It is seen from Table 5 that linear polystyrene shows a hydrophobic surface and the contact angle hysteresis, i.e., difference between the advancing and the receding contact angles, is small. The contact angle on the linear PEO could not be measured because water droplets were gradually sucked into the specimen. On the other hand, the heteroarm polymers and graft copolymers show more hydrophilic surfaces due to the presence of the hydrophilic PEO chains. In addition, the contact angle hysteresis is much larger than that of polystyrene and the hysteresis on the heteroarm polymers is larger than that on the graft copolymers. This indicates that the hydrophobic polystyrene chains existing as branch chains in the heteroarm polymers are able to more responsible to the environmental condition than that of the PSt backbone in the graft copolymer. The heteroarm architecture might provide more mobility to the polystyrene chain resulting in the large hysteresis behavior. This is consistent with the effect of polymer architecture on Xc.

Table 5.
Contact Angles toward Water on Cast Film Specimens [a]

Sample	Composition (wt%)		Contact Angle θ (degree)		
	PEO	PSt	Advancing θ_a	Receding θ_r	Hysterisis $\theta_a - \theta_r$
-Linear Homopolymers -					
PSt	0	100	83	80	3
PEO	100	0	—	—	—
- Heteroarm polymers -					
Poly[(MA-PSt8060)-co-(MA-PEO6100)]	22.3	77.7	71	12	59
Poly[(MA-PSt14000)-co-(MA-PEO4900)]	34.0	66.6	70	22	48
Poly[(MA-PSt14000)-co-(MA-PEO4900)]	86.3	13.6	65	—	—
-Graft copolymers -					
Poly[St-co-(MAPEO4900)]	44.7	55.3	61	24	37
Poly[St-co-(MAPEO4900)]	56.8	43.3	56	—	—

[a] Measured by Kyowa-Kagaku contact angle goniometer at Room Temperature.

4. CONCLUSION

Graft copolymers and heteroarm polymers composed of PEO chains and PSt chains were prepared by the macromonomer method to investigate the effect of polymer architecture on the properties. It is clarified that the polystyrene chain within the molecule influences the melting point, the crystallinity of PEO chains, and contanct angle hysteresis. The effect of polystyrene chain is suggested to be dependent on the polymer architecture. However, further study is necessary to draw the clear picture on the influence of polymer architecture including the co-polymerization behavior between unlike macromonomers and now in progress.

REFERENCES

1. H. Iatrou and N. Hadjichristidis, *Macromolecules*, **25**, 4649 (1992).
2. J. Allgaier, R. N. Young, V. Efstratiadis, and N. Hadjichristidis, *Macromolecules*, **29**, 1794 (1996).
3. A. Avgeropoulos, Y. Poulos, N. Hadjichristidis, and J. Roovers, *Macromolecules*, **29**, 6076 (1996).
4. T. Fujimoto, H. Zhang, T. Kazama, Y. Isono, H. Hasegawa and T. Hashomoto, *Polymer*, **33**, 2208 (1992).
5. H. Huchstadt, V. Abetz, and R. Stadler, *Macromol. Rapid Commun.*, **17**, 599 (1996).
6. O. Lambert, P. Dumas, G. Hurtrez and G. Riess, *Macromol. Rapid Commun.*, **18**, 343 (1997).
7. S. Sioula, Y. Tselikas and N. Hadjichristidis, *Macromolecules*, **30**, 1518 (1997).
8. E. Molau, Ed., *Colloidal and Morphological Behavior of Block and Graft Copolymers*, Plenum, New York (1971).
9. I. Goodman, Ed., *Developments in Block Copolymers*, Applied Science Publisher, London (1982).
10. V. Heroguez, Y. Gnanou, and M. Fontanille, *Macromolecules*, **30**, 4791 (1997).
11. K. Ishizu and X. Shen, *Polymer*, **40**, 3251 (1999).
12. M. Takatsuka, Y. Tsukahara and K. Kaeriyama, *Polym. Prepr. Jpn.*, **47**, 80 (1998).
13. K. Ito, K. Tanaka, H. Tanaka, G. Imai, S. Kawaguchi and S. Itsuno, *Macromolecules*, **24**, 2348 (1991).
14. Y. Tsukahara, K. Mizuno, A. Segawa and Y. Yamashita, *Macromolecules*, **22**, 1564 (1989).
15. Y. Tsukahara, in *Macromolecular Design: Concept and Practice*, Polymer Frontier International, New York (1995), pp. 161–227.
16. Y. Tsukahara, Y. Ohta and K. Senoo, *Polymer*, **36**, 3413 (1995).
17. Y. Tsukahara, K.Yai and K. Kaeriyama, *Polymer*, **40**, 729 (1999).
18. J. Brandrup and E. H. Immergut, *Polymer Handbook*, 3rd Ed., John Wiley & Sons, New York (1989).
19. A. M. Afifi-Effat and J. N. Hay, *J. Chem. Soc., Faraday Trans. 2*, **68**, 656 (1972).
20. I. C. Sanchez and R. K. Eby, *Macromolecules*, **8**, 639 (1975).
21. T. Nishi and T. T. Wang, *Macromolecules*, **8**, 909 (1975).
22. L. Mandelkern, *Chem. Rev.*, **56**, 903 (1956).

Simultaneous block copolymerization via macromolecular engineering. One-step, one-pot initiation of living free radical and cationic polymerization

DOTSEVI Y. SOGAH,* RUTGER D. PUTS, OREN A. SCHERMAN and MARC W. WEIMER

Baker Laboratory, Department of Chemistry & Chemical Biology, Cornell University, Ithaca, New York 14853-1301

Abstract—This paper describes a novel approach involving the use of multifunctional initiators to initiate living free radical and cationic polymerization either sequentially or concurrently. Cationic ring-opening polymerization of oxazolines was carried out concurrently with living free radical polymerization to provide a library of amphiphilic polymers. Polymers of varying block lengths, molecular weights, and compositions and low polydispersity indices were produced. Evidence for block copolymer formation was obtained from studies involving SEC, NMR, extractions, hydrolysis, and investigation of solution properties of the polymers.

Keywords: Multifunctional; living polymerization; copolymerization; poly(oxazoline); poly(ethylene imine).

1. INTRODUCTION

The major thrust of our research is the development of a simple, one-step, one-pot methodology for the synthesis of block copolymers from monomers that polymerize by different mechanisms. Polymers of controlled architectures such as block and graft copolymers are important technologically because of their utility in several applications. The synthesis of such controlled architectures is usually accomplished through living polymerization [1]. Living polymerizations minimize the number of dead chain ends, thus allowing efficient formation of block copolymers through sequential addition of monomers. The monomers, in this case, must polymerize by the same mechanism. However, the preparation of block copolymers from very dissimilar monomers, especially those that polymerize by different mechanisms, poses a special challenge since invariably either the mechanisms are inherently incompatible or the reaction conditions are significantly different.

---

* To whom correspondence should be addressed. E-mail: dys2@cornell.edu

Block copolymers can also be prepared through the coupling of two preformed polymers [1, 2]. However, this is generally inefficient due to the low availability and accessibility of the reactive chain ends, and sometimes contamination from difficult-to-remove homopolymers. Other strategies for block copolymer preparation include transformation of a living chain end in situ into an appropriate macroinitiator for polymerization of the second monomer [2, 3]. The efficiency of this approach depends on how well the amount of dead or dormant chain ends are controlled, the reactivity and accessibility of the living end, and the reactivity of the specific reagent, because any chain not quenched with the specific reagent would remain as a homopolymer contaminant. A one-step, one-pot block copolymerization involving monomers of vastly different reactivity ratios, whereby one monomer is essentially completely consumed before the second monomer polymerizes, is another strategy [1e, 4]. However, this approach is not applicable to monomers that polymerize by different mechanisms. Besides, the number of available monomer combinations is limited.

Recently techniques involving the use of multifunctional initiators **1** and **2** (Figure 1) containing orthogonal reactive sites that could be independently and selectively accessed either simultaneously or sequentially have been developed in our laboratories [5]. These initiators permitted block copolymer formation in a straightforward manner. The initiating sites survived the different polymerization methods employed leading to successful preparation of block and graft copolymers containing vinyl pyridine (VP), styrene (S), and 2-oxazolines (OZ) without resorting to intermediate steps or chain end transformations. The compatible reaction conditions for cationic ring-opening polymerization (CROP) of oxazolines and nitroxide-mediated living free radical polymerization (LFRP) of styrene render the one-pot, one-step block copolymerization feasible [6–8]. The versatility of multifunctional initiators is broad in that they possess not just initiating sites for LFRP and CROP but also groups that can be readily converted into anionic initiators for the polymerization of lactones, lactide, epoxides, and methacrylates [9]. Since the appearance of our report [5, 8, 9], others have reported a complementary approach involving simultaneous anionic ring opening polymerization (AROP)

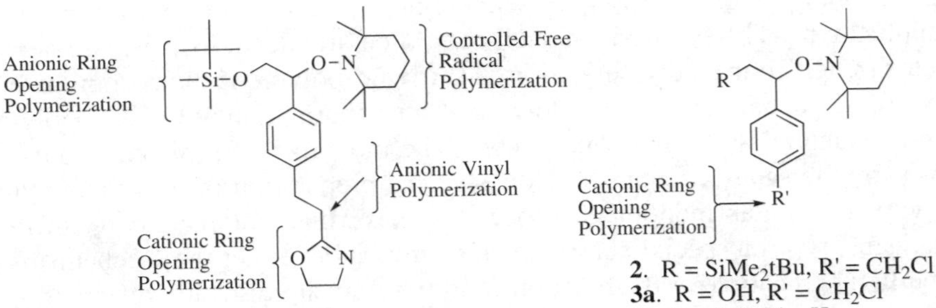

2. R = SiMe$_2$tBu, R' = CH$_2$Cl
3a. R = OH, R' = CH$_2$Cl
3b. R = OH, R' = H

Figure 1. Multifunctional initiators.

and either atom transfer radical polymerization (ATRP) or nitroxide-mediated LFRP to produce block and graft copolymers of ε-caprolactone (CL), S, methyl methacrylate (MMA), and 2-hydroxyethyl methacrylate (HEMA) [10]. The advent of simultaneous multiple polymerizations eliminates the need for intermediate end group transformations and difficult purification steps [9]. This novel technology is especially applicable to (1) polymerization mechanisms that do not interfere with each other in a way that deactivates living chain ends, (2) monomers that neither cross polymerize nor interfere adversely with each other to any appreciable extent, and (3) polymerizations whose reaction conditions, i.e. temperature, solvent, etc., are similar.

We provide herein further evidence that the one-step, one pot block copolymerization technology involving different polymerization reactions is, indeed, general. This is illustrated by the successful synthesis of poly[(2-phenyl-2-oxazoline)-*b*-styrene] (PPOZ-*b*-PS) using concurrent CROP and nitroxide-mediated LFRP. Convincing evidence for block copolymer formation was obtained from studies involving SEC, NMR, solvent extractions, and hydrolysis, and evaluation of solution properties of the resulting polymers.

2. RESULTS AND DISCUSSION

2.1. (PPOZ-b-PS) by CROP and LFRP

2.1.1. Synthesis. Although benzyl chloride initiator has been used for CROP of POZ, the reaction is usually slow due to the propagation occurring by a covalent mechanism [6]. Therefore, to polymerize POZ at an appreciable rate, we transformed in situ **4** in to the corresponding triflate (**5**) using AgOTf (Figure 2). The CROP of POZ and LFRP of styrene occurred to give **7** in moderate to excellent isolated yields. Peaks corresponding to both PS and PPOZ were seen in the ^{1}H NMR spectrum, confirming that both monomers underwent the desired polymerization.

Table 1 summarizes typical polymerization results. The observed molecular weights were moderate and slightly lower than the calculated values probably due to PS not being the best standard for the block copolymers. Even though the lengths of the two blocks were varied over a broad range, the polydispersity remained relatively low (1.27–1.40) in all cases. Furthermore, the composition of the block copolymers, determined from ^{1}H NMR, were in close agreement with that expected from monomer feed molar ratios, indicating control over composition and block length.

The question as whether or not the two polymerizations were concurrent was addressed by monitoring the conversion of each monomer as a function of time (Figure 3). The two curves revealed that the two processes were, indeed, concurrent. However, as expected, the CROP of POZ, an ionic reaction, was slightly faster than the LFRP of styrene. It is intriguing to note that both reactions reached very high conversions in less than 7 hours. In absence of each other these same

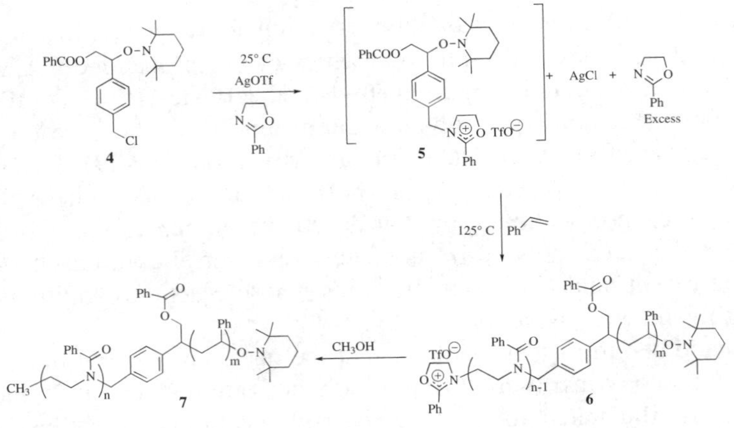

Figure 2. Concurrent polymerization of POZ and styrene.

Table 1.
Results of CROP of 2-Phenyl-2-oxazoline (POZ) and LFRP of Styrene (S)

Sample	M_n(SEC)[a] ($\times 10^{-3}$)	M_n(calc)[b] ($\times 10^{-3}$)	M_w/M_n (SEC)[a]	% Yield	f_{POZ}[c]	f_S[c]	F_{POZ}[d] (NMR)	F_S[d] (NMR)
PPOZ	26.8	14.0	1.3	99	1.00	0.00	1.00	0.00
7a	28.0	18.7	1.4	80	0.47	0.54	0.47	0.54
7b	24.3	11.6	1.3	45	0.15	0.85	0.12	0.88
7c	33.5	25.0	1.4	99	0.81	0.19	0.88	0.12

[a]SEC was performed in DMAC versus PS standards. [b]M_n calculated from monomer/initiator ratio and corrected for each monomer conversion. [c]Monomer mole fraction in feed. [d]Copolymer composition determined from NMR.

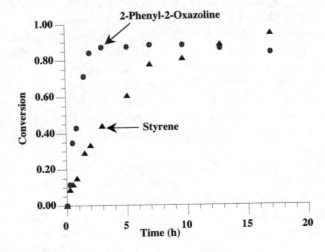

Figure 3. Conversion versus Time plots for PPOZ and PS.

polymerizations, under similar conditions, would typically have taken over 24 hours to reach a similar conversion [7]. Hence, one important unpredicted consequence of this one-step, one-pot concurrent process is the synergistic rate enhancing influence each reaction experiences.

2.1.2. Evidence for Block Copolymerization from SEC and Solvent Extraction Studies. The first evidence for the formation of block copolymers came from the fact that the SEC traces for all polymer samples were unimodal, even in case cases where the block lengths (degrees of polymerization, DP) were vastly different, such as **7b** ($DP_{POZ}:DP_S = 25:198$) and **7c** ($DP_{POZ}:DP_S = 207:29$). Further evidence was obtained from Soxhlet extractions of **7b** and **7c** with methanol and cyclohexane, respectively. Since methanol is a good solvent for PPOZ but not for PS, any PPOZ would be removed from the mixture. Likewise, cyclohexane, a θ-solvent for PS, but a poor one for PPOZ would remove any PS. After 20 hours of extraction the polymers remained unchanged in MW and composition. Hence, it can be concluded that little or no homopolymer formed during the polymerizations.

2.1.3. Evidence for Block Copolymerization from Solution Properties of Hydrolyzed Polymer. Amphiphilic block copolymers are known to form micelles in appropriate solvents. This is usually demonstrated through the behavior of the polymers in different solvents [11]. A recent example was provided by Meijer and co-workers for amphiphilic PS-dendritic poly(ethylene imine) (PEI) hybrid block copolymers [11a,b], while another one was reported by Frechet *et al.* involving solvent-responsive hybrid dendrimers [11c]. Poly(oxazoline)s can be hydrolyzed under acid conditions to give hydrophilic PEIs [12]. In the case of our polymers, the hydrolysis would produce an amphiphilic block copolymer only if the starting polymer was also a block copolymer. When a sample of the polymer produced by our process was subjected to acid hydrolysis, the product, presumed to have structure **8** (Figure 4), was isolated. The SEC traces of **8** from both refractive index (RI) and UV detectors were superimposable on each other, with no evidence of any low MW peak. This indicates that every polymer chain observed by the RI detector, including the non-UV active PEI, contained the UV active PS.

To determine if the hydrolyzed polymer was amphiphilic, the polymer with the shortest PS segment **8c** ($DP_{POZ}:DP_S = 207:29$), which dissolved completely to form a homogenous solution in both methanol and chloroform, was studied by ^{1}H NMR in CD_3OD and $CDCl_3$. The ^{1}H NMR spectrum of **8c** in CD_3OD showed no signals due to PS [8]. This suggests that in a very polar solvent such as methanol the hydrophilic PEI sequestered the hydrophobic PS so that it was not in direct contact with the polar methanol. This process should be further enhanced by hydrophobic interactions within the PS segments leading to their clustering together in the interior of the folded polymer chains. In contrast, the spectrum of **8c** in $CDCl_3$ showed signals due to both PEI and PS [8]. These results confirmed that the hydrolyzed polymer **8** was indeed amphiphilic, which is a direct evidence for the starting polymer being a block copolymer.

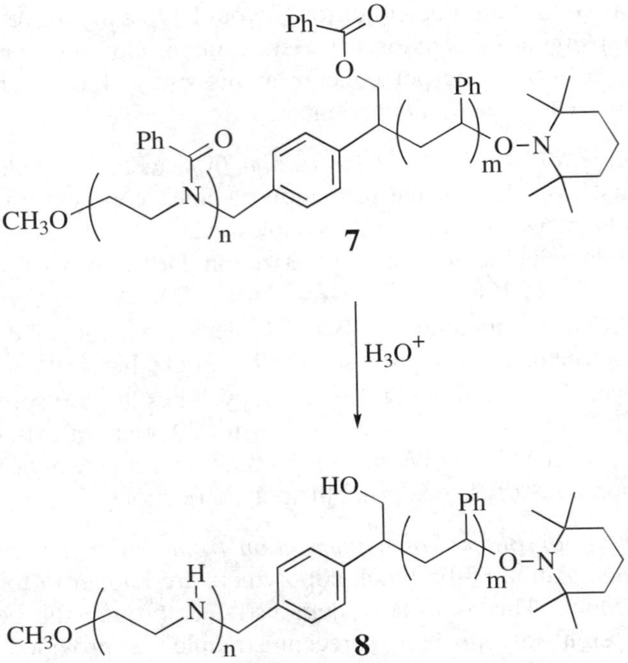

Figure 4. Hydrolysis of PPOZ-b-PS.

2.1.4. Evidence for Block Copolymerization from DSC Studies. Block copolymers containing sufficiently long incompatible segments usually exhibit separate thermal transitions for each block [1b]. The thermal data for the various samples are listed in Table 2. The unhydrolyzed polymers showed two transitions. The copolymer with approximately equal block lengths (**7a**) gave one transition at 104 °C corresponding to overlapped T_gs of both PS and PPOZ, and a second transition at 237 °C corresponding to the T_m of PPOZ. The independently prepared PPOZ gave T_g at 109 °C and T_m at 237 °C. Obviously, the T_gs for PS (100 °C) and PPOZ (109 °C) are too close to be resolved in the DSC. The copolymer with the shortest PPOZ block (**7b**) gave a T_g at 103 °C, which is essentially that of PS, and a T_m at 237 °C identical to T_m of PPOZ. Similarly, the polymer with the shortest PS block (**7c**) gave a T_g at 109 °C and T_m at 235 °C, consistent with its being mostly PPOZ. In the case of the hydrolyzed polymers, the block lengths are apparently too short to give separate T_gs. Thus, **8b** (mostly PS) showed a single T_g at 82 °C, and **8c** (mostly PEI) gave a single T_g at –2 °C. The reported T_g for PEI is –23.5 °C [13]. Calculation of the T_gs using values for the homopolymers and the relative block sizes gave 90 °C for **8b** and –7 °C for **8c**. These thermal analysis data support the conclusion that the reaction produced the expected block copolymer directly.

Table 2.
Thermal Properties of PPOZ-*b*-PS and PEI-*b*-PS from Differential Scanning Calorimetry (DSC) Studies

Sample No.	T_g (DSC) (°C)	T_g (calc) (°C)	T_m (DSC) (°C)
PPOZ	109		237
7a	104		237
7b	103		237
7c	109		235
8b	82	90	None
8c	–2	–7	None

3. CONCLUSIONS

We have demonstrated the feasibility of performing two independent polymerizations simultaneously from multifunctional initiators leading to the formation of block copolymers. This eliminates the need for unnecessary handling and intermediate transformations. In all cases, there was a synergistic rate-enhancing influence of one reaction on the other. This opens the door for the creation of new architectures and new materials. Work is in progress to determine the general applicability of the one-step, one-pot block polymerization methodology.

4. EXPERIMENTAL

Materials. Styrene (S) (99%) and 2-phenyl 2-oxazoline (POZ) (99%) (Aldrich), were purified by distillation under reduced pressure from CaH_2 and stored under argon. Benzoyl peroxide (97%), TEMPO (98%), Et_3Al (1.0 M in hexanes), and silver triflate were used as received from Aldrich. All solvents were of the highest grade from Fisher. THF and toluene were distilled from sodium benzophenone prior to use. Other solvents (Fisher) were used as received. All glassware was oven dried and cooled in a desiccator. Synthesis of initiators **5** and **6**, were done as reported in the literature [5, 10].

Methods. NMR spectra were taken on a Brucker WM 300 (300 MHz) spectrometer at 25 °C in $CDCl_3$ with chemical shifts relative to the solvent (7.24 ppm). Size-exclusion chromatography (SEC) in THF was performed on a SEC line consisting of an M510 pump, U6K universal injector, 486 tunable UV detector (all from Waters), and a differential refractive index (RI) detector, RefractoMonitor IV (Milton Roy). The separations were achieved on a bank of three 5 μm Polymer Standards Service columns with pore sizes 50Å, 1000Å and linear using a flow rate of 1 mL/min. Molecular weights were calculated against PS standards using Polymer Standards Service WINSEC software. SEC in DMAC was performed on-line consisting of a 222 pump from Viscotek, U6K injector from Waters, and a model 200 differential refractometer. Separations were preformed on two mixed bed linear and one 500Å 10 µm columns. A flow rate of 1 mL/min was used and calculations were preformed using SEC-PRO from Viscotek. Thermal analysis was performed on a Seiko Instruments 5200 thermal analysis station. TGA was performed under N_2 (60 mL/min) in a temperature range of 30–550 °C with a ramp of 20 °C/min. DSC was performed under N_2 with a temperature range of 25–150 °C heated and cooled at 10 °C/min. The second heating cycle was used for the calculation of the thermal transitions. Conversion was determined on HP 5890A GC equipped with HP 7673A autosampler and HP 3396A integrator. Runs for the PPOZ-b-PS kinetics were performed from 90–200 °C at 15 °C/min. Separations were performed on HP-1 cross-linked dimethylsilicone gum capillary column (25 m x 0.32 mm, 0.17 µm film) with a flow rate of 1.1 mL/min of He and FID detector.

Poly(2-Phenyl-2-Oxazoline) (PPOZ). A round bottom flask, fitted with a condenser and magnetic stirring bar, was charged with **5** (50.7 mg, 0.19 mmol) followed by AgOTf (32.3 mg, 0.13 mmol). The vessel was purged with argon for 15 min. POZ (1.5 mL, 1.67 g 11.4 mmol) was added via syringe followed by 1.5 mL of Toluene. The reaction mixture was placed in a thermally regulated bath preheated to 125 °C for 23 hrs. The solid polymer sample, once cooled, was dissolved in 20 mL of CH_2Cl_2 and treated with 1mL of MeOH to quench the living end. The polymer was isolated by precipitation into hexane to yield 2.18 g of material that contained some hexane. SEC in DMAC: M_n 26,800, PDI = 1.32 $M_{n(calc)}$, 14,000. T_g 109 °C, T_m 237 °C.

Typical One-Pot, One-Step CROP of POZ and LFRP of Styrene. A round bottom flask, fitted with a condenser and stir bar, was charged with **4** (54.0 mg, 0.13 mmol) and AgOTf (41.3 mg, 0.16 mmol). The vessel was purged 15 min with argon. POZ (1.5 mL, 1.67 g, 11.4 mmol) was added via syringe followed by styrene (1.5 mL, 1.36 g, 13.1 mmol). The reaction mixture was placed in a thermally regulated bath preheated to 125 °C and heated for 44 hours. The solid polymer sample once cool was dissolved in 20 mL of CH_2Cl_2 and treated with 1mL of MeOH to quench living end. The solution was filtered to remove AgCl and precipitated into a 10-fold excess of hexane. The sample was redissolved in CH_2Cl_2 and precipitated into hexanes/ ethyl acetate (9/1, v/v). The precipitate was dried under vacuum overnight to yield 2.47 g (80%) of polymer **7**. ^{1}H NMR (300 MHz, $CDCl_3$, ppm): δ 1.2-1.6 (CH_2 of PS), 1.6-2.2 (CH of PS), 2.5-4 (br, CH_2 of PPOZ), 6.3-6.8 (ArH for PS), 6.8-7.2 (ArH for both PS and PPOZ), 7.2-7.5 (ArH of PPOZ). Integration gave S:POZ ratio of 0.87. SEC in DMAC with PS standards: M_n 28,000, PDI 1.36, M_n(calc) 18,700. DSC: T_g 104 °C, T_m 237 °C.

Formation of PPOZ-b-PS as a Function of Time. To one round bottom flask was charged a stir bar and **4** (0.35 g, 0.82 mmol). The flask was fitted with a 3-way stopcock and evacuated. The process was repeated 3 times. POZ (10.0 mL, 11.18 g, 76.5 mmol) was added via syringe. To a second round bottom was charged AgOTf (0.23 g, 0.89 mmol) and a stir bar. The flask was fitted with a 3-way stopcock and evacuated. This was repeated 3 times. Styrene (10.0 mL, 9.09 g,

87.2 mmol) was added via syringe. Sixteen glass tubes fitted with stir bars and septa were purged with a flow of argon for 5 min. Then the tubes were filled with 0.5 mL of the styrene solution followed by 0.5 mL of the POZ solution. The AgOTf precipitated immediately upon the addition of the POZ. The tubes were degassed with three freeze-pump-thaw cycles and sealed under vacuum. They were then placed in a preheated bath (125 °C). The first tube was removed when the temperature inside an unsealed tube (control) reached 125 °C. This tube became the point for time zero. Tubes were removed at 0, 5, 15, 30, 50, 90 min, and at 2, 3, 5.08, 7, 9.7, 12.83, 17, 21, 25, 43.5 hours. Upon removal the tubes were quickly cooled in liquid nitrogen. The solution hardened at 90 min. The polymer began to turn a pinkish purple after 7 hours. Each tube was broken open and the polymer was dissolved in $CHCl_3$ (stabilized with ethanol) and diluted up to 50.00 mL. Aliquots were removed and injected into the GC to measure the conversion of the monomer.

Hydrolysis of **7c** *to* **8c**. A round bottom flask was charged with **7c** (1.52 g), 60 mL of water and 15 mL of conc. HCl. The mixture was refluxed for 48 hours. After cooling, the mixture was bought to pH 10 (pH paper) using 1M KOH and then dialyzed in a cellulose acetate bag with molecular weight cut off of 2000 daltons. The solution after dialysis had pH 7. The bulk of the water was removed under vacuum and the residue was lyophilized to dryness.

Hydrolysis of **7b** *to* **8b**. A round bottom flask was charged with **7b** (0.60 g), 100 mL of 1,4-dioxane and 20 mL of conc. HCl. The mixture was refluxed for 24 hours. After cooling the solution was brought to pH 13 (pH paper) with 1M KOH. The solvent was removed under vacuum and the residue was extracted with CH_2Cl_2.

REFERENCES

1. (a) Szwarc, M. *Carbanions, Living Polymers and Electron Transfer Processes*; Wiley-Interscience: New York, 1968. (b) Noshay, A.; McGrath, J. E. *Block Copolymers;* Academic: New York, 1977. (c) Velichkova, R. S.; Christova, D. C. *Prog. Polym. Sci.* 1995, **20**, 819. (d) Webster, O. W. *Science* 1991, **251**, 887. (e) Hertler, W. R.; Webster, O. W.; Cohen, G. M.; Sogah, D. Y. *Macromolecules* 1987, **20**, 1473. (f) Kobatake, S.; Harwood, H. J.; Quirk, R. P. *Macromolecules* 1997, **30**, 4238.
2. (a) Rempp, P. F.; Lutz, P. J. In *Comprehensive Polymer Chemistry*; Allen, G.; Ed.; Pergamon: Oxford, 1992, Chapter 12. (b) Rempp, P. F.; Franta, E. *Adv. Polym. Sci.* 1984, **58**, 3. (c) Yoshida, E.; Sugita, A. *Macromolecules* 1996, **29**, 6422. (d) Yoshida, E.; Ishizone, T.; Hirao, A.; Nakahama, S.; Takao, Y.; Endo, T. *Macromolecules* 1994, **27**, 3119. (e) Sogah, D. Y.; Kaku, M.; Shinohara, K.-I.; Rodriguez-Parada, J. M.; Levy, M. *Makromol. Chem., Macromol. Symp.* 1992, **64**, 49, and references cited therein.
3. Nomura, R.; Endo, T. *Chem. Eur. J.* 1998, **4**, No.9, 1605.
4. Hou, Z.; Tezuka, H.; Zhang, Y.; Yamazaki, H.; Wakatsuki, Y. *Macromolecules* 1998, **31**, 8650.
5. (a) Puts, R. D.; Sogah, D. Y. *Macromolecules* 1997, **30**, 7050. (b) Sogah, D. Y.; Puts, R. D.; Trimble, A.; Scherman, O. *Polym Prepr. (Am. Chem. Soc., Div. Polym. Chem.)* 1997, **38**(1), 731.

6. (a) Kobayashi, S. *Prog. Polym. Sci.* 1990, **15**, 751. (b) Bassiri, T. G.; Levy, A.; Litt, M. J. *Polym. Sci., Polym. Letters* 1967, **5**, 871. (c) Kobayashi, S.; Saegusa, T. *Ring Opening Polymerization*; Elsevier: New York, 1985, vol. 2. (d) Puts, R. D.; Sogah, D. Y. *Macromolecules* 1997, **30**, 6826. (e) Schulz, R. C. *Makromol. Chem., Macromol. Symp.* 1993, **73**, 103.
7. (a) Moad, G.; Rizzardo, E.; Solomon, D. H. *Polym. Bull.* 1982, **6**, 589. (b) Georges, M. K.; Veregin, R. P. N.; Kazmaier, P. M.; Hamer, G. K. *Macromolecules* 1993, **26**, 2987. (c) Kazmaier, P. M.; Moffat, K. A.; Georges, M. K.; Veregin, R. P. N.; Hamer, G. K. *Macromolecules* 1995, **28**, 1841. (d) Fukuda, T.; Tareuchi, T.; Goto, A.; Tsuji, Y.; Miyamoto, T.; Shimizu, Y. *Macromolecules* 1996, **29**, 3050. (e) Hawker, C. J. *Angew. Chem. Int. Ed. Engl.* 1995, **34**, 1456. (f) Hawker, C. J. *J. Am. Chem. Soc.* 1994, **116**, 11185. (g) Puts, R. D.; Sogah, D. Y. *Macromolecules* 1996, **29**, 3323.
8. Weimer, M. W.; Scherman, O. A.; Sogah, D. Y. *Macromolecules* 1998, **31**, 8425.
9. Sogah, D. Y.; Weimer. M. W.; Scherman, O. A. *Polym. Mater. Sci. Eng.* 1999, **80**, 86.
10. (a) Hawker, C. J.; Hedrick, J. L.; Malmström, K. E.; Trollsas, M.; Mecerreyes, D.; Moineau, G.; Dobois, P.; Jérôme, E. *Macromolecules* 1998, **31**, 213. (b) Mecerreyes, D.; Moineau, G.; Dobois, P.; Jérôme, E.; Hedrick, J. L.; Hawker, C. J.; Malmström, K. E.; Trollsas, M. *Angew. Chem. Int. Ed.* 1998, **37**, 1274.
11. (a) van Hest, J. C. M.; Delnoye, D. A. P.; Baars, M. W. P. L.; Gendenen, M. H. P.; Meijer, E. W. *Science* 1995, **268**, 1592. (b) van Hest, J. C. M.; Delnoye, D. A. P.; Baars, M. W. P. L.; Gendenen, M. H. P.; Meijer, E. W. Macromolecules 1995, 28, 6689. (c) Gitsov, I.; Frechet, J. M. J. *J. Am. Chem. Soc.* 1996, **118**, 3785, and references cited therein.
12. Tanaka, R.; Ueoka, I.; Takaki, Y.; Kataoka, K.; Saito, S. *Macromolecules* 1983, **16**, 849.
13. Reference 6(c), p 778.

Selection of polymeric materials for biomedical applications

MIROSLAWA EL FRAY

Technical University of Szczecin, Department of Chemical Fibers & Physical Chemistry of Polymers, Pulaskiego Str. 10, 70-322 Szczecin, Poland

Keywords: Sterilization methods; degradation of polymers; synthetic polyesters.

1. INTRODUCTION

Selecting a polymer for biomedical applications it should be specified which chemical and physical properties are required. Although the specific materials requirements will differ according to the nature of the application (degradable sutures, vascular grafts, drug delivery systems, permanent implants etc. [1, 2]), it is fundamental requirement in every case that the polymer should display adequate biocompatibility. Thus, it is very important to perform a detailed investigation of the characteristics of any synthetic polymer *in vitro* in order to establish structure-response relationship before using it as biomaterial.

Polymeric materials had been utilised in the medical practice many years ago when first synthetic polymers were obtained. Polyethylene (PE) display very good biocompatibility and is used as endoprosthesis, cathethers, etc [1, 2]. Polypropylene (PP) has found its application as surgical sutures, medical containers, grids etc. Polyamide is used for similar applications. Polyurethanes (PU) have found widespread application as heart valves, vascular grafts, elements of artificial hearts [3]. In recent years, synthetic triblock or segmented block (multiblock) copolymers exhibiting microheterogenous (hydrophobic-hydrophilic) structures have been paid much attention for biomedical applications [4]. Multiblock polyurethanes form the microphase separated structure composed of the two incompatible segments namely hard segments consisting of urethane or urea linkages, which are dispersed in the matrix of the soft segments. Polymers with such microdomain structure showed reduction of blood cell activation appearing on the polymer surface [5]. Block copolymers having hydrophobic-hydrophilic microdomain structures exhibit non-thrombogenicity due to suppression of the activation of adherent platelets.

Multiblock poly(ester-ether) elastomers with the soft segment composed of modified poly(ethylene oxide) were found to be degradable, biocompatible and bioerodible materials, with their end usage as an artificial tympanic membrane and bone graft substitute [6–10]. Other soft segments material, namely nonlinear hydrophobic dimer fatty acids (aliphatic dibasic acids) was successfully applied to a new class of biocompatible polyanhydride-based polymers developed for a drug carrying [11].

In biomedical applications, environmental stability tends to be particularly important in processing, sterilization and long-term implantation. Multiblock poly(ester-ester)s [12, 13] elaborated in our laboratory derived from a poly(butylene terephthalate) (PBT) and a dimer fatty acid (DFA) are stable during processing [13] and sterilization [14]. As will be demonstrated their hydrolytic stability (hydrolysis and oxidation are the only feasible degradation mechanisms in the physiological environment) depends on the aromatic (hard) and aliphatic (soft) segments content. In the present work the test methods for the detection of degradation of these multiblock poly(ester-ester)s is reviewed and some experimental work on the sterilization of material is discussed.

2. STERILIZATION OF POLYMERS

Several chemical and physical procedures can be chosen for sterilization purposes of biomedical devices. Use of ethylene oxide (EtO) and γ-irradiation are the two possible alternatives. The doses most commonly used for sterilization of the bulk materials for biomedical application lie between 10 and 30 kGy (1–3 Mrad). Electron-beam radiation successfully used as a bectericide, give an attempt to develop sterilization technique for different materials. The only disadvantage is that irradiation of polymers can promote chain scission, crosslinking or photooxidation reactions. Ethylene oxide gas treatment is also capable of degrading the polymer through hydrolysis, and presence of residual EtO may result in cytotoxicity after implantation.

The effect of two sterilization procedure on the physical and mechanical properties of selected multiblock polyester (26%-wt. of PBT hard segments) have been examined in our earlier work [14]. The electron-beam irradiation from the accelerator instead of izotope method was used.

The sterilization of multiblock copolymer can greatly influence the structure of the polymer, depending on the sterilization technique (gas or radiation treatment), as well as from the radiation dose. As we demonstrated, the ethylene oxide gas sterilization does not appear to alter significantly the multiblock polymer properties and structure. Beam-radiation, on the other hand significantly changed polymer structure and properties with increasing radiation doses (from 25 up to 75 kGy). As we also showed, the commonly recommended irradiation with a dose of 25 kGy as a suitable for sterilization of bulk material for biomedical application can be applied for studied polymer because this dose does not affect the polymer structure and mechanical properties.

3. DEGRADATION PROCESS OF POLYMERIC MATERIALS

In general all polymers are susceptible to degradation, but the conditions under which polymers degrade vary within wide ranges. Degraded materials changing their chemical, mechanical and biochemical properties.

Polymeric materials degrade in two ways: either by hydrolytic or by oxidative scissions of the polymer backbone releasing low molecular fragments, oligomers or monomers which might be toxic [15–17]. Other factors leading to degradation mechanism can be the simple incorporation of low molecular weigh compounds (water, lipids, organic acids) weaken the secondary bonds within the polymer structure like plasticizers. This enlarges the distance between the polymer chains and causes the polymer to swell. The consequences are reduced mechanical strength and increased flexibility and softness [18].

Biodegradation of polymers is predictable to a considerable degree based on *in vitro* experiments [19]. Especially simulation tests have been developed to get preliminary information on the relative stability of polymers. One of these is the so-called "boiling test" which must be considered as an *in vitro* screening test for accelerated degradation of polymers and useful method to preselect biomaterials [20]. The indication for the progressive degradation process is the decreasing average molecular weight and the altered molecular weight distribution obtained by the gel permeation chromatography (GPC) for soluble polymers. In case of polymers which cannot be dissolved in organic solvents the extent of degradation after the exposure to boiling water can only be identified by altered mechanical parameters (stress-strain behaviour).

Multiblock copolymers PBT-DFA containing different concentrations of the hard, crystallizable segments [13] varying from rigid plastic to soft elastomer (hard segments concentration was ranging from 26%-wt. up to 70%-wt.). Therefore they cannot be in general dissolved in common organic solvents (except the most flexible sample containing 26%-wt. of hard segments). The polymer samples were boiled under reflux conditions up to 100 hours and the change of mechanical properties and molecular weight were observed with respect to exposition to hot water. The decreasing viscosity of the polymer solution after time intervals (Fig. 1) as well as decreasing of ultimate tensile strength (Fig. 2) provide a sensitive indication of polymer degradation. A remarkable dependence of hydrolytic degradation appeared for sample containing the highest concentration of amorphous phase (26 wt.-% of hard segment). Increased elongation at break (Fig. 3) can indicate incorporation of low molecular fragments in the material, which are plasticizing the molecular structure. Additionally, a so-called half-life can be defined to quantify the effect of degradation, indicating the boiling time necessary before the tested material had lost 50% of its original average molecular weight (limiting viscosity number $[\eta]$ value can be considered as relative indicator of molecular weight). A comparison of the half-life of polymers is shown in Table 1. It can be seen from the Table 1 that polymer containing the lowest content of semicrystalline (PBT) hard segments degrade very fast (half-life comparable

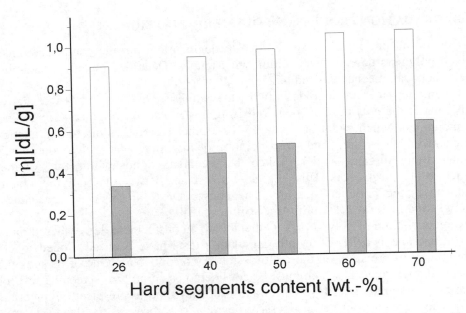

Figure 1. Changes of limiting viscosity number [η] after 100 hours boiling as a function of hard segments content (wt.-%). 1st column: before boiling test; 2nd column: after boiling test.

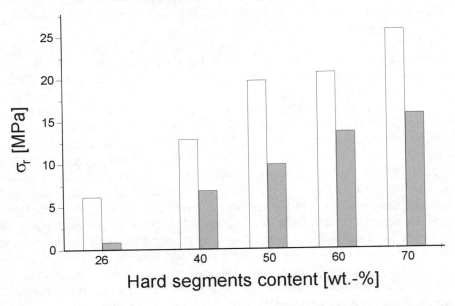

Figure 2. Tensile strength (σ_r) as a function of hard segments content (wt.-%) after boiling time of 100 hours. 1st column: before boiling test; 2nd column: after boiling test.

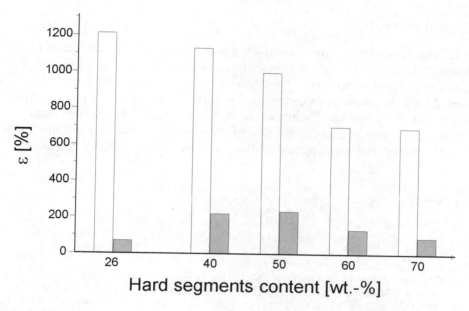

Figure 3. Elongation (ε) of polymers as a function of hard segments content (wt.-%) after boiling time of 100 hours. 1st column: before boiling test; 2nd column: after boiling test.

Table 1.
The degradation of polymer series in boiling water

Hard segments content in polyesters (wt.-%)	Half-life* (hours)
26	< 100
40	100
50	>100
60	>100
70	>100

* half-life estimated based on reduction of limiting viscosity number [η].

to polyurethanes [19]). Degradation *in vitro* is considered as a result of simple hydrolysis and it takes place preferentially in the amorphous phase of these semi-crystalline polymers, as was observed for other semicrystalline materials by Gilding [21] and Chu [22]. They argued that water is able to penetrate amorphous area more rapidly than crystalline areas. Therefore the amorphous regions are first removed by hydrolysis, and then next step of degradation involves the crystalline areas. The ester bonds are readily hydrolyzable, resulting from the primary attack of the hydroxyl ion on the positive carbonyl C-atom. In general, the aromatic polyestrs are less sensitive to moisture than the aliphatic polyesters because of the greater hydrophobicity of the aromatic parts.

4. CONCLUSIONS

As it was demonstrated the molecular weight, expressed by limiting viscosity number, and tensile strength decreases in proportion to the hard segment content in newly developed polymers. Degradation of polymers in the presence of hot water shows variations in their degradation profile according to changes of polymers composition (hard/soft segments i.e. hydrophobic/hydrophilic balance).

Results can suggest that degradation of these polymers can be regulated by using defined length of aromatic and aliphatic residues in the polymer backbone.

REFERENCES

1. J. M. Anderson, *Trans. Am. Soc. Artif. Intern. Organs*, **34**, 101–107 (1988).
2. J. J. Kroschwitz, *"Polymers: Biomaterials and Medical Applications"*, J Willey & Sons, New York 1989.
3. S. Mazurkiewicz, *Polimery* (Warsaw), **44**, 403–406 (1999).
4. S. Dumitriu, *"Polymeric Biomaterials"*, Marcel Dekker Inc., New York 1994.
5. T. Tsuruta, *"Biomedical Applications of Polymeric Materials*, CRC Press, Boca Raton 1993.
6. N. Yui, J. Tanaka, K. Sunai, N. Ogata, K. Kataoka, T. Okano, Y. Sakurai, *Polym. J.,* **16**, 119–126 (1984).
7. D. Bakker, C. A. van Blitterswijk, S. C. Hessling, W. Th. Daems, J. J. Grote, *J. Biomed. Mater. Res.* **24**, 277–293 (1990).
8. A. M. Radder, J. A. van Loon, G. J. Puppels, C. A. van Blitterswijk, *J. Mat. Sci. Mat. Med.*, **6**, 510–517 (1995).
9. E. A. Bakkum, J. B. Trimbos, R. A. J. Dalmeijer, C. A. van Blitterswijk, *J. Mat. Sci. Mat. Med.* **6**, 41–45 (1995).
10. R. M. van Haastert, J. J. Grotte, C. A. van Blitterswijk, *J. Mat. Sci. Mat. Med.* **5**, 764–769 (1994).
11. A. J. Domb, S. Amselem, R. Langer, M. Maniar, *"Biomedical Polymers, Designed-to-Degrade Systems"*, Ed. S. W. Shalaby, Hanser Publishers, New York 1994.
12. M. El Fray, P. Prowans, J. Słonecki, *Polish Patent pending No P 329794* (1998).
13. M. El Fray, J. Slonecki, *Angew. Makromol. Chem.* **234**, 103–109 (1996).
14. M. El Fray, A. Bartkowiak, P. Prowans, J. Slonecki, *J. Mater. Sci.-Mater. Med.*, in press.
15. W. Schnabel, *Polymer Degradation: Principles and Practice*, Carl Hanser Verlag, Wien 1981.
16. A. J. Domb, A. Bentolila, D. Teomin, *Acta Polym.*, **49**, 526–533 (1998).
17. P. J. A. int't Veld, P. J. Dijkstra, J. Feijen, *Adv. in Biomater.* **10**, 420–429 (1982).
18. S. Dawids, *Test Procedures for the Blood Compatibility of Biomaterials*, Kluwer Academic Publishers, Amsterdam 1993.
19. D. F. Williams, *J. Mater. Sci.*, **17**, 1233–1246 (1982).
20. LINI Test, *Meilliand Textilberichte*, **9**, 1008–1009 (1965).
21. D. F. Williams, Ed., *Biocompatibility of Clinical Implant Materials*, CRC Press, Boca Raton, Florida, USA, 1982.
22. C. C. Chu, *J. Appl. Polym. Sci.*, **26**, 1727–1736 (1981).

Synthesis and thermo-responsive properties of poly(N-vinyl caprolactam)/polyether segmented networks

Dedicated to the memory of Prof. Kirsh[†]

NATALYA A. YANUL,[1] YURY E. KIRSH,[1] SAM VERBRUGGHE,[2] ERIC J. GOETHALS[2] and FILIP E. DU PREZ[2,*]

[1]*Laboratory of Membrane Processes, Karpov Institute of Physical Chemistry, Vorontzovo pole 10, Moscow, 106034 Russia*
[2]*Department of Organic Chemistry, Polymer Division, University of Ghent, Krijgslaan 281 (S4), B-9000 Ghent, Belgium*

Abstract—Segmented networks, in which a bismacromonomer acts as macromolecular crosslinker for a polymer with a lower critical solution temperature, are introduced as a new type of hydrogel polymeric structure with thermo-sensitive properties. Thermo-responsive properties such as swelling or shrinking behavior and cloud point temperature of segmented networks based on poly(N-vinyl caprolactam), crosslinked with hydrophilic poly(ethylene oxide) or hydrophobic poly(tetrahydrofuran) α,ω-bis(meth)acrylates, were investigated. It was found that these properties can be regulated by the block-copolymer composition, crosslinking density and nature and molecular weight of the crosslinker.

Keywords: Poly(N-vinyl caprolactam); poly(ethylene oxide); poly(tetrahydrofuran); segmented network; thermo-responsive; LCST; bismacromonomer.

1. INTRODUCTION

In the last two decades, thermo-responsive polymer systems have become the subject of extensive investigations because of their wide range of potential applications in medicine, pharmacology, ecology and even in architecture [1–5]. One important aspect in these investigations is the control of the lower critical solution temperature (LCST) in aqueous solution and hydrogels [6–9]. During the last decade, many concepts to regulate the LCST of water-polymer systems, such as the introduction of additives, copolymerization, variation of polymer concentration and others, have been the subject of a large number of investigations. For the

---

[*] To whom correspondence should be addressed. E-mail: filip.duprez@rug.ac.be
[†] Prof. Kirsh deceased on January 19, 2000.

hydrogels that consist of crosslinked LCST-polymers, some attempts to control the water content by changing the gel polymeric structure have been reported. In order to govern the hydrophilic-hydrophobic balance of the gels, the introduction of a second component in the same network in the form of a randomly copolymerized hydrogel [10] or in a second network in the form of an interpenetrating polymer network (IPN) [11, 12] are two well-known strategies. In most of these studies, the influence of the crosslinkers on the water dependent properties of the hydrogels is not considered. However, as it is generally accepted that the temperature dependence of swelling of such hydrogels is closely related to the temperature dependence of polymer-water and polymer-polymer interactions, it is important to understand the effect of the degree of crosslinking and the nature of the crosslinker on the thermo-responsive properties.

In this work, these effects were investigated on segmented networks, a new type of hydrogel architecture with thermo-sensitive properties. In these AB-crosslinked polymers, a low molecular weight polymer, bearing polymerizable groups at both chain ends (bismacromonomer), acts as a macromolecular crosslinker for another polymer with LCST behavior. In this way, segments with desired philicity can be independently introduced into the polymer network without breaking the sequences of the thermo-sensitive polymer. Moreover, the physicochemical properties of these networks are expected to be refined by the nature, ratio and molecular weight of these crosslinkers. It is known from earlier studies that in such segmented networks not only the hydrophobic-hydrophilic balance but also the morphology of the material can be controlled by the composition and by the crosslink density of the network [13–16]. It was expected that also the thermo-responsive properties would be influenced by these parameters.

A well-known example of a polymer with LCST behavior is poly(N-vinyl caprolactam) (PVCL), which shows a "dissolving – precipitation" transition in water close to physiological temperatures (30°–40°C).

$$PVCL \quad \begin{array}{c} -[CH_2-CH]_n- \\ | \\ N \\ / \quad \backslash \\ CH_2 \quad C=O \\ | \quad \quad | \\ CH_2 \quad CH_2 \\ \backslash \quad / \\ CH_2-CH_2 \end{array}$$

Polymer gels based on PVCL have been described earlier and it was demonstrated that these materials show thermo-responsive properties, such as swelling, shrinking and change in opacity [17]. In the present work, we report on the synthesis of segmented networks based on PVCL and on the thermo-responsive properties of these networks in water. As bismacromonomer, we chose respectively for the hydrophilic poly(ethylene oxide) (PEO) and the hydrophobic poly(tetrahydrofuran) (PTHF). The segmented polymer networks are prepared by free radical initiated copolymerization of N-vinyl caprolactam (VCL) with the corresponding α,ω-bis(meth)acrylates of the bismacromonomer.

2. EXPERIMENTAL SECTION

2.1. Materials

Tetrahydrofuran (Acros 99%) was dried for 24h over calcium hydride and then refluxed over sodium wire until the addition of benzophenone generated a blue color. The dry THF was distilled into the reaction vessel under nitrogen prior to use. 2,2,6,6-Tetramethylpiperidine (Across 98%), trifluoromethane sulfonic anhydride (triflic anhydride) (Aldrich 98%), triethylamine (Aldrich 99%), methacryloyl chloride (Acros 99%), acrylic acid (Acros 99%) and N-vinyl caprolactam (Aldrich 98%) were purified by vacuum distillation before use. Poly(ethylene glycol) (Aldrich, M_n = 600, 2000, 3400 and 4600) was dried by azeotropic distillation in toluene and by further drying under vacuum during six hours at 80°C. Isopropyl alcohol (Aldrich, HPLC-grade), 2,2'-azobisisobutyronitrile (AIBN) (Aldrich) and tetra(ethylene glycol) dimethacrylate (Aldrich) were used without further purification.

2.2. Bismacromonomers and segmented networks

The synthesis of α,ω-acrylate terminated PEO (illustrated for M_n = 2000) starts from the commercially available PEG. Into a 500 ml three-necked flask, equipped with a magnetic stirrer, 0.05 mol dried PEG (100 g), 200 ml of dichloromethane and 0.11 mol of triethylamine (proton trap) were charged under nitrogen atmosphere. To this mixture, 0.11 mol of methacryloyl chloride was added dropwise at a temperature below 10°C. The esterification reaction proceeded overnight at room temperature. The solution was filtrated several times to remove the precipitated triethylamine hydrochloride salt. The clear liquid was washed with 5% NaOH and several times with distilled water until it became neutral. After treatment with $MgSO_4$ to remove water, the solution was filtered and CH_2Cl_2 was removed by rotary evaporation. Further drying was done under high vacuum at room temperature. A yield of approximately 85% was obtained.

The synthesis of PTHF α,ω-bisacrylate with M_n equal to 4000 is as follows: 0.207 ml (1.2 mmol) of freshly distilled triflic anhydride is added to 50 ml (44.45 g, 0.616 mol) of dry THF at 22°C in a two-necked flask provided with a magnetic stirring rod, an inlet for argon and a glass neck equipped with a rubber septum. After 11 minutes, 0.423 ml (6.2 mmol) of acrylic acid is introduced through the septum by means of a hypodermal syringe, immediately followed by 1.2 ml (7.1 mmol) of 2,2,6,6-tetramethylpiperidine. After stirring for 10 minutes, the polymer is precipitated in a cold (0°C) NaOH aqueous solution (0.1 M), filtered on a cooled glass filter, washed with ice water and dried by lyofilization for 24 hours. A yield of 3.8g of PTHF bisacrylate was obtained.

Segmented networks, in the form of 1 mm thick sheets, were obtained as follows: after mixing VCL, the bismacromonomer and a minimal amount (up to 15 wt.-%) of isopropyl alcohol, used as common solvent, in the desired ratio at 40°C for 2 minutes, 1 mol-% (relative to VCL) of AIBN is added. The mixture is degassed for 30 s under vacuum, injected between two hot glass plates (50°C) separated by a 1mm-silicone spacer and kept at 70°C for 16 hrs. After 3 hrs post-curing at 130°C, the film is removed and all soluble fractions are removed with CH_2Cl_2 in a soxhlet apparatus.

2.3. Methods of analysis

^{1}H NMR spectra were recorded in $CDCl_3$ on a Bruker AC360 FT-NMR apparatus. Molecular weights of the bismacromonomers were determined by gel permeation chromatography using a 60 cm 10^3 Å column from Toyo Soda Manufacturing Co and a Melz RI (LCD212) detector with chloroform as an eluent at a flow rate of 1.0 ml/min (polystyrene standards).

The swelling of the PVCL-PEO and PVCL-PTHF networks in distilled water was determined gravimetrically as a function of temperature. After immersion in water at a desired temperature, the samples were removed from the water and tapped with filter paper to remove excess water on the sample surface. The equilibrium weight of the swollen samples was determined after a weight change of less than 1 wt.-%. The degree of swelling was defined as $S = 100 \cdot (W_{sw} - W_o) / W_o$, where W_{sw} and W_o respectively denote the weight of the swollen and dried sample (vacuum, 60°C, 16 hours).

For cloud point temperature (T_{cp}) measurements, the swollen networks are put into test tubes and kept at 0°C for 24 hours. The T_{cp} was determined visually by transferring the test tubes into a water bath, the temperature of which is increased stepwise with 1°C increments every 10 min.

The T_{cp} could also be obtained from differential scanning calorimetry (DSC). A Perkin Elmer DSC7 apparatus with thermal analysis controller TAC7/DX was used. The samples, swollen to equilibrium, were kept into a hermetically closed sample pan at 0°C for 2 hours, then quenched to -100°C and finally heated to 100°C at a scanning rate of 5°C/min. The onset of the endothermal signal at 25–45°C is ascribed to the phase separation effect at the T_{cp} and corresponds quite well (within 2°C) with the temperatures obtained from the visual T_{cp} determination method.

3. RESULTS AND DISCUSSION

3.1. Synthesis of bismacromonomers and segmented networks

Bismacromonomers are preferred building blocks for the synthesis of well-defined segmented networks as they afford the possibility to separate the polymerization process from network formation. In order to obtain macromolecular network structures with control of structural parameters such as molecular weight between the crosslinks and to investigate the influence of the macromolecular crosslinkers on the swelling behavior of the segmented networks, a precise control of molecular weight, molecular weight distribution and functionality is required. Two different strategies were followed to obtain such hydrophobic and hydrophilic bismacromonomers.

Triflic anhydride has been described as an initiator which leads to bifunctionally living PTHF with low polydispersity and predictable molecular weight if the polymerization is carried out in bulk and if the monomer to initiator concentration ratio is properly chosen [18]. In this work, we used this initiator for the synthesis of polyTHF bisacrylate by end-capping of both growing chain ends with acrylic acid in the presence of 2,2,6,6-tetramethylpiperidine (TMP) (Fig. 1).

PEO bismethacrylate was prepared by reacting both OH-endgroups of commercial PEG with methacryloyl chloride in the presence of triethylamine.

The molecular weights, the molecular weight distributions and nomenclature of both kind of bismacromonomers are given in Table 1. The degree of functionalization (F), expressed as the number of (meth)acrylate endgroups per polymer chain, was determined from the integral ratio of the ^{1}H NMR peaks of the polymer chain and those of the (meth)acrylate endgroups. As shown in Table 1, this value was situated between 1.8 and 2.0 for all the samples.

The segmented polymer networks were obtained by free radical copolymerization of VCL with PTHF bisacrylate or PEO bismethacrylate at 70°C with AIBN as initiator. The soluble fraction of the networks, obtained after soxhlet extraction, varied between 3 and 10 wt.-%, depending on the ratio of the components and the molecular weight of the bismacromonomer used. These low values indicate that the crosslinking reaction between VCL and the bismacromonomers proceeds with high yields. The addition of small amounts of isopropyl alcohol as common solvent in the reaction mixture helped to further decrease the soluble fraction. The NMR analysis of the soluble fraction showed that it consisted mainly of linear PVCL. This could be explained by the combination of different parameters, such as differences in copolymerization parameters, viscosity of the reaction mixture, low concentration of polymerizable endgroups of the bismacromonomers, etc.

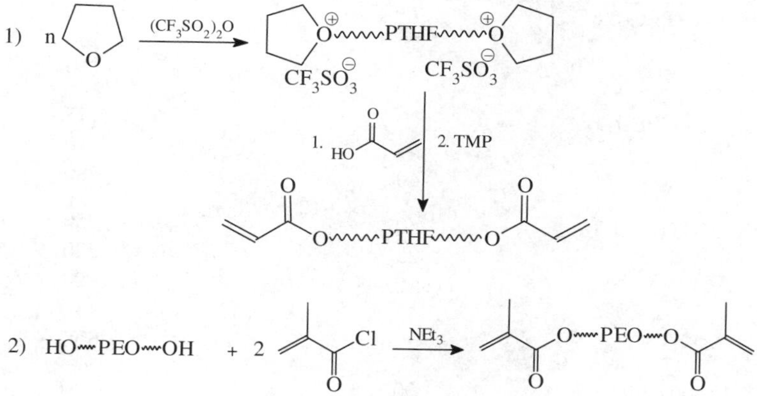

Figure 1. Reaction scheme of the synthesis of the PTHF (1) and PEO (2) bismacromonomers.

Table 1.
Nomenclature, molecular weight (M_n), polydispersity (M_w/M_n) and degree of functionalization (F) of bismacromonomers of PEO and PTHF

	M_n (PEO bis-methacrylate) g/mol	M_w/M_n	F		M_n (PTHF bisacrylate) g/mol	M_w/M_n	F
PEO-0.6	640	1.07	1.8	PTHF-2	1700	1.19	1.9
PEO-2	1960	1.17	2.0	PTHF-3	3200	1.09	2.0
PEO-3	3700	1.09	1.8	PTHF-6	5600	1.14	1.9
PEO-5	4600	1.13	1.9				

3.2. Swelling behavior of the segmented networks in water below the LCST

The degrees of swelling S of the segmented polymer networks were determined as a function of the composition and the molecular weight of the macromolecular crosslinker. The results are presented in Tables 2 and 3.

The data in Table 2 show that the swelling of the PVCL-PEO networks increases with increasing VCL content and with increasing molecular weight of PEO. Both parameters result in lower crosslink densities: the increase of the VCL content favors the formation of longer PVCL chain lenghts between the crosslinks whereas increase of M_n(PEO) leads to increased length of the PEO chains in the network. Thus, the swelling of the PVCL-PEO systems is mainly governed by the crosslinking density, which is the classical behavior of a polymer network in which all the segments swell in a similar way [19].

Table 2.
The influence of the fraction and the molecular weight of the crosslinker PEO on the degree of swelling (S) of the PVCL-PEO networks (at 18°C)

M_n (PEO)	PEO content in segmented networks (wt.-%)	S [a]
640 (PEO-0.6)	75	80
	50	100
	25	150
1960 (PEO-2)	80	175
	50	210
	40	280
3700 (PEO-3)	50	250
	25	320
	10	900
	5	1100
	2	1500
4600 (PEO-5)	50	330
	30	420

[a] The definition of S is given in the experimental part.

Table 3.
The influence of the fraction and the molecular weight of the crosslinker PTHF on the degree of swelling (S) of the PVCL-PTHF networks (at 18°C)

M_n (PTHF)	PTHF content in segmented networks (wt.-%)	S [a]
1700 (PTHF-1.7)	75	9
	50	26
	25	65
3200 (PTHF-3)	80	13
	50	37
	25	90
	10	150
	5	170
	2	250
5600 (PTHF-6)	75	14
	50	40
	25	95

[a] The definition of S is given in the experimental part.

The degrees of swelling (S) of the PVCL-PTHF networks (Table 3) are much lower than those of the comparable PVCL-PEO networks, even for small fractions of the bismacromonomer. For example, the introduction of 2 wt.-% of PTHF-3 leads to an S of 250%, whereas the corresponding network with 2 wt.-% of PEO-3 leads to an S of 1500%.

Another feature of the PVCL-PTHF networks is that M_n(PTHF) has only a small effect on the S-value compared with the PVCL-PEO systems. This is explained by assuming that the hydrophobic PTHF segments form microdomains within the water-swollen network that act as physical crosslinks and therefore lead to less swelling. A schematic comparison of both types of swollen networks is depicted in Fig. 2.

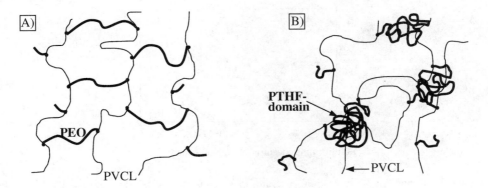

Figure 2. Schematic drawing of a PVCL-PEO (A) and a PVCL-PTHF (B) segmented network, swollen at equilibrium conditions below the T_{cp}.

3.3. Thermo-responsive shrinking effect

Temperature-composition diagrams of the swollen segmented networks are presented in Fig.3.

It is clear that the shrinking effect, i.e. the sharp volume transition of the crosslinked system from swollen to collapsed state by heating above the LCST, is only demonstrated for weakly crosslinked samples (Fig. 3, curves 1–2). For crosslinker contents higher than 10 wt.-%, a continuous contraction over a wide temperature range takes place (curves 3 in Fig. 3a and 3b).

The effect of the hydrophilicity-hydrophobicity of the macromolecular crosslinker on the shrinking behavior manifests itself in the higher water content of PVCL-PEO networks (Fig. 3a), in comparison to PVCL-PTHF networks (Fig. 3b), in the swollen as well as in the collapsed states. For example, the samples with 50 wt.-% of PEO or PTHF crosslinker (curves 3 in Fig. 3) contain 80 and 25 wt.-% of water at 20°C, and 59 and 14 wt.-% of water at 70°C, respectively. The PVCL-PEO networks also display more pronounced contractions. In a temperature range of 5°C, the concentration of water in the PVCL-PEO networks with 2 wt.-% of PEO (curve 1 in Fig. 3a) decreases 25%, while it only decreases 12% in the corresponding PTHF-containing networks (curve 1 in Fig. 3b). Again, this can be ascribed to the nature and hydrophilicity of the crosslinker. The existence of collapsed microdomains in PVCL-PTHF networks, already below the LCST, prevents the PVCL chains to demonstrate a large gel collapse phenomenon.

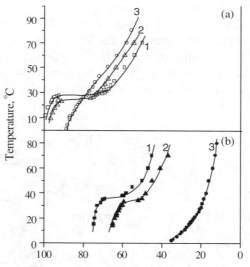

Figure 3. "Temperature-composition" diagrams of swollen PVCL-PEO (Fig. 3a) and PVCL-PTHF (Fig. 3b) networks at equilibrium. Fig. 3a: curve 1 - 2 wt.-% of PEO-3; curve 2 - 10 wt.-% of PEO-3; curve 3 - 50 wt.-% of PEO-3. Fig. 3b: curve 1 - 2 wt.-% of PTHF-3; curve 2 - 10 wt.-% of PTHF-3; curve 3 - 50 wt.-% of PTHF-3.

3.4. Cloud point temperatures (T_{cp}) of water-swollen systems

In Table 4, the T_{cp}-values of water-swollen networks are given as a function of the crosslinker content. These values were obtained by visual observation, although similar values could be found by DSC measurements (Fig. 4). It is clear that these values are determined by the nature of the crosslinker, the block-copolymer composition and the crosslink density and can be varied between 25 and 45°C.

Table 4.
Comparison of the cloud point temperatures (T_{cp}) of swollen PVCL-PEO and PVCL-PTHF networks as a function of their composition

Weight fraction of PEO-3 or PTHF-3 in dry PVCL-PEO and PVCL-PTHF networks /wt.-%	T_{cp} of the swollen PVCL-PEO networks /°C	T_{cp} of the swollen PVCL-PTHF networks /°C
0 [a]	37	37
2	28	36.5
5	26.5	36.5
10	25.5	37
25	32	42
50	33	43.5
75	34	45

[a] Slightly crosslinked PVCL-network, 0.05 mol-% of tetra(ethylene glycol) dimethacrylate as crosslinker.

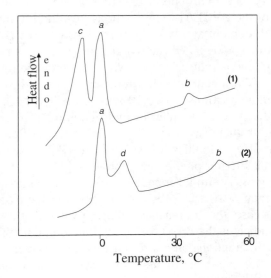

Figure 4. DSC spectra of a PVCL-PEO (1) and PVCL-PTHF (2) segmented network, swollen at equilibrium: a) melting peak of bulk water; b) endothermic peak of phase separation at T_{cp}; c) melting peak of PEO-H$_2$O complex; d) melting peak of crystalline PTHF domains (PEO-3/PVCL and PTHF-3/PVCL = 50/50).

One remarkable observation is that the T_{cp}'s of the PVCL-PEO networks are considerably lower than those of the PVCL-PTHF networks. As the T_{cp} mostly increases with the hydrophilicity of the polymer-water system, the opposite observation could be expected. However, the influence of the introduction of hydrophilic units in thermo-responsive polymer-water systems is not well understood. In some cases, this was found to lead to an increase of T_{cp} [6, 8], in other cases a decrease was reported [7, 9]. It was shown before that in the PVCL/H_2O binary system, the T_{cp} is mainly determined by a competition between PVCL/H_2O hydrogen bond interactions and the hydrophobic interactions between the PVCL sidegroups that respectively will be broken and formed at the T_{cp} [17, 20]. On the other hand, the hydrophilic PEO-segments in the PVCL-PEO networks are also assumed to form hydrogen bond interactions with water, leading to the formation of a PEO-H_2O complex [21–23]. The existence of such a complex in the swollen PVCL-PEO networks could be evidenced by calorimetric measurements. In the DSC spectra, a melting peak is observed at -15°C, clearly separated from the melting peak of bulk water at 0°C (curve 1, Fig. 4). Similar observations have been found in literature [21, 22].

The lower T_{cp}-values of the PVCL-PEO networks could therefore be explained by assuming that the introduction of PEO-segments, that are forced by the network structure to be in the neighborhood of the PVCL chains (Fig. 2), leads to a water-swollen system in which the competition between the PVCL/H_2O and the PEO/H_2O interactions causes weaker PVCL/H_2O interactions and thus water expulsion from the PVCL segments at lower temperatures. In the PVCL-PTHF networks, on the contrary, the hydrophobic PTHF segments do not influence the number and strength of the PVCL/H_2O interactions, resulting in the independence of the T_{cp} on the PTHF content at low weight fractions (2–10 wt.-%, Table 4). In fact, the DSC spectra of PVCL-PTHF networks (curve 2, Fig. 4) show a melting peak of PTHF at 10°C, indicating that the driving force for the microdomain formation in the water environment is still strong enough to form crystalline domains. For PTHF weight fractions above 10 wt.-%, the T_{cp}-value increases continuously. Probably, the size and number of hydrophobic PTHF domains at these contents become so high that they lower the mobility of the PVCL chains. This results in a restriction of the globular formation of the PVCL segments, which is an essential step in the thermo-precipitation process [24, 25], and demands for higher activation energies. A similar increase of the T_{cp} beyond a certain content of the bismacromonomer (> 10 wt.-%) is also observed in the case of the PVCL-PEO networks. In this case, the T_{cp} reaches a plateau value that is close to the T_{cp} of pure PVCL. This indicates that PEO forms separate domains that do not influence the PVCL/water interactions anymore but that are flexible enough to allow the normal aggregation phenomena of the PVCL segments. A more detailed investigation of the phase morphology of the swollen networks and the comparison of the data with the phase behavior of the ternary PVCL-PEO-H_2O and PVCL-PTHF-H_2O mixtures are currently in progress to confirm the above mentioned assumptions.

Acknowledgements

F. Du Prez thanks the F.W.O. (Fund for Scientific Research Flanders) for a postdoctoral fellowship. The INTAS-project (97–1676) and the Belgian Programme on Interuniversity Attraction Poles (IUAP P4/11) initiated by the Belgian State, Prime Minister's office, are acknowledged for financial support.

REFERENCES

1. R. Dinarvand, A. D'Emanuele, *J. Control. Release*, **36**, 221 (1995).
2. M. Okubo, H. Ahmad, *J. Polym. Sci., Polym. Chem.*, **A 36**, 883 (1998).
3. J. E. Chung, M. Yokoyama, T. Aoyagi, Y. Sakurai, T. Okano, *J. Control. Release*, **53**, 119 (1998).
4. E. A. Markvicheva, N. E. Tkachuk, S. V. Kuptsova, T. N. Dugina, S. M. Strukova, Y. E. Kirsh, V. P. Zubov, L. D. Rumsh, *Appl. Biochem. and Biotech.*, **61**, 75 (1996).
5. D. P. Gundlach, K. A. Burdett, *J. Appl. Polym. Sci.*, **51**, 731 (1994).
6. D. L. Taylor, L. D. Cerankowski, *J. Polym. Sci.*, **13**, 2551 (1975).
7. S. H. Yuk, S. H. Cho, S. H. Lee, *Macromolecules*, **30**, 6856 (1997).
8. H. Feil, Y. H. Bae, J. Feijen, S. W. Kim, *Macromolecules*, **26**, 2496 (1993).
9. B. S. Shin, M. S. Jhon, H. B. Lee, S. H. Yuk, *Eur. Polym. J.*, **34**, 171 (1998).
10. Y. Kaneko, R. Yoshida, K. Sakai, Y. Sakurai, T. Okano, *J. Membr. Sci.*, **101**, 13 (1995).
11. S. H. Gehrke, *Adv. Polym. Sci.*, **110**, 81 (1993).
12. T. Aoki, M. Kawashima, H. Katono, K. Sanui, N. Ogata, T. Okano, Y. Sakurai, *Macromolecules*, **27**, 947 (1994).
13. F. E. Du Prez, E. J. Goethals, in: *Ionic Polymerizations and Related Processes*, J. E. Puskas (Ed.), pp. 75–98. Kluwer Academic Publishers, Netherlands (1999).
14. F. E. Du Prez, E. J. Goethals, R. Schué, H. Qariouh, F. Schué, *Polym. Int.*, **46**, 117 (1998).
15. Y. Tezuka, Y. Murakami, T. Shiomi, *Polymer*, **39**, 2973 (1998).
16. B. Iván, J. P. Kennedy, P. W. Mackey, in: *Polymer Drugs and Delivery Systems*, R. L. Dunn and R. M. Ottenbrite (Eds), pp. 194–212. ACS Symp. Book Series 469, Washington, DC (1991).
17. Y. E. Kirsh, *Water Soluble Poly-N-vinylamides,* John Wiley & Sons, New York (1998).
18. F. D'Haese, E. J. Goethals, *Br. Polym. J.*, **20**, 103 (1988).
19. J.-P. Queslel, J. E. Mark, in: *Comprehensive Polymer Science*, vol. 2, C. Booth and C. Price (Eds), pp. 271–309. Pergamon Press (1989).
20. L. M. Mikheeva, N. V. Grinberg, A. Y. Mashkevich, V. Y. Grinberg, L. T. M. Thanh, E. E. Makhaeva, A. R. Khokhlov, *Macromolecules*, **30**, 2693 (1997).
21. N. B. Graham, M. Zulfigar, N. E. Nwachuku, A. Rashid, *Polymer*, **30**, 528 (1989).
22. B. Bogdanov, M. Mihailov, *J. Polym. Sci., Part B: Polym. Phys.*, **23**, 2149 (1985).
23. D. Eagland, N. J. Crowther, C. J. Butler, *Polymer*, **34**, 2804 (1993).
24. K. Otake, H. Inomata, M. Konno, S. Saito, *Macromolecules*, **23**, 283 (1990).
25. Y. E. Kirsh, N. A. Yanul, K. K. Kalninsh, *Eur. Polym. J.*, **35**, 305 (1999).

Poly(rotaxane)s as building blocks for the preparation of cyclodextrin-containing membranes

LAURENT DUVIGNAC and ANDRÉ DERATANI[*]

Laboratoire des Matériaux et Procédés Membranaires, UMR 5635 CNRS-ENSCM-Université Montpellier 2, 2, place Eugène Bataillon, 34095 Montpellier cedex, France

Abstract—Poly(rotaxane)s were synthesised by capping with 2,4-dinitrofluorobenzene the corresponding pseudopoly(rotaxane)s consisting of α-cyclodextrin (α-CD) threaded onto two end-aminated poly(ethyleneglycol) (PEG, Mw 2000 and 3500), with the aim to be used for nanoscale building blocks of facilitated transport membranes. The number of α-CD threaded in poly(rotaxane)s obtained proved to remain constant whatever the initial molar ratio ethylene oxide units: α-CD under our conditions of capping. The resulting polymers had a similar polymolecularity index as the parent PEGs. Thermal properties and light scattering analysis in DMSO showed that inclusion complexation in α-CD stiffens the PEG chains rendering them as semi rod-like macromolecules. A strong tendency to auto-association was observed which is attributed to hydrophobic interactions between the blocking end-groups. Films were prepared by crosslinking poly(vinylalcohol) (PVA) and poly(rotaxane)s with hexamethylenediisocyanate. After removing the end-stoppers under mild alkaline conditions and releasing PEG from the CD cavities, membranes were studied in a water/methanol pertraction system. Large increases in the permeation rate of toluene were observed by replacing α-CD with an α-CD molecular assembly. This effect was interpreted by the cooperativity of CD cavities in the toluene permeation across channels formed inside the membrane.

Keywords: Poly(rotaxane); cyclodextrin; membrane; facilitated transport; pertraction.

1. INTRODUCTION

Cyclodextrins (CDs) are ring-shaped oligosaccharides made of glucopyranoses (Fig. 1) linked together so that the molecules have an internal cavity on the order of a nanometer, the size of which depends on the numbers of units (Table 1) [1]. CDs are polyfunctional with two kinds of hydroxyl groups, each located on a different side of the ring. The primary hydroxyls on C-6 atom are the more nucleophilic whereas the secondary C-2 OH groups are the more acidic. A network of hydrogen bonds easily forms between the hydroxyl groups resulting in a poor solubility of CDs (water and such polar solvents as DMSO, DMF, pyridine). Chemical modification even incomplete disorganises the hydrogen bonding increasing the solubility in a large extent.

---

[*] To whom correspondence should be addressed. E-mail: deratani@crit.univ-montp2.fr

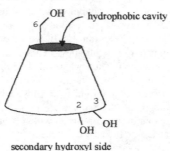

Figure 1. α-Cyclodextrin structure.

Table 1.
Structural characteristics of cyclodextrins

	Number of units	Number of hydroxyl groups	Molar mass	Size of the inner cavity [1]	
				Diameter (nm)	Height (nm)
α–Cyclodextrin	6	18	973	0.47–0.53	0.79
β–Cyclodextrin	7	21	1135	0.60–0.65	0.79
γ–Cyclodextrin	8	24	1297	0.75–0.83	0.79

[1] from W. Saenger, *Angew. Chem. Int., Ed. Engl.*, **19**, 344 (1980).

Considerable interest in CDs comes from their ability to accommodate hydrophobic guests in water. This binding property originates from the amphiphilic nature of CDs, hydrophilic outside – hydrophobic inside. Requirements of a geometric fitting in size and shape between the CD cavity and the molecule included (or a part of it) must be fulfilled to the inclusion take place. As non-covalent bonds are formed, the binding constant is mainly related to the strength of Van der Waals forces and hydrophobic interactions. As a consequence, molecular recognition can be achieved and CDs found numerous applications in separation science. For example, geometric and optical isomers can be readily discriminated using liquid [2] and gas chromatography [3] with supports bearing grafted CDs. Advantage can be also taken of this property to perform extractions of valuable compounds or organic pollutants from aqueous solutions. Several techniques for removal and recovery by CD-complex formation have been proposed such as selective precipitation [4], liquid-liquid extraction [5] and sorption onto CD-immobilised polymer network [6].

CDs have been also introduced as fixed site carriers in facilitated transport membranes [7–10]. In this case, CDs were either embedded or covalently bound to the polymer matrix of non porous films. The proposed uptake mechanism, schematically shown in Figure 2, is mediated by the CD cavities. It consists of three different steps: (a) formation of an inclusion compound at the feed side of the membrane, (b) diffusion by hopping from a complexing site to another inside the film, (c) complex dissociation at the permeate side.

Lee, in a pioneering work [7], prepared the first membranes containing CDs physically entrapped in a cellulosic polymer for pervaporation of aromatic compounds. Incorporation of α-CD resulted in a strong decrease of the membrane permeability. In the same time, a selectivity was observed in extraction of geometric isomers for disubstituted benzenes (for dichlorobenzenes, selectivity values for *p*- over *m*- and *p*- over *o*- were of 3.2 and 2, respectively). It was recently found that poly(vinylalcohol) (PVA), acts as a good polymer matrix for CD-containing membranes likely due to its hydrophilic character and the presence of hydroxyl groups enabling a very high compatibility between both species by hydrogen bonding [8–10]. Yamasaki *et al.* [8] prepared pervaporation membranes for dehydration of C2-C3 alcohols by crosslinking of a mixture PVA and

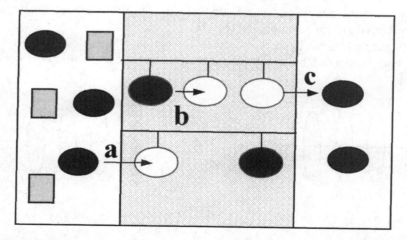

Figure 2. Liquid-liquid extraction mediated by cyclodextrin in facilitated transport membranes.

β-CD/epichlorohydrin prepolymer [11] with glutaraldehyde. The increase of selectivity for water observed was accounted for a decrease of the diffusion of alcohol because of inclusion in β-CD. Miyata *et al.* [9] showed that β-CD embedded in PVA films gave better selectivity values for separation of propanol and *p*- over *o*- xylene isomers by vapour permeation than by pervaporation. Our recent studies [10] have focused on the design of membranes prepared by covalent bonding of CDs to PVA films with epichlorohydrin and hexamethylene diisocyanate (HMDI). They were successfully applied to the water/methanol pertraction of toluene. Pertraction is a continuous liquid-liquid extraction technique based on a membrane process to overcome loss of solvent due to emulsifying tendencies of systems [12]. Pertraction fluxes which are significantly higher than those observed by pervaporation or vapour permeation, varied with the CD content and the cavity size. The β-CD 38 % membrane exhibited a selectivity of 6 for toluene over ethylbenzene and 1.5 for *p*- over *o*-xylene [13].

In the case of the large majority of facilitated transport membranes, the complexing sites are more or less randomly distributed into the polymer matrix. It seems a promising way to design membranes from nanoscale structures constructed by self-assemblies to better control the permeation properties. Macrocycles are especially suitable for this purpose. It is now well recognised that crown ethers and CDs can thread onto a linear polymeric chain to form pseudopoly(rotaxane)s by host-guest complexation (Fig. 3) [14]. Stable supramolecular structures, poly(rotaxane)s are obtained after capping of the terminal extremities by moieties bulky enough to prevent the dethreading of the macrocycles.

Pseudopoly(rotaxane)s can be easily created by inclusion complexation of poly(alkylene oxide) through the cavity of CDs, for instance poly(ethylene glycol) Mw 400-10,000 (PEG) and poly(propylene glycol) Mw 400-4,000 (PPG) with α-CD and β-CD, respectively [15]. Poly(rotaxane)s consisting of many CDs threaded on PEG and triblock copolymer of PEG and PPG were recently synthesised by reacting amine end-terminated chains of pseudopoly(rotaxane)s with large protective groups as stoppers [16, 17].

Poly(rotaxane)s have been proposed as molecular nanoscale devices for electronics and biodegradable materials for pharmaceutical applications [14, 17, 18].

(a) (b)

Figure 3. Schematic structure of (a) pseudopoly(rotaxane) and (b) poly(rotaxane).

Our concern is to investigate how these molecular assemblies can be utilised to form channels of fixed site carriers thereby to increase the permeability by cooperativity in facilitated transport membranes.

We now report on the preparation of such CD-containing membranes. Poly(rotaxane)s formed by threading α-CDs onto PEGs capped by 2,4-dinitrofluorobenzene [16], were attached to PVA using HMDI to yield crosslinked polymeric films. The end-stoppers were then removed under mild alkaline conditions involving the release of PEG chains from the film. Facilitated transport experiments in water/methanol pertraction of toluene were performed on the resulting α-CD containing membranes to validate the concept.

2. EXPERIMENTAL SECTION

2.1. Materials

Bis(amino) PEG (Mw 3350) was purchased from Sigma. O,O'-bis(2-aminopropyl) PEG (Mw 2000), PVA (degree of polymerisation 500, degree of hydrolysis 99.5 mol.-%), DMF and DMSO were obtained from Fluka. 2,4-Dinitrofluorobenzene, hexamethylenediisocyanate and DMSO-$d6$ were purchased from Aldrich. α-CD was kindly supplied from Orsan (France).18 MΩ MilliQ water was used for the preparation of aqueous solutions.

2.2. Measurements

The chromatographic system operating at 70°C, consisted of a Waters 515 pump, a Styragel HR-4 column (Waters), a Dawn™-DSP (Wyatt Technology Corporation) and a Wyatt Optilab DSP. DMSO at a flow rate of 0.3 mL/mn was used as mobile phase. NMR spectra were recorded with a Brüker AC 250 and UV-Vis absorption were measured on Philips PU 8710 spectrophotometer. Thermal properties of polymers were determined on a Perkin-Elmer DSC4 apparatus at a heating rate of 20°C/min.

2.3. Pseudopoly(rotaxane)s 2 [16]

An aqueous solution of end-aminated PEG **1** (1g in 7 mL) was added to a saturated aqueous solution of α-CD (10 g in 69 mL). The mixture was stirred overnight and then freeze-dried. The resulting pseudopoly(rotaxane) was recovered as a white powder and used without further purification.

In the study of the influence of the molar ratio of reactants, the amount of **1a** added was adjusted to the desired fraction.

2.4. Poly(rotaxane)s 3 [16]

A fiftyfold excess of 2,4-dinitrofluorobenzene to end-amino groups of PEG was poured into a 15% solution of **2** in DMF. The reaction mixture stirred overnight at room temperature was then heated at 80°C for 4 hrs. The reaction mixture was precipitated by addition of diethyl oxide. The precipitate was filtered and washed with successively, methanol (3 times) and water (2 times). The crude product was then precipitated twice in pyridine. The resulting yellow solid was dried under vacuum. The overall yields based on the starting PEG were 25-30 %.

3a. ^{1}H NMR (DMSO-$d6$) (250 MHz): δ 3.28 (m,C(2)H α-CD), 3.38 (m,C(4)H α-CD), 3.50 (s,CH$_2$ PEG), 3.56 (m,C(5)H α-CD), 3.63 (m,C(6)H$_2$ α-CD), 3.73 (m,C(3)H α-CD), 4.44 (m,O(6)H α-CD), 4.79 (d,C(1)H α-CD), 5.51 (m,O(3)H α-CD), 5.66 (m,O(2)H α-CD), 7.28 (d,C(6)H phenyl), 8.27 (d,C(5)H phenyl), 8.87 (s,C(3)H phenyl), 7.48-8.92 (NH)

^{13}C NMR (DMSO-$d6$) (62.5 MHz): δ 59.49 (C(6)H$_2$ α-CD), 69.34-69.69 (CH$_2$ PEG), 71.51 (C(2)H α-CD), 72.01 (C(5)H α-CD), 73.16 (C(3)H α-CD), 81.64-81.98 (C(4)H α-CD), 101.87 (C(1)H α-CD)

3b. ^{1}H NMR (DMSO-$d6$) (250 MHz): δ 3.28 (m,C(2)H α-CD), 3.38 (m,C(4)H α-CD), 3.50 (s,CH$_2$ PEG), 3.56 (m,C(5)H α-CD), 3.63 (m,C(6)H$_2$ α-CD), 3.73 (m,C(3)H α-CD), 4.44 (m,O(6)H α-CD), 4.79 (d,C(1)H α-CD), 5.51 (m,O(3)H α-CD), 5.66 (m,O(2)H α-CD), 7.28 (d,C(6)H phenyl), 8.27 (d,C(5)H phenyl), 8.87 (s,C(3)H phenyl), 7.48-8.92 (NH)

^{13}C NMR (DMSO-$d6$) (62.5 MHz): δ 59.38-59.85 (C(6)H$_2$ α-CD), 69.38-69.68 (CH$_2$ PEG), 71.49 (C(2)H α-CD), 71.98 (C(5)H α-CD), 73.14 (C(3)H α-CD), 81.62-81.95 (C(4)H α-CD), 101.86 (C(1)H α-CD)

2.5. Preparation of membranes

0.2 g poly(rotaxane) (0.17 mmol α-CD) and 0.2 g PVA (4.54 mmol repeat unit) were dissolved into 4 mL dry DMSO at room temperature. 0.061 mL HMDI (0.38 mmol) were then added and the solution stirred for 1 min. The free standing film was immediately cast on a glass plate using a doctor blade technique with a gap of 200 µm and dried in an oven at 60°C overnight. The weight ratio poly(rotaxane)/PVA was adjusted according to the chosen composition of the final film. The molar ratio HMDI/total OH groups was kept to 1/10.

2.6. Removing of end stoppers

The yellow films obtained were soaked successively in NaOH 0.1% water/acetone (10/90 v/v) solution up to the complete discoloration of the film (6–8 hrs) and in water which was exchanged several times to remove all residual reactants. The transparent resulting films were ca 20 µm thick. The absence of defects (crack, pin hole) was checked by Scanning Electron Microscopy and gas permeation.

2.7. Pertraction experiments

Liquid/liquid extraction were carried out at 25°C using a cell in which the aqueous and the organic (methanol) phases were separated by the membrane, each solution being continuously recirculated by two peristaltic pumps. The toluene concentration was determined spectrophotometrically at 254 nm.

3. RESULTS AND DISCUSSION

3.1. α–CD/PEG poly(rotaxane)s

We prepared pseudopoly(rotaxane)s **2** from bis aminated PEG (**1a**, Mw 3350) and O,O'-bis(2-aminopropyl) PEG (**1b**, Mw 2000) (Fig. 4). Contacting the PEG samples with a saturated aqueous solution of α-CD at room temperature yielded **2** as white powders. Poly(rotaxane)s **3** were then synthesised in DMF at 80°C using 2,4-dinitrofluorobenzene according to the procedure previously described by Harada et al. [16], modified in a way that the crude product was precipitated in pyridine to eliminate unthreaded α-CD and low molecular adducts instead of to be purified by gel chromatography on Sephadex G-50.

The purity of samples **3** was checked by GPC coupled with a multi angle laser light scattering photometer for the determination of molecular weight and size distribution and a refractive index detector for concentration measurement. As depicted in Fig. 5 for poly(rotaxane) **3b**, the final product was detected by refractive index as a single peak. It must be noted that **2** eluted under the same operating conditions gave rise only to the peak at the longer retention time corresponding to low molecular weight compounds. This demonstrates the dethreading of α-CD from the PEG chains for pseudopoly(rotaxane) compounds resulting in the instability of these molecular assemblies in DMSO.

^{13}C and ^{1}H NMR spectra of **3** showed a considerable broadening of all the peaks assigned to α-CD and PEG and a significant upfield shift (PEG and C3, C5 and C6 of α-CD) which is consistent with the inclusion complexation of ethylene oxide (EO) units by α-CD and a strong decrease of the molecular mobility. The temperature of the glass transition (Tg) is indicative of possibilities of chain

$H_2N\text{-}(CH_2CH_2O)_n\text{-}CH_2CH_2NH_2$ **1a** Mw 3350

$H_2N\text{-}CH(CH_3)CH_2\text{-}(CH_2CH_2O)_n\text{-}CH_2CH(CH_3)NH_2$ **1b** Mw 2000

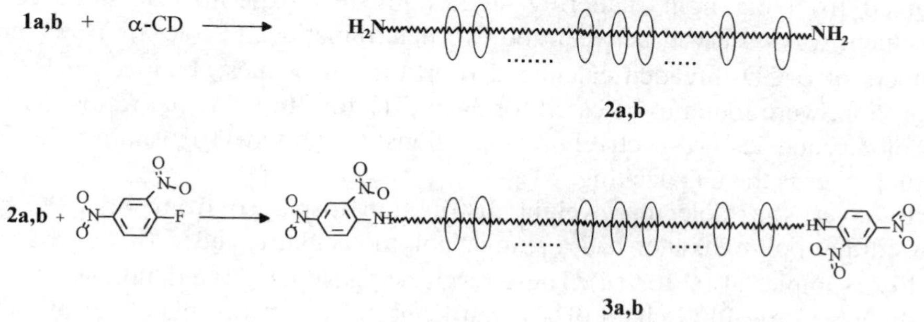

Figure 4. Synthesis of poly(rotaxanes).

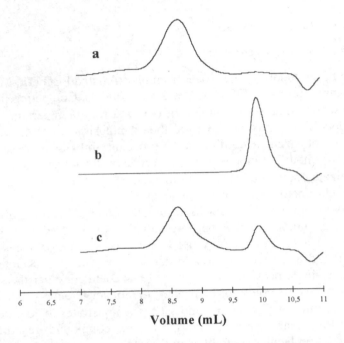

Figure 5. GPC traces of poly(rotaxane) **3b** (refractive index response): (a) product purified by precipitation in pyridine, (b) pyridine soluble fraction, (c) crude product.

movements and related to the stiffness of chains. The introduction of α-CD to form pseudopoly(rotaxane) and poly(rotaxane) inclusion compounds drastically alters the thermal properties of the starting PEG as shown in Table 2. The inclusion complexation is again well exemplified by the disappearance of the melting point and the higher temperature of degradation of PEG chains in compounds **2a** and **3a**. Moreover, the large increase of Tg observed for **2a** and **3a** implies much more rigid structure of the resulting molecular assemblies.

The number-average molecular weight Mn was determined by UV-Vis titration of the end groups of **3** at 363 nm. The weight-average molecular weight Mw was obtained from the light scattering signals in GPC experiments. An excellent agreement was observed between the two determinations (Table 3). The average numbers of α-CD threaded calculated from the Mn values obtained by UV-Vis absorption, were found to be ca 22 for **3a** and 17 for **3b**, i.e. molar ratios EO unit: CD of 3.6 and 2.7, respectively. As previously reported [16], shorter the PEG chain, higher is the CD packing.

Moreover, the molecular weight distribution displayed from the GPC peak showed that polyrotaxanes had a similar polymolecularity index (Ip) as the starting PEG samples (1.02 for **1a**). These results suggest that a fixed number of α-CD are threaded on a PEG chain of a given length. To confirm this observation, the molar ratio EO unit: CD was varied in the range 2 to 15 in the preparation of **2a**.

Table 2.
Thermal properties of pseudopoly(rotaxane) **2a** and poly(rotaxane) **3a** compared to those of PEG **1a**

	1a	α–CD	2a	3a
Glass transition (°C)	-32	–	49	44
Melting point (°C)	63	–	–	–
Degradation (°C)	200	300	300	320

Table 3.
Molecular weight of poly(rotaxane)s **3** and calculated number of α-CD threaded

	$Mw^{1)}$	$Mn^{2)}$	Ip	Number of α-CD threaded[2]	Molar ratio EO : α-CD[2]
3a	25,200	24,700	1.02	22	3.5 : 1
3b	20,000	17,900	1.12	16	2.7 : 1

[1] Determined by light scattering in GPC of **3**.
[2] Calculated from UV-Vis spectra of **3**.

The average number of α-CD threaded calculated for the corresponding poly(rotaxane) proved to be almost constant whatever the initial molar ratio (Table 4). It is believed that the driving force for the formation of such a self assembly of CDs is mainly due to intermolecular H-bonding between neighbouring CDs providing that the conditions of steric fittings are achieved [16, 19]. Our findings and the previously reported results indicate that pseudopoly(rotaxane) precipitates for a given stoichiometry (determined to be 2 EO units per α-CD [15]). During the step of end blocking, a partial dethreading of CD cavities from the polymer chain occurs due to a decrease of intermolecular forces between neighbouring CDs in DMF as above shown in DMSO by GPC. The dethreading, evidenced by the change of the molar ratio EO unit: CD on going from pseudopoly(rotaxane) to poly(rotaxane), is kinetically controlled [20]. As above shown, it depends on the length of the PEG chain and on the blocking reaction conditions.

Table 4.
Molecular weight of poly(rotaxane) **3a** and calculated number of α-CD threaded as a function of the initial molar ratio EO : α-CD

Entry	Initial molar ratio EO : α-CD	$Mn^{1)}$	Number of α-CD threaded[1]	Final molar ratio EO : α-CD[1]
1	2 : 1	24,700	22	3.5 : 1
2	4 : 1	24,500	21	3.7 : 1
3	8 : 1	21,600	19	4.0 : 1
4	15 : 1	21,800	19	4.0 : 1

[1] Calculated from UV-Vis spectra of **3a**.

Table 5 presents the macromolecular characteristics of **3** in DMSO determined from light scattering data. The dn/dc value was found to be close to that generally found for polysaccharides in DMSO (0.062). Figure 6 shows the refractive index and light scattering (90°) trace for **3b**. The light scattering chromatogram exhibits three peak maxima, while only one was recorded by the refractive index detector. The peak at the lower elution volumes on the light scattering chromatogram can be explained by the presence of supermolecular aggregates, the concentration of which being very low. Aggregation was dramatically enhanced when DMSO contained 5 % of water. These results imply that poly(rotaxane)s **3** present a tendency to auto-association (**3b** more than **3a**) even in a such good solvent as DMSO. From our findings, it appears that the phenomenon of aggregation probably occurs by hydrophobic interactions between the protective end-groups. It can be seen on the refractive index profile a shoulder on the lower elution volume side pointing out a beginning of aggregation corresponding to a dimer (Mw 45,000– 60000).

Table 5.
Macromolecular characteristics of poly(rotaxane) **3** in DMSO

	A_2	$<r_g^2>^{1/2}$ [1] (nm)	L rod-like[2] (nm)	$<L^2>^{1/2}$ random coil[2] (nm)	Number of α-CD threaded[3]	CD length[4] (nm)	Fraction of EO units included[5] (%)
3a	$3 \cdot 10^{-3}$	7.5	26	19	22	17.6	57
3b	–	5.5	18.5	13.5	16	12.8	71

[1] Estimated from the elution volume in GPC experiments.
[2] Calculated (see text).
[3] Calculated from UV-Vis spectra.
[4] Calculated from the number of α-CD cavities (0.8 nm height) threaded.
[5] Assuming a stoechiometric ratio EO units : α-CD of 2 : 1.

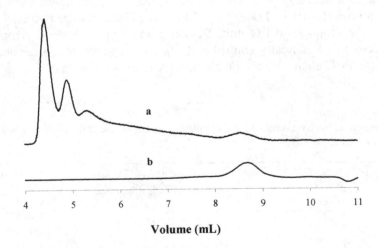

Figure 6. GPC traces of poly(rotaxane) **3b**: (a) 90° light scattering, (b) refractive index.

The elution volume of poly(rotaxane)s in DMSO at 40°C were found to be equivalent to those of poly(styrene) samples Mw ca. 50,000 (**3a**) and 30,000 (**3b**) in THF. From these values, the root mean square radii $<r_g^2>^{1/2}$ may be estimated to be ca. 7.5 and 5.5 nm for **3a** and **3b**, respectively. This means that the conformation of poly(rotaxane)s is more expanded than the random coil of polystyrene in THF. Calculations of the macromolecular sizes using equations giving the relationship between the root mean square radius and the size of simple geometric shapes ($<r_g^2> = L^2 / 12$ for a rod and $<r_g^2> = <L^2> / 6$ for a coil) suggest that poly(rotaxane)s **3** behave in DMSO under an intermediate conformation between the random coil and the rod-like macromolecule (Table 5). In fact, the calculated dimensions of the end-to-end arrangement of α-CD cavities prove to have a similar size as that calculated for a random coil. As can be seen in Table 5, only 60 to 70 % of EO units were included assuming a stoichiometric molar ratio EO units : α-CD of 2 : 1 [16]. As the hydrogen bonding between neighbouring CDs is weakened in DMSO, it can be assumed that CD cavities move along the chain depending on the nature of solvent and temperature, switching from an assembled (insoluble) to a dispersed (soluble) state.

3.2. Facilitated transport membranes

α-CD containing membranes were prepared according to the schematic route presented in Figure 7. PVA was used as the polymer matrix to ensure the mechanical stability of films. Only brittle films can be prepared with more than 80 wt.-% of **3** or 60 wt.-% of α-CD in the starting mixture. The isocyanate chemistry was employed to graft α-CD onto PVA and to crosslink the material instead of epichlorohydrin to avoid the activation of hydroxyl groups in an alkaline medium which can hydrolyse the protective end-groups of poly(rotaxane)s. The casting of DMSO solutions onto glass plates produced yellow films, the colour of which were related to the presence of dinitrophenyl groups.

Cleavage of the end-stoppers was carried out in a mild alkaline medium to prevent the breaking of urethane links. Transparent free-standing membranes were then obtained from **3** with a CD content approximatively the same as in the starting mixture (Table 6). Membranes were prepared under analogue conditions with α-CD/PVA and PVA alone for comparison.

3.3. Pertraction of toluene

In the liquid-liquid pertraction technique using CD-containing films, it is expected that a molecule having a high affinity for CDs can be transferred to the permeate side more rapidly than others, providing that its diffusivity into the membrane are not too low. In this study, performances of membranes were evaluated using toluene as a model molecule and methanol as the organic solvent. It should be noted that the organic phase can be miscible to water in this process since the membrane acts as a separator.

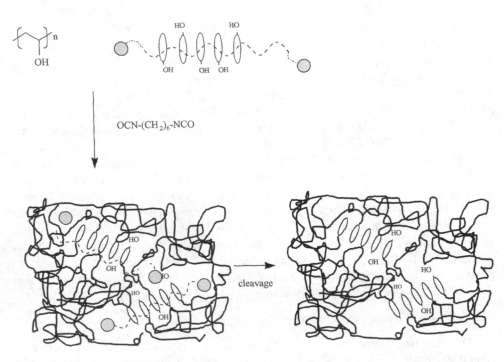

Figure 7. Preparation of membranes.

Table 6.
Facilitated transport membranes prepared from poly(rotaxane) **3a**

Starting weight ratio α-CD / APV / HMDI	25 / 60 / 15	40 / 49 / 14	49 / 34 / 17	61 / 25 / 14
Weight ratio α-CD in membrane (%)[1]	24.2	36.8	46.5	57.3
Thickness (μm)	20–22	17–18	13–16	15–16

[1] Determined by quantitative analysis after acidic hydrolysis [11].

Figure 8 presents plots of the concentration of toluene extracted by α-CD (24 wt.-%) containing membranes compared to PVA membrane. As shown, the transport of toluene is facilitated by the presence of CD cavities inside the membrane demonstrating that the complex binding to CD can accelerate the transfer of molecules to the permeate side of the membrane. As expected, the efficiency of the toluene pertraction was strongly enhanced by using poly(rotaxane) instead of native α-CD as a starting material for the preparation of membrane. This means that the attached α-CD cavities are organised in more or less interconnected channels inside the membrane which improves the hopping from one site to another.

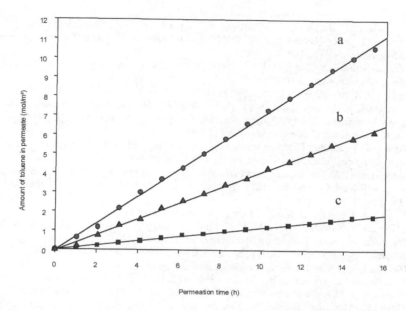

Figure 8. Pertraction of toluene: (a) α-CD (20 wt. %) containing PVA membrane prepared from poly(rotaxane) **3a**, (b) α-CD (20 wt. %) containing PVA membrane, (c) PVA membrane.

4. CONCLUSION

Facilitated transport membranes were synthesised using α-CD/PEG poly(rotaxane)s as molecular assembly of complexing sites. The use of these structures as building blocks gave rise to an organisation of CDs at the nanoscale level in the membranes allowing the permeation rate to increase in comparison with membranes in which CDs are randomly distributed.

Acknowledgement

We are grateful to the Ministry of Scientific Research and Education for the grant of L. Duvignac.

REFERENCES

1. J. Szejtli, *Cyclodextrin Technology,* Kluwer Academic Publishers, The Netherlands (1988); J. L. Atwood, J. E. D. Davies, D. D. MacNicol, J.-M. Lehn, J. A. Ripmeester and F. Vögtle (Eds), *Comprehensive Supramolecular Chemistry,* Vol. 3, Pergamon, Oxford (1996).
2. See for example: T. J. Ward, D. W. Armstrong, *J. Liquid Chomatogr.*, **9**, 407 (1986); A. M. Stalcup, S. Chang, D. W. Armstrong, J. Pitha, *J. Chromatogr.*, **513**, 181 (1990); N. Thuaud, G. Lelièvre, A. Deratani, B. Sébille, *Europ. Polymer J.*, **33**, 1015 (1997).
3. See for example: W. A. König, S. Lutz, P. Mischnik-Lübbecke, B. Brassat, G. Wenz, *J. Chromatogr.*, **447**, 193 (1988); W. A. König in: *New Trends in Cyclodextrins and Derivatives,* D. Duchêne (Ed.), p. 551. Editions de Santé, Paris (1991).

4. E. Y. Zhou, G. L. Bertrand, D. W. Armstrong, *Sep. Sci. Technol.*, **30**, 2259 (1995).
5. I. Uemasu, *J. Incl. Phenom.*, **13**, 1 (1992); J. Andreaus, J. Draxler, R. Marr, A. Hermetter, *J. Colloid Interface Sci.*, **193**, 8 (1997).
6. D. Warner-Schmid, Y. Tang, D. W. Armstrong, *J. Liquid Chromatogr.*, **17**, 1721 (1994); Q. Li, M. Ma, *Chemtech,* **may**, 31 (1999).
7. C. H. Lee, *J. Appl. Polym. Sci.*, **26**, 489 (1981).
8. A. Yamasaki, K. Mizogushi, *J. Appl. Polym. Sci.*, **53**, 1669 (1994); A. Yamasaki, K. Ogasawara, K. Mizogushi, *J. Appl. Polym. Sci.*, **54**, 867 (1994); A. Yamasaki, T. Iwatsubo, T. Masuoka, K. Mizogushi, *J. Membr. Sci.*, **89**, 111 (1994).
9. T. Myata, T. Iwamoto, T. Uragami, *J. Appl. Polym. Sci.*, **51**, 2007 (1994); T. Myata, T. Iwamoto, T. Uragami, *Macromol. Chem. Phys.*, **197**, 2909 (1996).
10. P. Baldet-Dupy, A. Deratani in *Proceedings of the 9th International Symposium on Cyclodextrins*, J. J. Torres-Labandeira (Ed.), Kluwer Academic Publishers, The Netherlands (1999) p. 193.
11. E. Renard, A. Deratani, B. Sébille, *Europ. Polymer J.*, **33**, 49 (1997).
12. S. K. Ray, S. B. Sawant, J. B. Joshi, V. G. Pangarkar, *Sep. Sci. Technol.*, **32**, 2669 (1997).
13. P. Baldet-Dupy, *thesis*, Université de Montpellier, France (1999).
14. D. B. Amabilino, I. W. Parsons, J. F. Stoddart, *Trends Polym. Sci.*, **2**, 146 (1994).
15. A. Harada, J. Li, M. Kamachi, *Macromolecules,* **26**, 5698 (1993).
16. A. Harada, J. Li, M. Kamachi, *Nature,* **356**, 325 (1992); A. Harada, J. Li, T. Nakamitsu, M. Kamachi, *J. Org. Chem.*, **58**, 7524 (1993); A. Harada, *Acta Polymer.*, **49**, 3 (1998).
17. T. Ooya, H. Mori, M. Terano, N. Yui, *Macromol. Chem. Rapid Commun.*, **16**, 259 (1995); H. Fujita, T. Ooya, N. Yui, *Macromolecules,* **32**, 2534 (1999).
18. C. Gong, H. W. Gibson, *Angew. Chem. Int. Ed. Engl.*, **36**, 2331 (1997); S. Anderson, R. T. Aplin, T. D. W. Claridge, T. Goodson III, A. C. Maciel, G. Rumbles, J. F. Ryan, H. L. Anderson, *J. Chem. Soc., Perkin Trans. 1*, 2383 (1998).
19. J. Pozuelo, F. Mendicuti, W. L. Mattice, *Macromolecules,* **30**, 3685 (1997).
20. M. Ceccato, P. Lo Nostro, P. Baglioni, *Langmuir,* **13**, 2436 (1997).

Design, synthesis, and uses of phosphazene high polymers

HARRY R. ALLCOCK,[1,*] JAMES M. NELSON,[1] CHRISTINE R. deDENUS[1] and IAN MANNERS[2]

[1] *Department of Chemistry, The Pennsylvania State University, University Park, Pennsylvania 16802, USA*
[2] *Department of Chemistry, The University of Toronto, Ont., Canada*

Abstract—Polyphosphazenes form one of the most diverse classes of macromolecules. They have a backbone of alternating phosphorus and nitrogen atoms and bear two organic or organometallic side groups linked to each phosphorus. Most of the known examples are prepared via the replacement of chlorine atoms in the macromolecular intermediate $(NPCl_2)_n$ by the use of organic or organometallic nucleophiles. The different polymers generated by the introduction of various side groups span the range of properties from elastomers to glasses, and from hydrogels to bioerodible polymers. Uses are being developed in biomedicine, aerospace technology, fire-resistance, fuel cell membranes, and in solid polymer electrolyte batteries.

However, until recently, the only access route to $(NPCl_2)_n$ was via the thermal ring-opening polymerization of the cyclic trimer, $(NPCl_2)_3$, which yields polymers with broad molecular weight distributions, and little or no means for molecular weight control or access to block copolymers. A new method for the synthesis of polyphosphazenes is described here. It involves the room-temperature, living cationic condensation polymerization of $Me_3SiN=PCl_3$ catalyzed by PCl_5, which gives $(NPCl_2)_n$ with precise molecular weight control and access via living mechanisms to block copolymers, stars, and telechelic systems. The recent synthesis of phosphazene-organic block copolymers by these methods is a major advance toward the commercialization of polyphosphazenes. The impact of these developments on the properties and uses of polyphosphazenes is also mentioned.

Keywords: Polymers; polyphosphazenes; synthesis; living polymerization; block copolymers; stars.

1. INTRODUCTION

Polyphosphazenes are a large group of unusual macromolecules based on the general formula shown in structure **1**.

$$\left[-N = \underset{\underset{R}{|}}{\overset{\overset{R}{|}}{P}} - \right]_n$$

1

---

[*] To whom correspondence should be addressed. E-mail: hra@chem.psu.edu

The inorganic backbone, combined with the organic or organometallic side groups, R, generates a wide range of property combinations that cannot be achieved with classical polymers [1]. For example, the backbone confers fire resistance or fire retardance, and an unusual degree of chain flexibility. Membranes, elastomers, hydrogels, fibers, films, and tailored surfaces can be produced. The biomedical attributes of polyphosphazenes are of considerable interst, as are their their uses as high refractive index glasses, liquid crystalline, and non-linear optical materials. Two areas of increasing focus are the use of polyphosphazenes in solid ionic conductors for rechargeable lithium batteries and fuel cell membranes. The first examples of these polymers were synthesized in our program in the 1960's [2–5], and commercial materials were first developed in the 1970's and 1980's [6–9].

Perhaps the most important advantage of polyphosphazenes is the ease with which different organic side groups can be introduced by the process of macromolecular substitution. This is illustrated in Scheme 1. The key to the effectiveness of this process is the existence of high molecular weight ($M_w = 1 \times 10^6$) poly(dichlorophosphazene) (**3**), a reactive polymeric intermediate that is produced normally by the thermal, molten-state, ring-opening polymerization of the cyclic trimer shown as **2**. A wide range of nucleophilic reagents can be used to replace the chlorine atoms in **3**, including alkoxides, aryloxides, primary and secondary

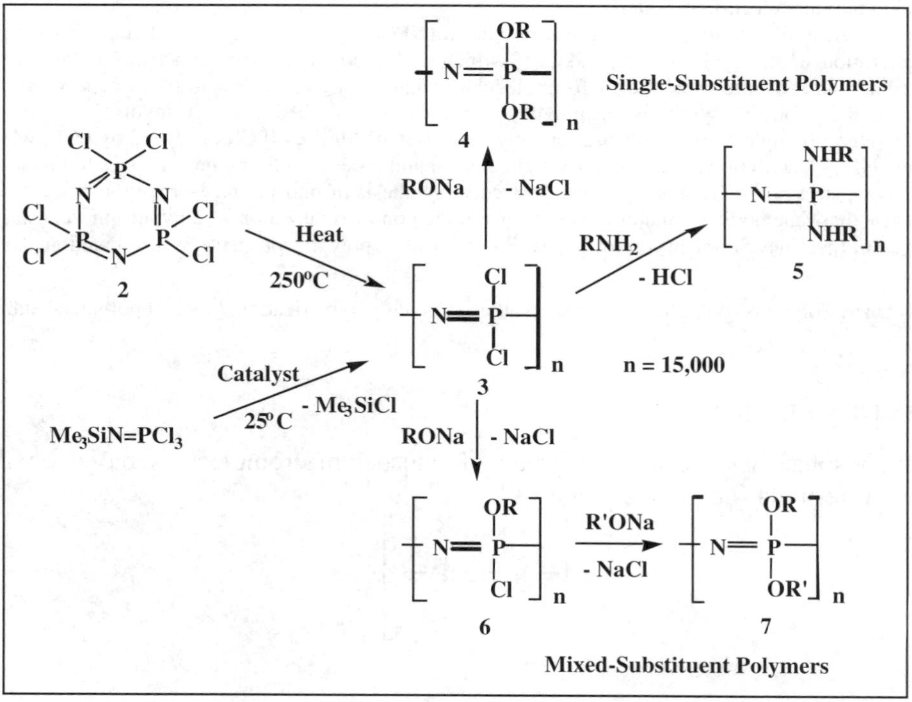

Scheme 1. Macromolecular Substitution with Polyphosphazenes.

amines, and organometallic reagents. By mid-1999 more than 250 different reagents had been utilized in this process. Because two or more different types of side groups can be introduced along the same polymer chain by sequential or simultaneous substitution, the potential number of different polymer structures accessible is very large. To date, more than 700 different polyphosphazenes have been described in the literature, with a large proportion of these being synthesized first in our laboratory at Penn State.

Poly(dichlorophosphazene) is such a crucial component in this process that its method of preparation is a limiting factor in the wider development of this field. The polymer produced by the molten-state, ring-opening polymerization of **2** has a broad molecular weight distribution, and control of the polymer chain length is a challenge. Reproducible polymerization rates are difficult to achieve, especially on a small scale. Solution-state reactions, which might alleviate some of these problems, are difficult to carry out because of the high reactivity of the cyclic trimer with many organic solvents at elevated temperatures. Finally, the ring-opening polymerization process does not lend itself to the production of block copolymers or telechelic macromolecules. This report summarizes recent work in our program to overcome these problems.

It has been known for several years that polyphosphazenes can also be formed by condensation reactions. For example, a report by Flindt and Rose in 1977 [10], a broad series of papers by Wisian-Neilson and Neilson from 1980 to the present [11–13], and reports by Matyjaszewski and Montague [14] demonstrated the direct formation of medium molecular weight poly(*organo*phosphazenes) with alkyl, aryl, or alkoxy side groups via the bulk thermal condensation of phosphoranimines of type $Me_3SiN=PR_2X$, where R = alkyl, aryl, or fluoroalkoxy units. The driving force for this reaction is the elimination of Me_3SiX. Other condensation reactions are known that involve the loss of nitrogen from diorganophosphorus azides [15]. However, no reports existed of the preparation of poly(di*chloro*phosphazene) by these methods. Attempts have been made to prepare this polymer by the elimination of hydrogen chloride from a mixture of ammonia and PCl_5 [16] and by the elimination of $POCl_3$ from $Cl_3PN=POCl_2$, [17] but these are highly corrosive reactions which yield low to medium molecular weight species with very broad molecular weight distributions.

2. THE NEW SYNTHESIS METHOD FOR $(NPCL_2)_n$

We have found that the phosphoranimine, $Cl_3P=NSiMe_3$, polymerizes *at room temperature* in the bulk or solution state in the presence of catalytic amounts of PCl_5 to give poly(dichlorophosphazene) [18]. The polymer molecular weight can be controlled by variations in the monomer to PCl_5 ratios. The molecular weight distribution is narrow. The polymerization is a living process, and this allows access to phosphazene-phosphazene and phosphazene-organic block copolymers. The following sections give details about this process.

3. THE GENERAL REACTION

The overall process is shown in Scheme 2. The monomer, $Cl_3P=NSiMe_3$, can be synthesized by two routes. The first involves the reaction of PCl_5 with $LiN(SiMe_3)_2$. A disadvantage of this process is the formation of $ClN(SiMe_3)_2$ as a side product. This is a powerful inhibitor of the polymerization reaction, and can be removed only with difficulty [19]. The alternative monomer synthesis, from PCl_5 and $N(SiMe_3)_3$, does not yield this impurity and is, thus, preferred [20].

Monomer **7** polymerizes in the bulk (liquid) state at room temperature when treated with small amounts of PCl_5. The proposed initiation and propagation mechanisms are shown in Scheme 2 [19]. However, the bulk state reaction is a heterogeneous process because the $ClSiMe_3$ formed by condensation is immiscible with poly(dichlorophosphazene), and the molecular weight distributions are quite broad (~1.8). The use of solvents such as methylene chloride or cyclohexane gives a homogeneous system and reduces the polydispersity index to 1.06-1.2. At 25°C the polymerization in methylene chloride is complete within 4 hours, while the reaction in cyclohexane is slower, requiring ~24 hours for completion. Other initiators, such as PBr_5, $Ph_3C^+ PF_6^-$, $SbCl_5$, or VCl_4, are also effective.

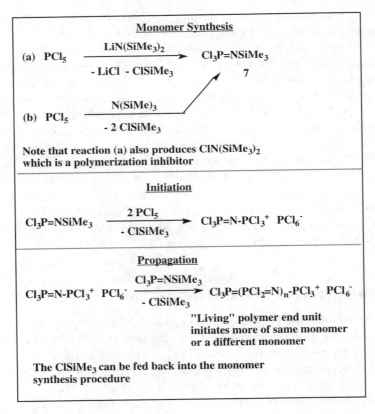

Scheme 2.

In each case, the addition of more monomer restarts the process and gives higher molecular weight polymer. Variations in the monomer to PCl_5 ratio, for example from 4.6 to 1 through 70 to 1 or higher, allow the chain length to be controlled. The molecular weights show a linear increase with conversion. Moreover, the reactions show pseudo first order kinetics. All these facts support the idea that this is a living cationic polymerization.

4. POLYMERIZATION OF ORGANOPHOSPHORANIMINE MONOMERS

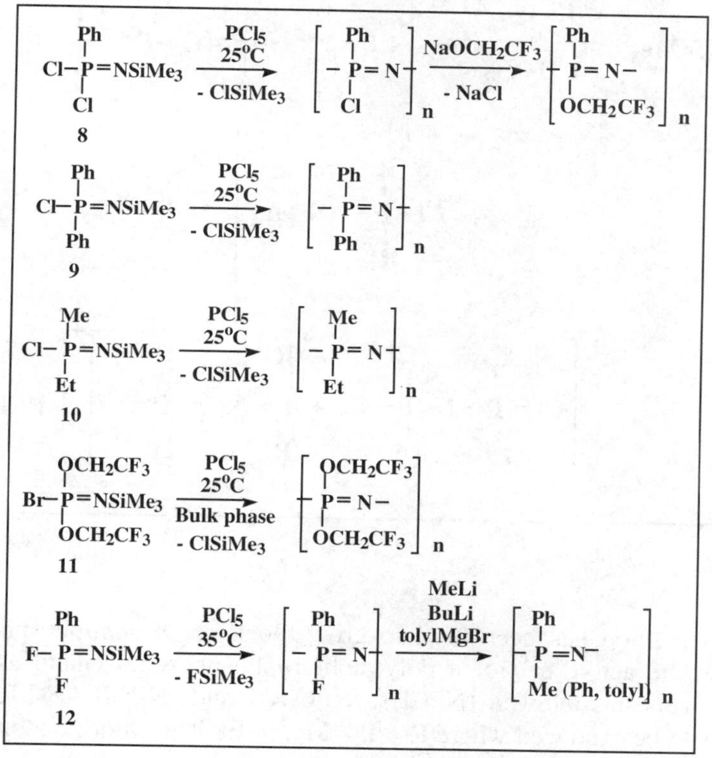

Scheme 3.

Scheme 3 shows a number or *organo*phosphoranimines that also polymerize via the PCl_5-induced cationic process [21]. In general these reactions are slower than those of the trichloro monomer, but they provide access to polymers that can be obtained only with difficulty from poly(dichlorophopsphazene) by reactions with organometallic reagents. Monomers **8–12** also polymerize at elevated temperatures in the absence of initiators [11–13] or with anionic initiation [14]. The polymers formed from **8** and **12** have reactive halogen atoms that can be replaced by organic groups to further increase the number of accessible polymer structures.

5. BLOCK COPOLYMERIZATION

Two methods normally employed for the synthesis of block copolymers are: (1) use of the living end units of a homopolymer to initiate growth of a different monomer, and (2) the use of end-functionalized homopolymers for coupling to initiator units from which a second block can be grown. Both approaches have been developed in our program for polyphosphazenes [22–24].

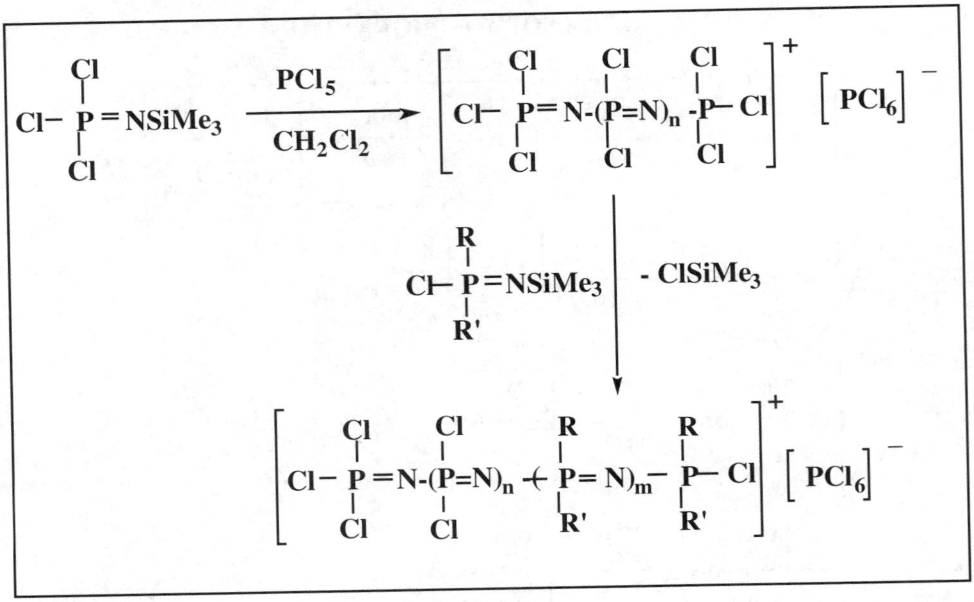

Scheme 4.

The first method has been used to grow blocks of *organo*phosphazene segments from the active end of a poly(dichlorophosphazene) chain, as shown in Scheme 4. This has allowed $(NPCl_2)_n$-$(NPClR)_m$ and $(NPCl_2)_n$-$(NPRR')_m$ block copolymers to be produced where R = Ph, Me, or Et. The chlorine atoms can then be replaced by macromolecular substitution.

The second approach, via telechelic polymers, allows phosphazene-organic block copolymers to be synthesized. An example is shown in Scheme 5, in which an amino-terminated poly(ethylene oxide) is linked to a phosphoranimine initiator unit [23]. Treatment with PCl_5 and exposure to $Cl_3P=NSiMe_3$ then induces phosphazene polymerization. Because poly(ethylene oxide) is available with either one or two terminal amino units, this provides access to di- or tri-block coploymers.

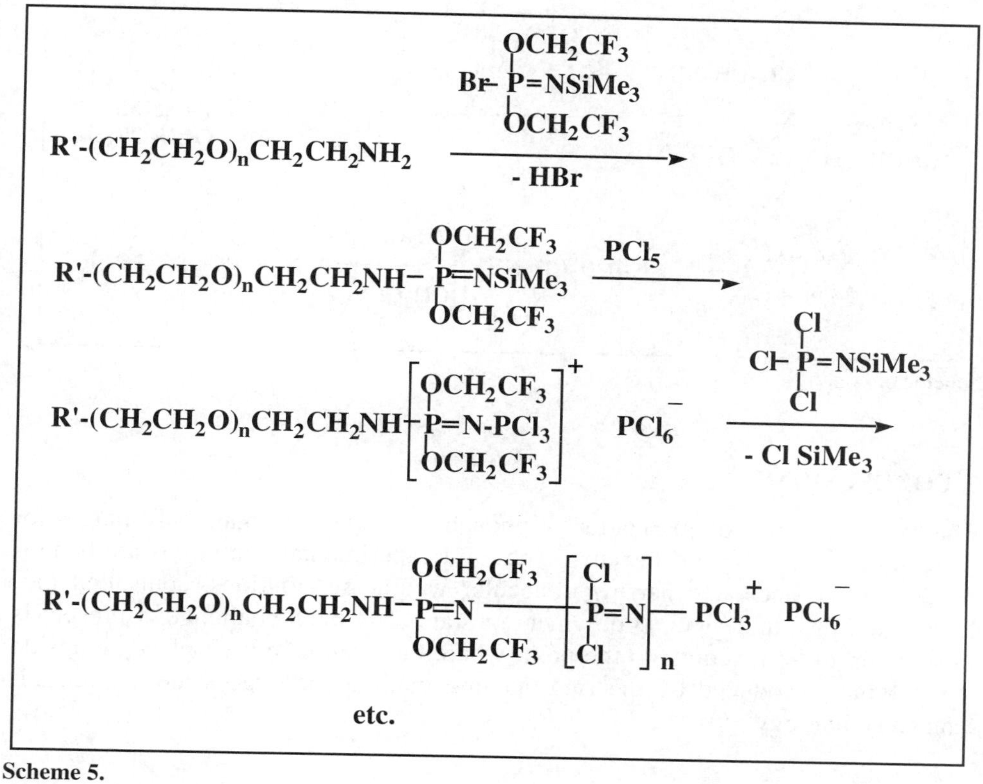

Scheme 5.

$R'-(CH_2CH_2O)_nCH_2CH_2NH_2$ $\xrightarrow[-HBr]{Br-P(OCH_2CF_3)_2=NSiMe_3}$

$R'-(CH_2CH_2O)_nCH_2CH_2NH-P(OCH_2CF_3)_2=NSiMe_3 \xrightarrow{PCl_5}$

$R'-(CH_2CH_2O)_nCH_2CH_2NH-[P(OCH_2CF_3)_2=N-PCl_3]^+ \; PCl_6^- \xrightarrow[-ClSiMe_3]{Cl-P(Cl)_2=NSiMe_3}$

$R'-(CH_2CH_2O)_nCH_2CH_2NH-P(OCH_2CF_3)_2=N-[P(Cl)_2=N]_n-PCl_3^+ \; PCl_6^-$

etc.

6. STAR-GEOMETRY POLYMERS

Scheme 6 illustrates how a tri-functional amine has been used as the core for the synthesis of tri-star polyphosphazenes [25]. As in the case of the linear block copolymers mentioned above, the first step is the linkage of a phosphoranimine to each amine functionality. Living polymerization is then induced by treatment with PCl$_5$ and a phosphoranimine monomer. The properties of the star polymers can, of course, be controlled by the introduction of different organic groups in place of the chlorine atoms. If the chlorine atoms in the star polymer are replaced by –OCH$_2$CH$_2$OCH$_2$CH$_2$OCH$_3$, units the product molecules are useful species for coordination to and solid state transport of lithium ions in energy storage devices [26].

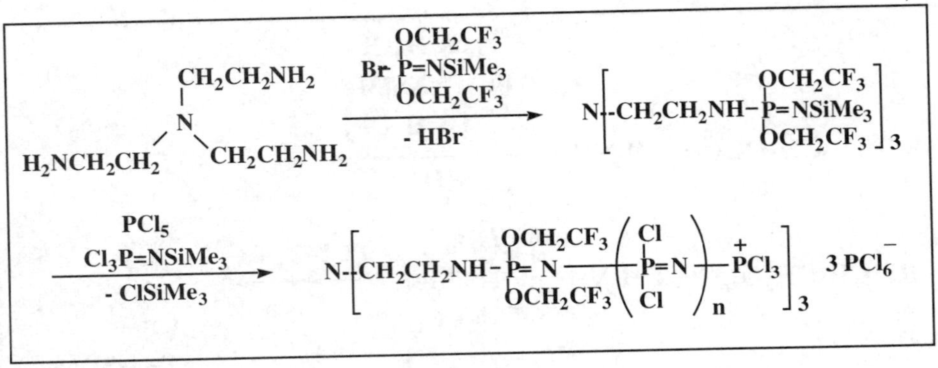

Scheme 6.

7. CONCLUSIONS

The living cationic polymerization of phosphoranimines has many advantages for the synthesis of polyphosphazenes. The mild experimental conditions are beneficial, as is the access to narrow molecular weight distributions, controlled molecular weights, and block copolymers or stars. The main challenge at present is to scale up these reactions to the point where they can take their place alongside the polymers produced by the ring-opening method for large scale uses in advanced technology.

Acknowledgments

We acknowledge the contributions to this work by our coworkers; C. H. Honeyman, C. T. Morrissey, S. D. Reeves, C. A. Crane, T. J. Hartle, M. A. Olshavsky, A. P. Primrose, and R. Ravikiran. This work was supported mainly by grants from the U.S. National Science Foundation through the NSF Polymers Program, the U.S. Federal Aviation Administration, and the Natural Sciences and Engineering Research Council of Canada.

REFERENCES

1. Mark, J. E.; Allcock, H. R.; West, R. *Inorganic Polymers;* 1st ed.; Prentice Hall: Englewood Hills, New Jersey, 1992, pp 272.
2. Allcock, H. R.; Kugel, R. L., *J. Am. Chem. Soc.* 1965, **87**, 4216–4217.
3. Allcock, H. R.; Kugel, R. L.; Valan, K. J., *Inorg. Chem.* 1966, **5**, 1709–1715.
4. Allcock, H. R.; Kugel, R. L., *Inorg. Chem.* 1966, **5**, 1716–1718.
5. Allcock, H. R.; Mack, D. P., *J. Chem. Soc. D* 1970, **11**, 685.
6. Rose, S. H., *J. Polym. Sci., Part B* 1968, **6**, 837–9.
7. Singler, R. E.; Hagnauer, G. L.; Schneider, N. S.; LaLiberte, B. R.; Sacher, R. E.; Matton, R. W. *J. Polym. Sci., Polym. Chem. Ed.* 1974, **12**, 433–44.
8. Tate, D. P., *J. Polym. Sci., Polym. Symp.* 1974, **48**, 33–45.
9. Penton, H. R., *Kautsch. Gummi, Kunstst,* 1986, **39**, 301–4.
10. Flindt, E. P.; Rose, H. *Z. Anorg. Allg. Chem.* 1977, **428**, 204–8.

11. Neilson, R. H.; Wisian-Neilson, P., *J. Macromol. Sci., Chem., A16* 1981, **1**, 425–39.
12. Neilson, R. H.; Wisian-Neilson, P., *Chem. Rev.* 1988, **88**, 541–62.
13. Gruneich, J. A.; Wisian-Neilson, P. *Macromolecules,* 1996, **29**, 5511–12.
14. Montague, R. A.; Matyjaszewski, K., *Polym. Prepr.* 1990, **31**, 679–80.
15. Matyjaszewski, K.; Franz, U.; Montague, R. A.; White, M. L., *Polymer* 1994, **35**, 5005–11.
16. Hornbaker, E. D.; Li, H. M., *Eur. Pat. Appl.* 1980, **18**.
17. De Jaeger, R.; Potin, P., *Phosphorus, Sulfur Silicon Relat. Elem.* 1993, **76**, 483–6.
18. Honeyman, C. H.; Manners, I.; Morrissey, C. T.; Allcock, H. R., *J. Am. Chem. Soc.* 1995, **117**, 7035–6.
19. Allcock, H. R.; Crane, C. A.; Morrissey, C. T.; Nelson, J. M.; Reeves, S. D.; Honeyman, C. H.; Manners, I. *Macromolecules* 1996, **29**, 7740–47.
20. Allcock, H. R.; Crane, C. A.; Morrissey, C. T.; Olshavsky, M. A. *Inorganic Chem.* 1999, **38**, 280–83.
21. Allcock, H. R.; Nelson, J. M.; Reeves, S. D.; Honeyman, C. H.; Manners, I. *Macromolecules* 1997, **30**, 50–56.
22. Allcock, H. R.; Reeves, S. D.; Nelson, J. M.; Crane, C. A.; Manners, I. *Macromolecules* 1997, **30**, 2213–15.
23. Nelson, J. M.; Primrose, A. P.; Hartle, T. J.; Allcock, H. R.; Manners, I. *Macromolecules* 1998, **31**, 947–49.
24. Nelson. J. M.; Allcock, H. R.; Manners, I. *Macromolecules,* 1997, **30**, 3191–96.
25. Nelson, J. M.; Allcock, H. R. *Macromolecules*, 1997, **30**, 1854–56.
26. Allcock, H. R.; Sunderland, N. J.; Ravikiran, R.; Nelson, J. M. *Macromolecules* 1998, **31**, 8026–35.

Design, synthesis and thermal behavior of liquid crystalline block copolymers by using transformation reactions

I. E. SERHATLI,[1] Y. HEPUZER,[1] Y. YAGCI,[1] E. CHIELLINI,[2] A. ROSATI[2] and G. GALLI[2]

[1]*Istanbul Technical University, Chemistry Department, 80626 Maslak-Istanbul, Turkey*
[2]*University of Pisa, Department of Chemistry and Industrial Chemistry, 56126 Pisa, Italy*

1. INTRODUCTION

Academic interest in the development of liquid crystalline (LC) polymers has focused largely on synthesizing and characterizing these systems and understanding their structure-property relationships [1]. LC polymers are in fact excellent structural materials for engineering applications, such as excellent chemical and heat resistance, high mechanical properties in the direction of orientation, high dimensional stability, low viscosity during processing, low thermal shrinkage, optical and rheological properties.

On the other hand, ferroelectric LC polymers exhibit polarization bistability and electro-optic response in the range of microseconds, therefore being functional materials of great potential for optical switches, light valves, display and storage devices.

More recently, much attention has been devoted to LC block copolymers. These systems are able to microphase-separate in ordered morphologies, because of the incompatibility of the different polymer segments. In particular, when a polymer system shows LC behavior, the order at molecular level as well as at supramolecular level can be studied in a single copolymer structure. LC block copolymers may play a role as interface active additives and shear flow modulators in the processing of engineering thermoplastic blends. In addition, they can provide fundamental elements in the elucidation of intriguing aspects of polymer physics. Block copolymers with LC segments can in fact be taken as models to improve the comprehension of the relationships between molecular properties, including block dimensions and their relative compositions, and the morphology, dimension and shape of the domain structure. The linking of an LC polymer with a conventional polymer in the form of an LC block or an LC graft copolymer allows a wide range of possible structure-property combinations. The macromolecular engineering of block and graft copolymers via various synthetic routes has resulted in the realization of very diverse copolymer structures.

To develop ordered morphologies it is necessary to polymerize block copolymers with a narrow molecular weight distribution which is normally realized by living anionic polymerization. For LC side chain polymers this polymerization is not suitable because most mesogens possess groups which react with the living anion. Moreover, the electronic structure of the monomers should allow these monomers to be polymerized preferably by an anionic mechanism. Therefore, polymer-analogous reactions on preformed block copolymers were performed to obtain the block copolymers with an LC block [2, 3]. Subsequently, a number of different routes to synthesize LC block and graft copolymers have been reported, e.g, group transfer polymerization [4], cationic polymerization [5, 6], combined anionic and photopolymerization [7], metathesis [8], direct anionic polymerization [9] and polycondensation [10–13].

In our work, we have developed a different synthetic strategy using "transformation reactions". This reaction scheme involves a synthetic route in which two or more mutually exclusive polymerization mechanisms are sequentially combined [14–18].

The transformation approach allows different propagating species and involves multiple combination of monomers [19–22]. Thus, monomer A is polymerized by a mechanism I to produce a polymer with a functional group F (by initiation or termination) that is capable of initiating a new polymerization by a different mechanism II, (Scheme 1). The polymer formed by mechanism I is used as a macroinitiator for the polymerization of other monomers using mechanism II. Indirectly one polymerization mechanism is transformed into another in order to incorporate into the copolymer combinations of blocks or grafts which could not be done by one single mechanism.

Structurally suitable azo initiators can be used as transformation agents, since they are able to combine other polymerization routes with radical polymerizations (Scheme 2). Bifunctional azo initiators with groups enabling them to participate in condensation or addition reactions can be classified as condensation-radical and addition-radical transfer agents, respectively.

2. TRANSFORMATION OF CATIONIC POLYMERIZATION TO RADICAL POLYMERIZATION

Until very recently, only tetrahydrofuran polymerization fulfilled the living conditions which is a critical requirement in the preparation of macroinitiators for a quantitative introduction of functional groups. Therefore, most of the reported cation to radical transformations involved poly(tetrahydrofuran) as a cationic segment. This type of transformation was the introduction of a common radical initiator group, such as the azo group, into the center of poly(tetrahydrofuran).

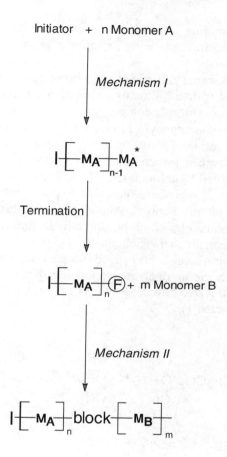

Scheme 1.

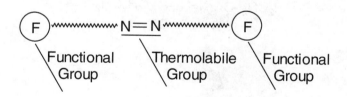

Scheme 2.

Expectedly, subsequent thermolysis would yield to polymeric radicals (Scheme 3). These polymers were used to initiate the polymerization of suitable LC monomers to produce AB and ABA block copolymers [23–25]. For instance, AB block copolymers were constituted by a poly(tetrahydrofuran) block and an LC polymethacrylate block. The latter was built up from either one of two methacrylate monomers containing a variously substituted azobenzene mesogenic unit, which had proved to form nematic, or smectic and nematic, mesophases in the relevant homopolymers.

The block copolymers were synthesized via a two-step cationic and free radical initiation process starting from a bifunctional low molar mass initiator containing one azo group and two acyl chloride end groups.

The α,α'-azobis(4-cyanopentanoyl chloride) 1 is highly efficient in initiating the free radical polymerization of a variety of monomers, very much like the most typical initiator AIBN, even if incorporated into a polymer chain. When reacted with a silver salt constituted by a low nucleophilic counterion in tetrahydrofuran, 1 produced diacyl cations 2 which are then able to polymerize tetrahydrofuran at both ends by an addition mechanism. Accordingly, in this reaction stage the macroinitiator 3 was obtained based on poly(tetrahydrofuran) and possessing one reactive azo group in the main chain (Scheme 3).

Subsequently, macroinitiator 3 was used to generate poly(tetrahydrofuran) macroradicals 4 through the thermal decomposition of the azo group at 70 °C, which in turn initiated the free radical polymerization of the LC methacrylate or acrylate monomers (Scheme 4).

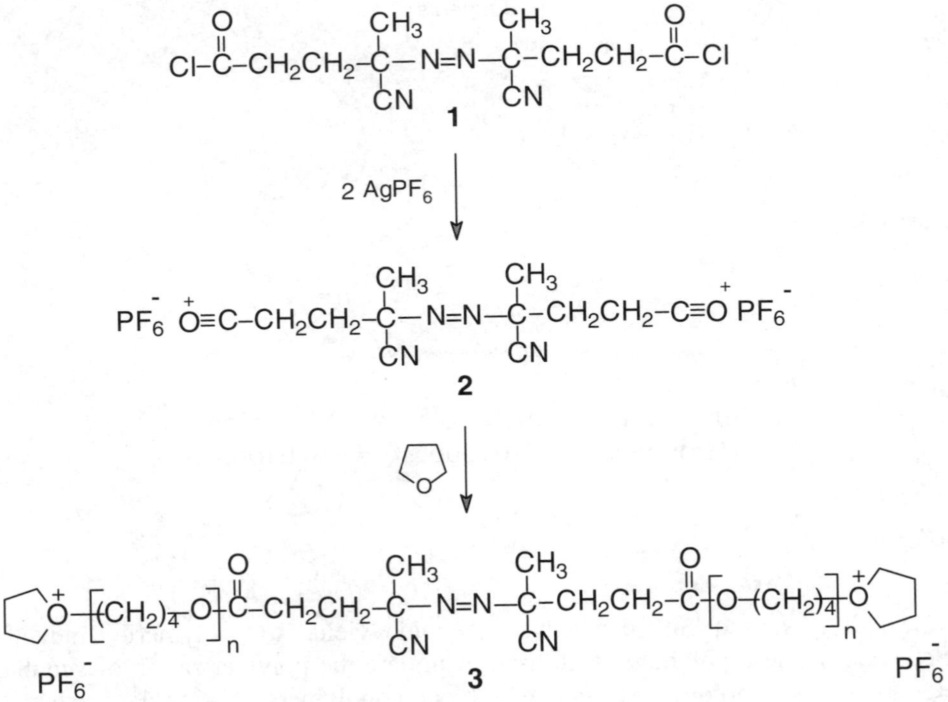

Scheme 3.

It is well established that free radical polymerization of methacrylate and acrylate monomers terminates by a disproportionation and combination process, respectively which in the present polymerization system results in the formation of AB and ABA block copolymers, respectively. Copolymers 5a-c and 6a-b were obtained with different molecular weights and block compositions by using different monomer ratios in the feed mixture and reaction times.

Living cationic polymerization enables control of the molecular weight distribution of the end groups in the resulting polymer. Matyjaszewski and Webster have reported a new initiating system for the synthesis of living alkyl vinyl ether polymers [26, 27]. In this system, a mixture of a strong protonic acid (trifluoromethanesulfonic acid or methyl triflouromethanesulfonate) and a Lewis base (tetrahydrothiophene), which stabilizes the active chain ends, produces the living polymerization (Scheme 5).

Amorphous-LC poly(vinyl ether) block copolymers were synthesized via a two-step procedure involving living cationic and free radical polymerization systems [28]. In the first stage an LC poly(vinyl ether) macroinitiator 7 was formed (Scheme 6). This was then use to polymerize various monomers, like methyl methacrylate or styrene, which resulted in the incorporation of a second, amorphous block component in the copolymers (Scheme 7).

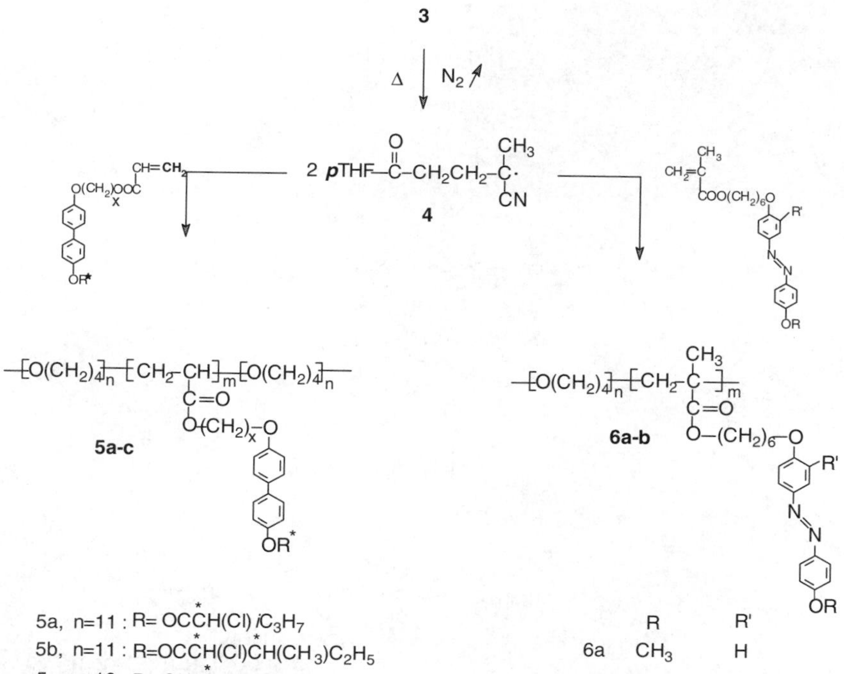

5a, n=11 : R= OCCH(Cl) iC_3H_7
5b, n=11 : R=OCCH(Cl)CH(CH_3)C_2H_5
5c, n=10 : R= CH_2CH(CH_3)C_2H_5

	R	R'
6a	CH_3	H
6b	$(CH_2)_5CH_3$	CH_3

Scheme 4.

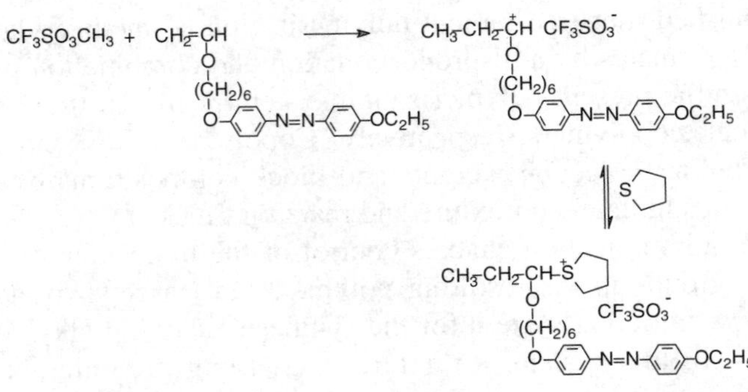

Scheme 5.

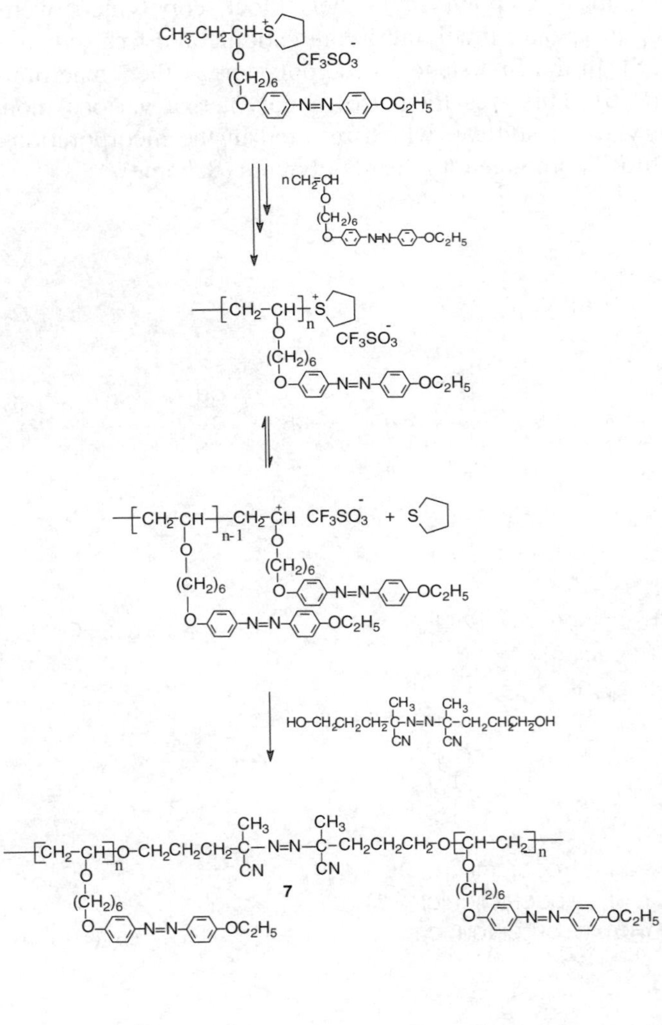

Scheme 6.

AB Block Copolymer

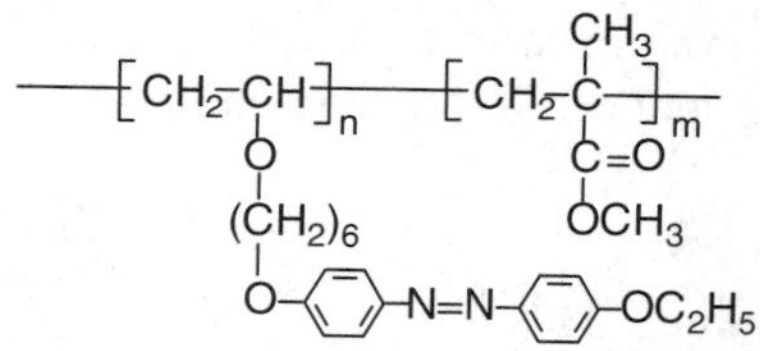

$$-[CH_2-CH]_n-[CH_2-C(CH_3)]_m-$$
with side chains: $O-(CH_2)_6-O-C_6H_4-N=N-C_6H_4-OC_2H_5$ on the first block and $C(=O)OCH_3$ on the second block.

8

ABA Block Copolymer

$$-[CH_2-CH]_n-[CH_2-CH]_m-[CH_2-CH]_n-$$
with side chains: $O-(CH_2)_6-O-C_6H_4-N=N-C_6H_4-OC_2H_5$ on the outer blocks and phenyl on the middle block.

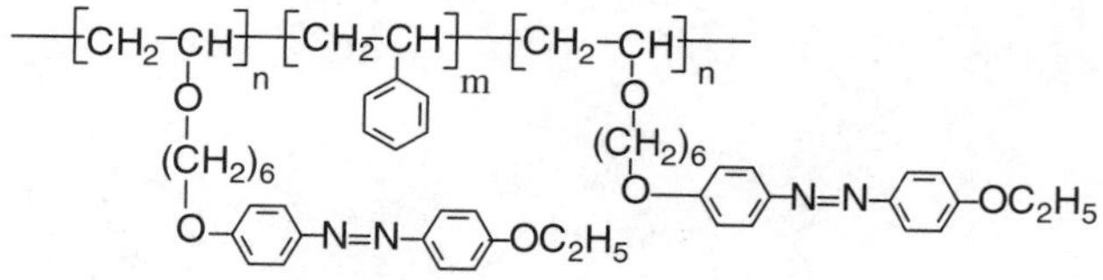

9

Scheme 7.

Block copolymers of poly(vinyl ether)-poly(methyl methacrylate) 8 and poly(vinyl ether)-poly(styrene) 9, were obtained with different compositions and conversions using different monomer ratios in the feed mixture. It should be pointed out that the termination mode of the vinyl monomer used in the second stage plays an important role in the structure of the resultant block copolymers.

A block of ABA type is obtained if the termination of macroradicals consisting of vinyl monomers occurs by combination, as in the case of styrene. On the contrary, block copolymers of the AB type will be formed if the termination of macroradicals occurs by disproportionation, as in the case of methyl methacrylate.

A typical example for the transformation from cationic to radical reactions, involves the use of macroinimers having both photoinitiator and polymerizable end functional groups, or macroinitiators containing both thermal and photoactive functional groups. Recently, thermally and photochemically active poly(tetrahydrofuran) macroinimers, which are essentially polymeric molecules containing both initiator and polymerizable end groups, were used for thermal and UV curing of formulations containing acrylate monomers in the absence of added low-molar mass photoinitiator [29, 30]. According to the macroinimer and macroinitiator concept [31], new classes of thermotropic block and graft copolymers consisting of LC polymer segments were synthesized [23, 32–35]. According to this approach, poly(tetrahydrofuran) macroinitiators and macroinimers possessing the appropriate thermal and photochemical functionality were used as precursors for the subsequent reactions of block and graft formation respectively.

The macroinimer 12 and the macroinitiator 17 were synthesized essentially by regulating the initiation and termination steps of the cationic polymerization of tetrahydrofuran. The living polymerization of tetrahydrofuran initiated by the

182 I. E. Serhatli et al.

methacryloyl chloride-silver salt system and by the α,α'-azobis (4-cyanopentanoyl chloride)-silver salt system was terminated with 2-picoline-N-oxide 11 to prepare the corresponding macroinimer 12 (Scheme 8) and macroinitiator 17 (Scheme 9), respectively.

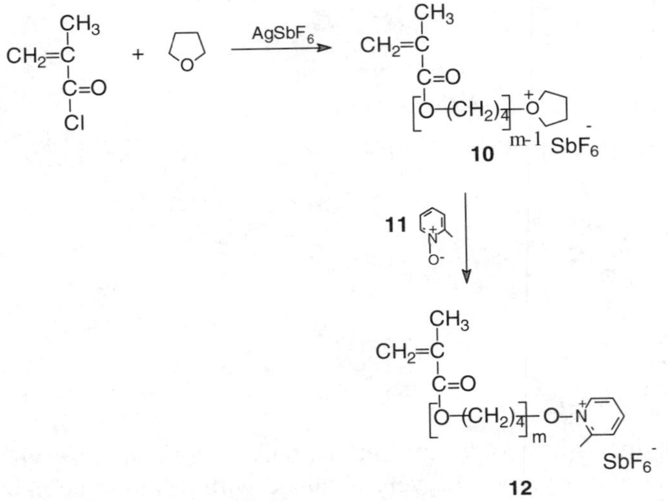

Scheme 8.

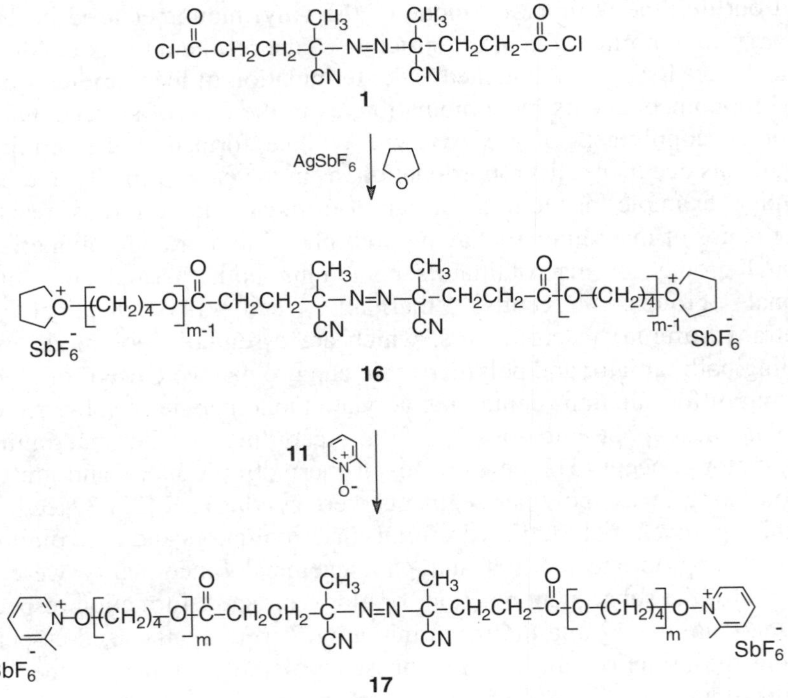

Scheme 9.

Macroinimer 12 and macroinitiator 17, having pyridinium groups were used in indirect photochemical polymerization of MMA to yield side and main-chain block copolymers, respectively. Detailed laser flash photolysis studies on this type of polymerization has been reported previously [36]. Upon irradiation of solutions containing polymers 13a-b and 18, and anthracene, macroradicals are

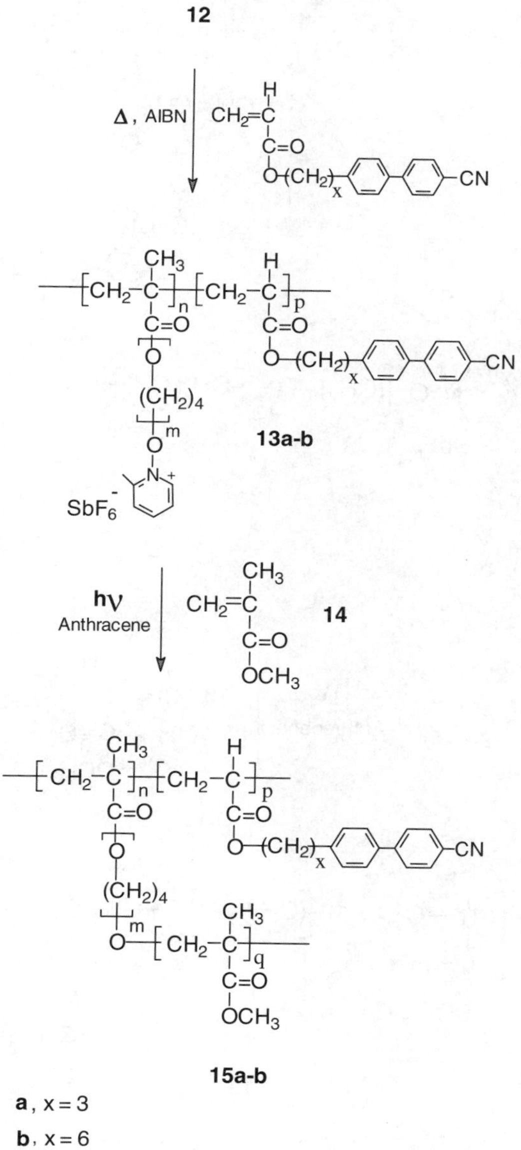

a, x = 3
b, x = 6

Scheme 10.

formed via electron transfer from the electronically excited sensitizer to the terminal pyridinium ion. Thus, a sensitizer radical cation and a pyridinyl radical are formed [37]. The latter rapidly decomposes to give pyridine and a polymeric radical capable of initiating MMA polymerization. Thus, block copolymers 15a-b and 19 are formed.

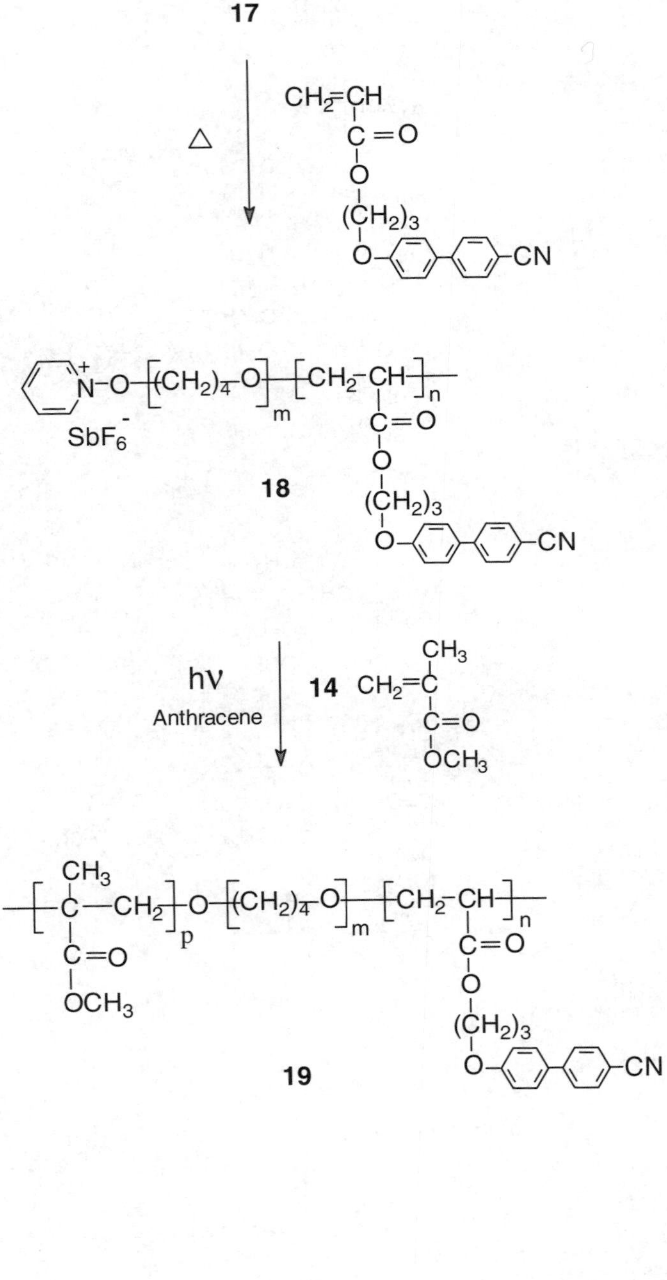

Scheme 11.

3. TRANSFORMATION OF PROMOTED CATIONIC POLYMERIZATION TO RADICAL POLYMERIZATION

It is well known that oxidation of electron donor radicals to corresponding cations may be used to initiate cationic polymerization of compounds such as epoxides, cyclic ethers and alkyl vinyl ethers. The overall process may be presented by the following reaction and useful oxidants include onium salts [38] and pyridinium salts [39].

Initiators like 20 possess two chromophoric groups, namely the azo and the benzoin methyl ether moities, that differ significantly in thermal activity and photoactivity. UV-light irradiation of the benzoin group provides free radicals (Scheme 12) and the oxidation of the electron donor radical to the corresponding carbocation may be conveniently employed to promote the cationic polymerization of epoxides, cyclic ether and vinyl ethers [40]. Irradiation of benzoin terminated polymers in conjunction with a pyridinium salt in the presence of cyclohexene oxide as a cationically polymerizable monomer can lead to the formation of block copolymers of different chemical nature [23, 41].

Following this scheme, block copolymers 26a-b were synthesized by a two-step procedure (Scheme 13). Photoinitiated polymerization of cyclohexene oxide using 20 in conjunction with 22 gave poly(cylohexene oxide) 23 incorporating one azo group per polymer chain. At the irradiation wavelength, $\lambda = 350$ nm, most of the emitted light is absorbed by benzoin chromophores ($\varepsilon = 225$ M^{-1}cm^{-1}), as the azo groups have a much weaker absorption ($\varepsilon = 20$ M^{-1}cm^{-1}). Poly(cyclohexene oxide) having functional azo group in the main chain was then used as a source of macroradicals at 70°C to polymerize LC acrylates 25a-b to yield block copolymers 26a-b.

In the synthetic scheme adopted, a macroinitiator is obtained by promoted cationic polymerization. The macroinitiator incorporates one central labile azo group per polymer chain. This generates macroradicals that then polymerize a mesogenic acrylate monomer by a free-radical mechanism leading to block copolymers [42].

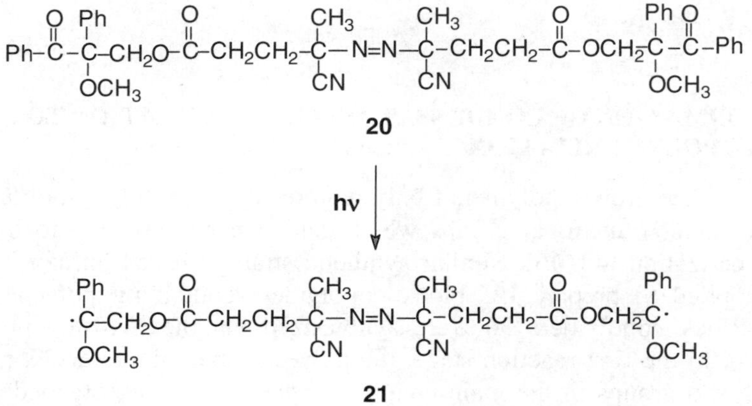

Scheme 12.

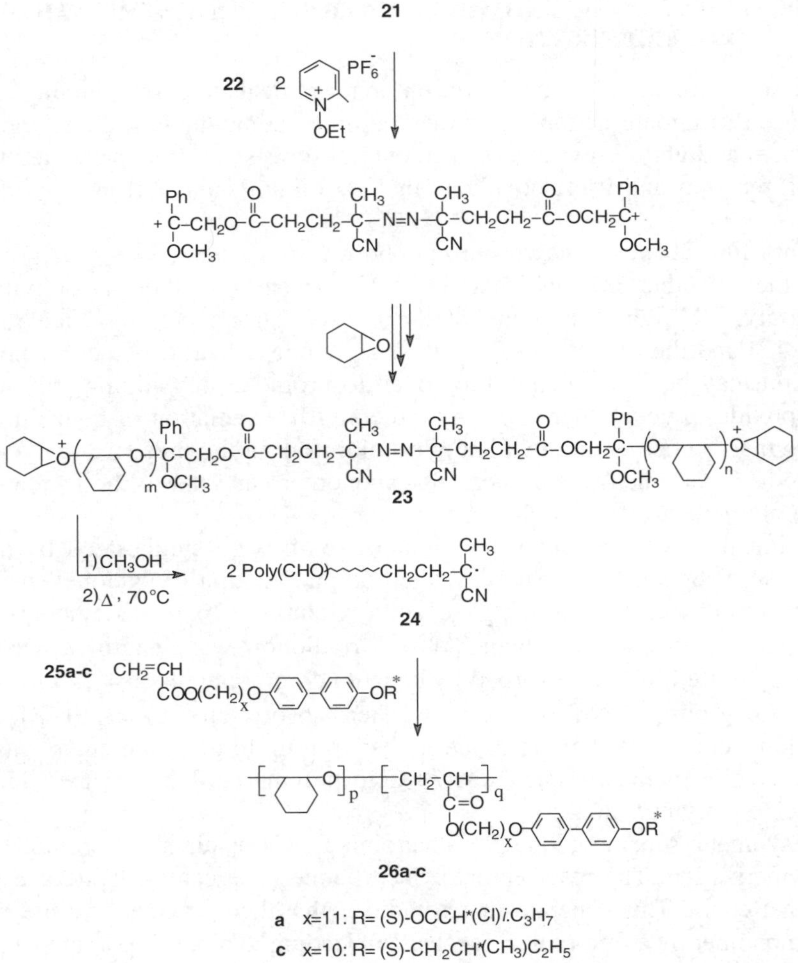

Scheme 13.

4. TRANSFORMATION OF CONDENSATION POLYMERIZATION TO RADICAL POLYMERIZATION

Preparation of step growth polymers firstly requires the incorporation of terminal, lateral, or in-chain functional groups which can then be activated to initiate a chain polymerization [43–46]. Similar synthetic strategy based on azo initiators was also applied to prepare LC block copolymers containing polycondensate segments. Block copolymers 30 were synthesized via the two-step procedure (Scheme 14). In the first reaction stage, the polyester macroinitiator 29, possessing reactive azo groups in the main chain, was prepared by a polycondensation reaction under phase transfer conditions between equimolar amounts of the diacyl chlorides 1 and 27 and sodium salt of the diphenol 28 in the presence of a catalytic amount of benzyltributylammonium bromide. The macroinitiator 29 was then used to initiate the free radical polymerization of styrene through the thermal decomposition of the azo group at 70°C. The copolymers 30 were prepared by using different concentrations of styrene in the feed mixture, and the resulting copolymers exhibited varying compositions of the main chain LC block. In this polymerization system initiation from macroinitiator chains containing one reactive azo group leads to the formation of a triblock ABA structure consisting of a central poly(styrene) block with two lateral polyester blocks.

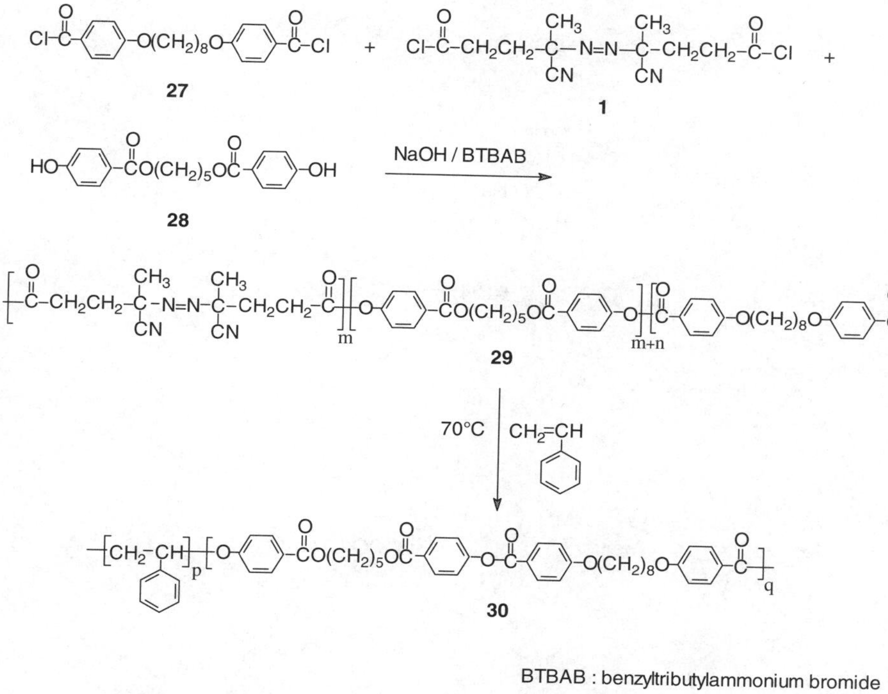

Scheme 14.

5. GRAFT COPOLYMERS

A two-step procedures involving a sequence of cationic and free radical polymerizations can also be applied to obtain LC graft copolymers as exemplified in Schemes 15 and 16. Tetrahydrofuran was polymerized using methyl triflate as a cationic initiator to produce polymer 31, which was end capped by treating it with tetrahydrothiophene. The resulting thiolanium salt 32 was isolated and then modified with sodium methacrylate which yielded methacrylate-terminated poly(tetrahydrofuran) 33 [47].

Finally, this macromonomer was copolymerized free radically in the presence of AIBN with the LC side chain monomer 34 forming graft copolymer 35 containing poly(tetrahydrofuran) grafts.

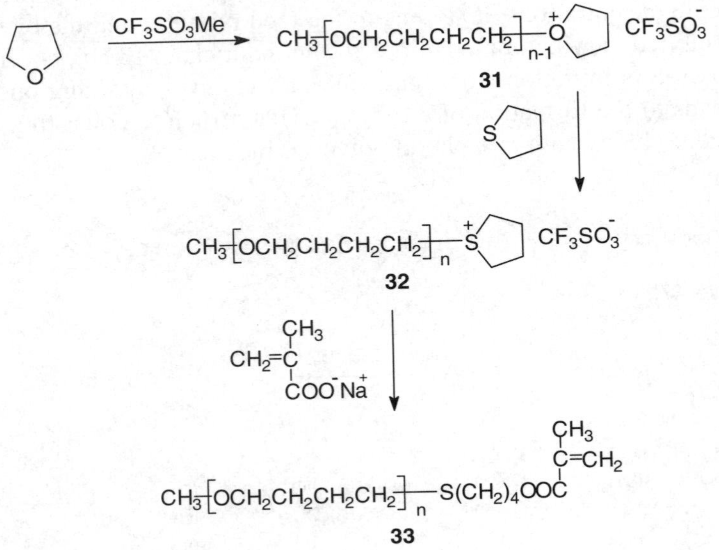

Scheme 15.

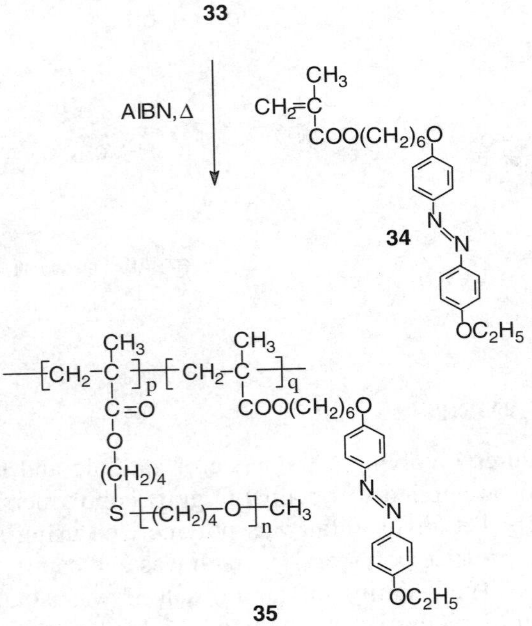

Scheme 16.

6. INTERPENETRATING NETWORKS (IPN's)

The molecules in an LC polymer possess an orientational order in the LC phases. These phases exist as intermediary but thermodynamically stable phases between the crystalline and the isotropic state and mainly arise by the form and dipolar anisotropy of the molecular components. If one of the components of a block or graft copolymer is a liquid-crystalline polymer, the structure of the polymers resulting may be influenced by the immiscibility of the components, which should cause a phase separation as well as by the nematic director of the mesogens in the LC sub-phase, which should force a spontaneous orientation of the mesogens in the sub-phase.

The parameters which control the degree of heterogeneity in the interpenetrating polymer networks (IPN's) have been investigated intensively during last two decades [48–50]. The combination of two or more polymers in network form, at least one of which is synthesized in the immediate presence of the other [51–53], results in IPN's formation. In the ideal case, there are no chemical links between both networks and the interpenetration of the network chains occurs on a molecular scale. Generally, the polymers involved are immiscible due to the low entropy of mixing, which results in a phase separation between the two networks.

Recently, Goethals *et al.* [54] investigated whether the presence of block copolymer structures in the first network of the IPN, in which one of the blocks is chemically identical to the component in the second network, increases the miscibility of the two networks due to the emulsifying effect of the block copolymer.

The applicability of this method was extended to the preparation of IPN's having LC blocks using poly(tetrahydrofuran) bifunctional macromonomer and LC monomer. The overall synthetic procedure is depicted in Scheme 17.

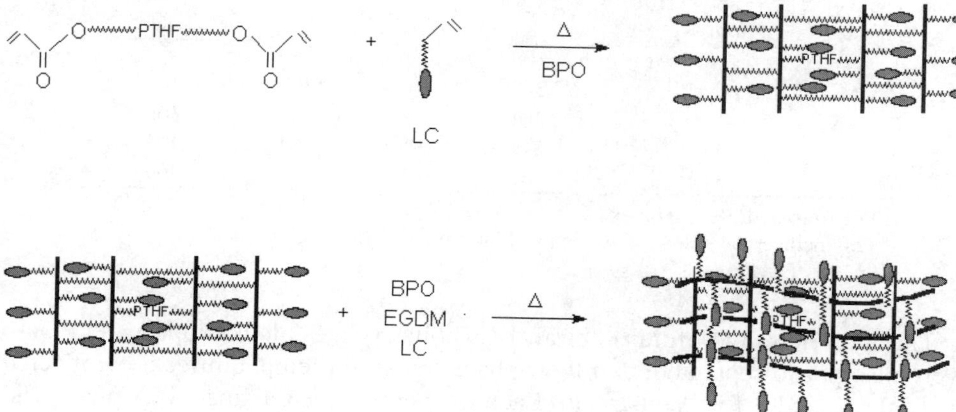

Scheme 17.

7. LIQUID CRYSTALLINE PROPERTIES OF BLOCK COPOLYMERS

The phase transition temperatures and relevant thermodynamic parameters of block copolymers 5a-c, 6a-b and macroinitiator 3 are collected in Table 1. In the DSC heating curve of the macroinitiator 3 one endothermic peak is centered at 300 K corresponding to the melting transition.

The homopolymer (PLCM-b) corresponding to the LC methacrylate block of 6b showed a nematic-isotropic transition at 368 K. This transition was preceded by the glass transition at 364 K. Block copolymer 6b exhibited the glass transition at 364 K and two endothermic transitions centered at 302 and 369 K. Comparison with the transitions of macroinitiator 3 and polymethacrylate revealed that the lower temperature transition was associated with the melting of the poly(tetrahydrofuran) block, whereas the higher temperature transitions were associated wth the glass transition and the nematic-isotropic transition of the polymethacrylate block. An analogous behavior was also shown by the block copolymer 6a. Additionally, the homopolymer (PLCM-a) corresponding to the LC polymethacrylate block of 6a formed an intermediate smectic phase that transformed into a nematic phase at 364 K. Such behavior was also detected in block copolymer 6a with a smectic-nematic transition occurring at 367 K.

Thus, the phase transition temperatures of the copolymers were very similar to those of the corresponding homopolymers, suggesting that the poly(tetrahydrofuran) and polymethacrylate blocks were strongly segregated and underwent their distinct transitions.

Table 1.
LC behavior of poly(tetrahydrofuran) 3, block copolymers 6a-b and polymethacrylates PLCM-a,b

Sample	poly(THF)[a] (wt.%)	T_g[b] (K)	T_m[a] (K)	T_{SN}[b] (K)	T_{NI}[b] (K)
3	–	300	–	–	–
PLCM-a	–	348	–	367	409
PLCM-b	–	364	–	–	368
6a	33	347	298	364	403
6b	48	364	302	–	369

[a] Poly(tetrahydrofuran) block.
[b] Polymethacrylate block. S: smectic; N: nematic; I: isotropic.

The poly(tetrahydrofuran)-chiral LC polyacrylate block copolymers 5a-c were also phase separated, and their phase transition temperatures are collected in Tables 2 and 3 for comparison. Each of them formed a smectic A phase and an ordered smectic phase, an additional monotropic smectic C phase being formed by 5c.

Table 2.
LC behavior of block copolymers 5a-b, relevant homopolymers PLCA-a,b and macroinitiator 3

Sample	poly(THF)[a] (wt.%)	Tm[a] (K)	TX-A[b] (K)	TAI[b] (K)
5a	9	297	370	387
5b	3	nd	360	369
PLCA-a	0	–	371	407
PLCA-b	0	–	345	382
3	100	302	–	–

[a] Poly(tetrahydrofuran) block.
[b] Polyacrylate block. X: semicrystalline or ordered smectic; A: smectic A; I: isotropic.

Table 3.
LC behavior of block copolymer 5c, relevant homopolymer PLCA-c and macroinitiator 3

Sample	poly(THF)[a] (wt.%)	Tm[a] (K)	TX-C[b] (K)	TCA[b] (K)	TAI[b] (K)
5c	13	300	362	369	387
PLCA-c	0	–	376	375[c]	399
3	100	302	–	–	–

[a] Poly(tetrahydrofuran) block.
[b] Polyacrylate block. X: semicrystalline or ordered smectic; C: smectic C; A: smectic A; I: isotropic.
[c] Monotropric transition.

The thermal behavior of LC poly(vinyl ether)-poly(methyl methacrylate) 8 and poly(vinyl ether)-poly(styrene) 9 block copolymers is summarized in Table 4. In the poly(vinyl ether) homopolymer 7 the nematic mesophase formed upon cooling of the isotropic melt and was, therefore, monotropic or metastable. For microphase separated block copolymers, the glass transitions of both blocks should be detected. However, the glass transition of the poly(methyl methacrylate) phase, which usually takes place at temperatures between 383 K and 398 K, was not revealed. The block copolymer with poly(styrene) block revealed the glass transition of the poly(styrene) phase at 378 K, whereas the glass temperature of the LC phase was not observed. The melting, isotropic-nematic and crystallization temperatures increased with an increase in the molar mass of the block copolymers [55].

The poly(vinyl ether) block easily crystallized on cooling with some 30 K of undercooling. It is well known that the present poly(vinyl ether)s tend to have a more stable crystalline phase than the structurally analogous polyacrylates.

Graft copolymers 13a-b and block copolymer 18 were microphase separated and the LC blocks clearly exhibited their phase transitions. The melting transition

Table 4.
Thermal behavior of polymer 7, and block copolymers 8 and 9

Sample	$T_m^{a)}$ (K)	$T_i^{a)}$ (K)	$T_c^{a)}$ (K)
7	411	404[b]	382
8	424	419[b]	393
9	424	422[b]	393

[a] Poly(vinyl ether) block. m: melting; i: isotropization; c: crystallization.
[b] Monotropic transition.

of poly(tetrahydrofuran) block was not detected probably because of a morphological effect of the LC block on either the poly(tetrahydrofuran) graft copolymers or the poly(tetrahydrofuran) block copolymers. The thermal behavior of the graft and block copolymers is reported in Table 5.

The graft copolymers 15a-b and multiblock copolymer 19 containing poly(methyl methacrylate) segments did not exhibit a mesophase because of the high content of poly(methyl methacrylate) resulting in an excessive dilution of the LC mesophase. However, the glass temperatures of the poly-LC component and the polymethyl methacrylate component were clearly revealed. The results are shown in Table 6.

Table 5.
LC behavior of graft copolymers 13a-b and block copolymer 18

Sample	poly(THF) (mol %)	$T_g^{a)}$ (K)	$T_{N-I}^{a)}$ (K)
13a	47	334	361
13b	39	307	394
18	30	338	361

[a] Polyacrylate block. N: nematic; I: isotropic.

Table 6.
Thermal behavior of graft copolymers 15a-b and block copolymer 19

Sample	Composition (mol %)			$T_g^{a)}$ (K)	$T_g^{b)}$ (K)
	PTHF	PLC	PMMA		
15a	2	2	96	330	391
15b	14	15	71	320	391
19	2	2	96	320	393

[a] LC block.
[b] Poly(methyl methacrylate) block.

Copolymer 26c gave rise to two mesophases in the thermal range between the melting and the isotropization temperatures, while copolymer 26a presented one mesophase (Table 7).

26c exhibited smectic C and A phases similar to its homopolymer but the C phase was enantiotropic in the copolymers and monotropic in the homopolymer. 26a formed one smectic A phase in agreement with its homopolymer. Thus, block copolymers 26a,c were essentially microphase-separated, with minor effects of the semicrystalline poly(cyclohexene oxide) block on the overall mesophase behavior. However, in each copolymer sample the mesophase transition temperatures were significantly lower than those of the corresponding LC homopolymers. Furthermore, the isotropization transition extended over a wider temperature range, and the isotropic and smectic phases coexisted over a broad thermal range. This suggests that some compatibility between the blocks may exist, especially in 26c comprising the rather short segments of the LC monomer units. As a result, a morphological effect appears to be completely inhibited for the poly(cyclohexene oxide) blocks in 26a,c.

Table 7.
Thermal behavior of macroinitiator 23, block copolymers 26a,c and relevant LC homopolymers

Sample	poly(CHO)[a] (wt.%)	$Tm^{a)}$ (K)	$Tm^{b)}$ (K)	$TC\text{-}A^{b)}$ (K)	$TA\text{-}I^{b)}$ (K)
23	100	340	–	–	–
25a	35	nd[c]	361	–	375
25b	35	nd[c]	356	363	376
PLCA-a	0	–	361	–	407
PLCA-c	0	–	376	375[d]	399

[a] Poly(cyclohexene oxide) block.
[b] Polyacrylate block. C: smectic C; A: smectic A; I: isotropic.
[c] Not detectable by DSC.
[d] Monotropic transition.

The LC behavior of block copolymers 30 and polyester 29 is summarized in Table 8.

Table 8.
LC behavior of block copolymers 20a-d and relevant homopolymer 29

Sample	polyester[a] (wt.%)	$Tg^{a)}$ (K)	$Tg^{b)}$ (K)	$TSN^{a)}$ (K)	$TNI^{a)}$ (K)
30a	32	306	381	400	451
30b	38	306	377	401	451
30c	59	305	368	402	451
30d	84	304	365	399	451
29	100	304	–	397	447

[a] Polyester block. S: smectic, N: nematic, I: isotropic.
[b] Poly(styrene) block.

All of the block copolymers exhibited the smectic-nematic and nematic-isotropic transitions of the main chain block around 400 and 450 K, respectively. The glass transition of poly(styrene) block was partly superposed to the smectic-nematic transition of the polyester block. However, the glass transitions of both blocks were clearly revealed by dynamic mechanical analysis. The glass transition temperature of the poly(styrene) block decreased progressively from 30a and 30d, probably due to the combined effects of a decrease in molecular weight of this block and the parallel growth of a diffuse interdomain region between the different blocks. The smectic-nematic and nematic-isotropic transition temperatures of block copolymers 30a-d were constant within the whole composition range and very similar to those of the related polyester 29.

Acknowledgment

This work was performed with the financial support from the Turkish TUBITAK and the Italian CNR. One of the authors (Y.H.) would like to thank Tubitak for Ph.D scholarship through BDP program.

REFERENCES

1. For a review see Galli, G. and Chiellini, E. in "Structure and Transport Properties in Organized Polymeric Materials", Chiellini, E., Giordano, M. and Leporini, D. Eds, World Scientific, Singapore, 1997, p.35.
2. Adams, J. and Gronski, W., Makromol. Chem., Rapid Commun., 1989, **10**, 553.
3. Saenger, J. Gronski, W., Macromol. Chem. Phys., 1998, **199**, 555.
4. Hefft, M., Springer, J., Makromol. Chem., Rapid Commun., 1990, **11**, 397.
5. Percec, V., Lee, M., J. Macromol. Sci., Pure Appl. Chem., 1992, **A29**, (9), 723.
6. Omenat, A., Lub, J., and Fischer, H., Chem. Mater, 1998, **10**, 518.
7. Kodaira, T. and Mori, K., Makromol. Chem., 1992, **193**, 1331.
8. Komiya, Z. and Schrock, R. R., Macromolecules, 1993, **26**, 1393.
9. Bohnert, R. and Finkelmann, H., Makromol. Chem. Phys., 1994, **195**, 689.
10. Moment, A., Miranda, R. Hammond, P. T., Macromol. Rapid Commun., 1998, **19**, 573.
11. Yamada, M., Iguchi, T., Hirao, Nakahama, S. and Watanabe, J., Polym. J. (Tokyo), 1998, **30**, 23.
12. Yamada, M., Itoh, T., Nakagawa, R., Hirao, A., Nakahama, S. and Watanabe, J., Macromolecules, 1999, **32**, 282.
13. Reichelt, N., Schulze, U. and Schmidt, H. W., Macromol. Chem. Phys., 1997, **198**, 3907.
14. Richards, D. H., Dev. Polym., 1979, **1**, 1.
15. Richards, D. H., Br. Polym. J., 1980, **12**, 89.
16. Quirk, R. P., Kinning, D. J. and Fetters, L. J., in Comprehensive Polymer Sci., vol 7 (Allen, G. and Bevington, J. C., Eds) Permagon, Oxford, 1989, Chapter 1.
17. Rempp, P., Franta, E. and Herz, J. E., Adv. Polym. Sci., 1988, **86**, 145.
18. Schue, F., in Comprehensive Polymer Sci., vol 7 (Allen, G. and Bevington, J. C., Eds.) Permagon, Oxford, 1989, Chapter 10.
19. Abadie, M. J. M. and Richards, D. H., Int. Chim., 1980, **208**, 135.
20. Richards, D. H., ACS. Symp. Ser., 1981, **166**, 343.
21. Richards, D. H., ACS. Symp. Ser., 1985, **286**, 87.
22. Abadie, M. J. M. and Qurahmoune, D., Br. Polym. J., 1987, **19**, 247.
23. Serhatli, I. E., Galli, G., Yagci, Y. and Chiellini, E., Polym. Bull., 1995, **34**, 539.

24. Galli, G., Chiellini, E., Yagci, Y., Serhatli, I. E., Laus, M., Bignozzi, C. M. and Angeloni, A. S., Macromol. Chem., Rapid Commun., 1993, **14**, 185.
25. Chiellini, E., Galli, G., Serhatli, I. E., Yagci, Y., Laus, M. and Angeloni, A. S., Ferroelectrics, 1993, **148**, 311.
26. Lin, H. and Matyjaszewski, K., Polym. Prepr. Am. Chem. Soc., Div. Polym. Chem., 1990, **31** (1), 599.
27. Cho, C. G., Feit, A. and Webster, O. W., Macromolecules, 1980, **23**, 1918.
28. Serhatli, I. E., Serhatli, M., Turk. J. Chem., 1998, **22**, 279.
29. Denizligil, S., Baskan, A. and Yagci, Y., Macromol. Rapid. Commun., 1995, **16**, 387.
30. Serhatli, I. E., Yagci, Y., Tomida, I., Suzuki, M. and Endo, T., J. Macromol. Sci. Pure Appl. Chem., 1997, **A34**(2), 383.
31. Nuyken, O. and Weidner, R., Adv. Polym. Sci., 1985, **73/74**, 145.
32. Galli, G., Chiellini, E., Laus, M., Angeloni, A. S., Bignozzi, M. C. and Francescangeli, O., Mol. Cryst. Liq. Cryst., 1994, **154**, 429.
33. Galli, G., Chiellini, E., Laus, M., Bignozzi, M. C. and Francescangeli, O., Makromol. Chem. Phys., 1994, **195**, 2247.
34. Chiellini, E., Galli, G., Angeloni, A. S. and Laus, M., Trends Polym. Sci., 1994, **2**, 244.
35. Laus, M., Bignozzi, M. C., Angeloni, A. S., Francescangeli, O., Galli, G. and Chiellini, E., Polym. J., 1995, **27**, 993.
36. Yagci, Y., Lukac, I. and Schnabel, W., Polymer, 1993, **34**, 1130.
37. Hızal, G., Yagci, Y. and Schnabel, Polymer, 1994, **35**, 20.
38. Yagci, Y. and Onen A., J. Macromol. Sci. Chem., 1991, **A28**, 129.
39. Yagci, Y. and Schnabel, W., Makromol. Chem., Macromol. Symp., 1988, **13–14**, 161.
40. Yagci, Y. and Onen, A., Macromolecules, 1991, **24**, 4620.
41. Yagci, Y. and Ledwith, A., J. Polym. Sci., Polym. Chem. Ed., 1988, **26**, 1911.
42. Komitov, L., Lagerwall, S. T., Stebler, B., Chiellini, E., Galli, G. and Strigazzi, A., Modern Topics in Liquid Crystals., A. Buka (Ed.), World Scientific, Singapore, 1993, 301.
43. Yagci, Y. and Mishra, M. K., in Macromolecular Design: Concept and Practise, edited by Mishra, M. K., (Polymer Frontiers Int., Inc., New York, 1994, Chap. 10).
44. Rempp, P. F., Lutz, P. J., in Comphrehensive Polymer Science, edited by Eastmond, G. C., Ledwith, A., Russo, S., and Sigwalt, P., (Pergamon Press, New York, 1989), 403.
45. Ueda, A. and Nagai, S., in Macromolecular Design: Concept and Practise, edited by Mishra, M. K., (Polymer Frontiers Int., Inc., New York, 1994, Chap. 7).
46. Nuyken, O. and Voit, B., Macromolecular Design: Concept and Practice, edited by Mishra, M. K., (Polymer Frontiers Int., Inc., New York, 1994, Chap. 8).
47. Asami, R. and Takaki, M., Macromol. Chem., Suppl. 1985, **12**, 163.
48. Klempner, D., Frisch, K. C., Lipatov, Y. S., "Advances in Interpenetrating Polymer Networks", Technomic, Lancaster PA, 1989, vol. I, 261.
49. Lipatov, Y. S., Grigor'yeva, O. P., Kovernik, Shilov, V. V. and Sergeyeva, L. M., Macromol. Chem., 1985, **186**, 1401.
50. Nevissas, V., Widmaier, J. M. and Meyer, G. C., J. Appl. Polym. Sci., 1988, **36**, 1467.
51. Sperling, L. H., "Interpenetrating Polymer Networks and Related Materials" Plenum, New York 1981.
52. Manson, A. J., Sperling, L. H., "Polymer Blends and Composites" Plenum, New York 1976.
53. Klempner, D., Sperling, L. H., Utracki, L. A., Eds, "Interpenetrating Polymer Networks" Adv. İn Chem. Series no.239, ACS Books, Am. Chem. Soc., Washington DC 1994.
54. Prez, F. and Goethals, E. J., Macromol. Chem. Phys., 1995, **196**, 903.
55. Portugall, M., Ringsdorf, H. and Zentel, R., Makromol. Chem., 1982, **183**, 2311.

Application of chiral polybinaphthyl-based Lewis acid catalysts to the asymmetric organozinc additions to aldehydes

LIN PU*

Department of Chemistry, University of Virginia, Charlottesville, VA 22901, USA

Abstract—A systematic study of novel rigid and sterically regular chiral binaphthyl polymer-based Lewis acid catalysts has been carried out. Through this study, a polymeric chiral Lewis acid catalyst with the most general enantioselectivity for the reaction of aldehydes with diethylzinc has been obtained. This polymer-based catalyst also shows high enantioselectivity for the reaction of other alkyl or aryl zinc reagents with aldehydes. The polymer ligands used in these reactions can be easily recovered after the reaction and the recovered ligands show the same catalytic properties as the original polymers. This work not only provides highly enantioselective methods to prepare optically active secondary alcohols, but also establishes new strategies to design and synthesize polymeric chiral catalysts. The sterics and electronics of the catalytic sites in these rigid and sterically regular polymers can be systematically modified. The rigid polymer structure can also be used to preserve the catalytic properties of monomer catalysts provided that the catalytically active species are monomeric rather than monomer aggregates.

Keywords: 1,1'-bi-2-naphthol; chiral polybinaphthyls; asymmetric catalysis; diethylzinc; diphenylzinc; aldehydes; chiral secondary alcohols.

1. INTRODUCTION

Lewis acid catalysis is an important synthetic method in organic chemistry and has found very broad applications in natural product and drug syntheses [1]. In recent years, enormous progress has been made on the use of chiral ligand-based Lewis acid catalysts for asymmetric organic reactions [1–3]. However, compared to some late transition metal-based catalysts, Lewis acid catalysts are generally used with a much larger quantity in a reaction. In a number of cases, a stoichiometric or even excess amount of Lewis acid catalysts is necessary. The larger amount required is due to the high sensitivity of Lewis acid complexes towards air, moisture and impurity. Because it is often quite expensive to prepare the chiral ligands used in asymmetric Lewis acid catalysis, recovery and reuse of these compounds are highly desirable. Therefore, using easily recoverable polymer-

---

* E-mail: lp6n@virginia.edu

based chiral catalysts becomes a very attractive process for the asymmetric Lewis acid catalysis.

Traditionally, polymer-based chiral catalysts are prepared by attaching chiral monomer ligands to flexible and sterically irregular polymers such as polystyrene [4–6]. Although a few good polymeric chiral catalysts have been obtained in this way, in many cases, a significant decrease in enantioselectivity has been observed when a monomer catalyst is converted into a polymer catalyst. Because of the sterically irregular polymer structure, it is very difficult to systematically modify the microenvironment of the catalytic sites in the traditional polymer catalysts to achieve the desired catalytic properties.

Recently, our laboratory has used rigid and sterically regular chiral polybinaphthyls such as (R)-1 to develop a new class of enantioselective polymeric Lewis acid catalysts [7–15]. Due to the rigidity and stereoregularity of these polymers, their catalytic sites have much better-defined microenvironment than the traditional polymeric chiral catalysts. Therefore, the catalytic sites in the polybinaphthyls can be systematically modified to carry out enantioselective organic reactions. By using this strategy, we have discovered highly enantioselective polymeric Lewis acid catalysts for the asymmetric reaction of organozincs with aldehydes. Herein, our recent work on this subject is reviewed.

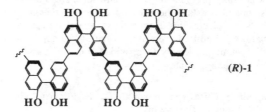

2. DISCUSSION

Starting from optically pure (R)-1,1'-bi-2-naphthol [(R)-BINOL], we have prepared chiral binaphthyl monomers (R)-2 that contain two bromine atoms at the 6,6'-positions and various protecting groups R on the 2,2'-oxygens. (R)-2 were polymerized in the presence of either a stoichiomeric amount of Ni(1,5-cyclooctadiene)$_2$ or a catalytic amount of NiCl$_2$ and excess Zn (Scheme 1) [7, 8]. The R groups of the resulting polymers were removed to generate polybinaphthol (R)-1. This polymer is soluble in N,N-dimethylforamide, benzyl alcohol, basic water solutions and DMSO, but insoluble in THF, chloroform and benzene. The hydroxyl groups in (R)-1 will allow the introduction of various Lewis acid metal centers for the preparation of novel polymeric chiral catalysts.

We have first examined the use of a polymeric aluminum complex (R)-3, prepared from the reaction of (R)-1 with diethylaluminum chloride, to catalyze the Mukaiyama aldol reaction (Scheme 2) [7, 8]. Although we found this polymeric aluminum complex having greatly enhanced catalytic activity over the aluminum

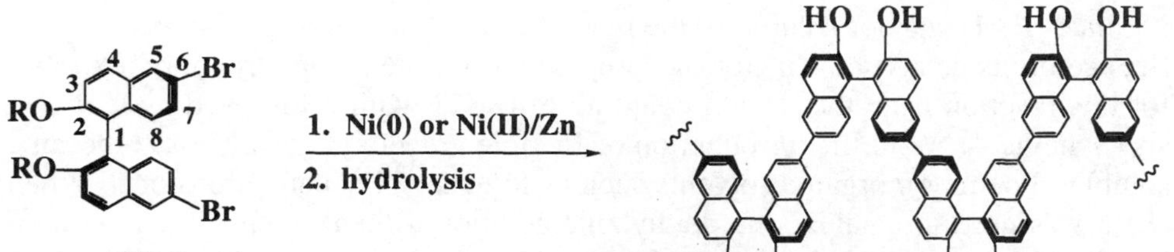

(*R*)-2
(*R*)-2a: R = Ac
(*R*)-2b: R = CH_2OCH_3
(*R*)-2c: R = *n*-C_6H_{13}

(*R*)-1

Scheme 1. Synthesis of the Optically Active Poly(1,1′-bi-2-naphthol) (*R*)-**1**.

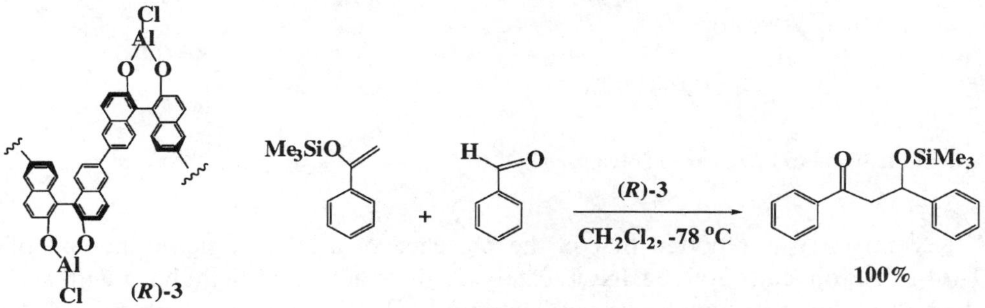

Scheme 2. A Mukaiyama Aldol Condensation Catalyzed by the Polybinaphthyl Aluminum Complex (*R*)-**3**.

complex made from the monomer ligand BINOL, both the polymer and monomer catalysts showed no enantioselectivity for this reaction.

We then applied polymer (*R*)-**1** to catalyze the reaction of benzaldehyde with diethylzinc. In the presence of this polymer, diethylzinc reacted with benzaldehyde at room temperature in methylene chloride to give a chiral alcohol product **4** and a side product benzyl alcohol with a ratio of 53:47 and an *ee* of 13% for **4** (Scheme 3) [9, 10].

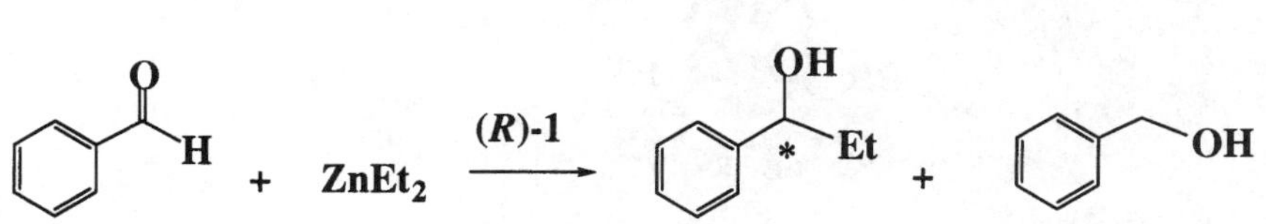

4

Scheme 3. Reaction of Benzaldehyde with Diethylzinc Catalyzed by (*R*)-**1**.

Since (R)-**1** was not soluble in the reaction media which might have strong influence on its catalytic behavior, we prepared a soluble binaphthyl polymer (R)-**6** for this reaction from the Suzuki coupling of (R)-**2b** with **5** followed by hydrolysis (Scheme 4). With the introduction of flexible hexyloxyl groups, (R)-**6** became soluble in common organic solvents such as toluene, THF and chloroform. When (R)-**6** was used to catalyze the diethylzinc addition to benzaldehyde, chiral alcohol **4** and benzyl alcohol were produced with a 71:29 ratio and the ee of **4** was 40% [9, 10].

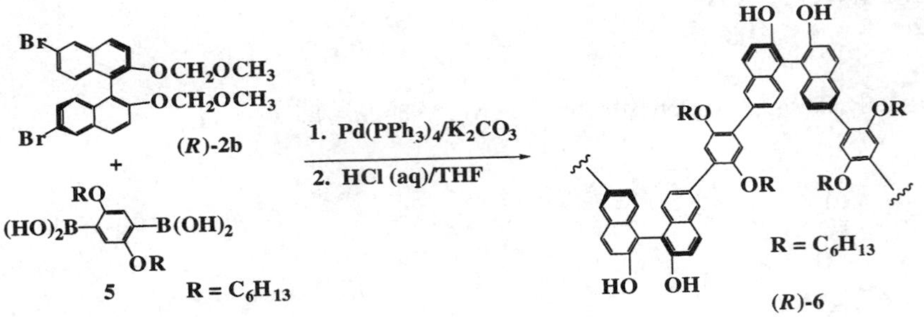

Scheme 4. Synthesis of a Soluble Polybinaphthyl (R)-**6**.

Recently it was reported that in the presence of a stoichiometric amount of Ti(O-i-Pr)$_4$, optically active BINOL catalyzed the reaction of diethylzinc with aryl aldehydes with very good enantioselectivity [16]. We thus used polymer (S)-**6**, prepared from the polymerization of (S)-**2b** with **5**, to catalyze the diethylzinc addition in the presence of Ti(O-i-Pr)$_4$. The reactions were carried out at 0 °C for 5 h in toluene by using 20 mol% of (S)-**6** and 1.4 equiv of Ti(O-i-Pr)$_4$. The observed ee for the reaction of benzaldehyde with diethylzinc was 86% and for the reaction of 1-naphthaldehyde 92% [11]. The catalytic properties of the polymer was very similar to those of the monomer. The catalytically active species for this reaction might be a polybinaphthyl titanium complex (S)-**7** generated from the reaction of (S)-**6** with Ti(O-i-Pr)$_4$.

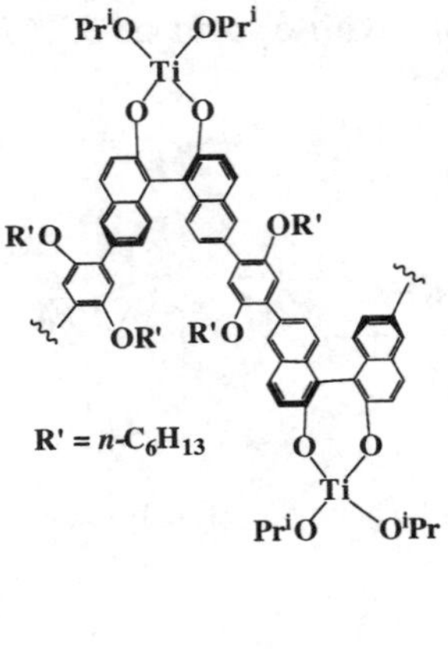

Although the titanium-based polybinaphthyl catalyst (S)-**7** showed good enantioselectivity for the reaction of diethylzinc with aryl aldehydes, it required the use of excess amount of Ti(O-i-Pr)$_4$ over the substrates. In order to avoid the large amount of Ti(O-i-Pr)$_4$ and to develop a more efficient catalytic process for the diethylzinc addition, we further modified the structure of the binaphthyl polymers. Going from (R)-**1** to (R)-**6**, only the solubility of the polymers was adjusted but not the structure of their catalytic sites. In the diethylzinc addition to benzaldehyde catalyzed by both (R)-**1** and (R)-**6**, the catalytically active species should be the Lewis acidic zinc centers generated from the reaction of diethylzinc with the hydroxyl groups of the binaphthyl units in the polymers. Therefore, in order to improve the catalytic properties of the polymers, it is necessary to modify the sterics and electronics of their Lewis acid centers.

We decided to introduce substituents to the 3,3'-positions of the binaphthyl units for better steric control at the catalytic sites of the polymers. Through the Suzuki coupling of (R)-**8** with **5** followed by hydrolysis, polymer (R)-**9** was produced with a molecular weight of Mw = 6,700 (PDI = 1.5) or Mw = 24,000 (PDI = 2.5) by using either Pd(OAc)$_2$/tris(o-tolylphosphine) or Pd(PPh$_3$)$_4$ as the catalyst (Scheme 5) [9, 10]. (R)-**9** was soluble in common organic solvents and displayed very sharp and well-resolved NMR signals. In the presence of 5 mol% of (R)-**9** at 0 °C in toluene solution, the diethylzinc addition to benzaldehyde proceeded with excellent enantioselectivity and complete chemical selectivity. It produced (R)-**4** with 92% ee. This polymer also showed high enantioselectivity for the reaction of para-substituted benzaldehydes and cinnamaldehyde (90–94% ee), and up to 83% ee for the reaction of aliphatic aldehydes [9, 10]. After completion of the reaction, (R)-**9** was recovered conveniently by precipitation with methanol. The catalytic properties of the recovered polymer were very similar to those of the original polymer and they were independent of the method of the polymer preparation, the molecular weight of the polymer and the molecular weight distribution.

In order to study the effect of the size of the alkyl groups in polymer (R)-**9** on the catalytic properties, we have prepared polymer (R)-**11** from the Suzuki coupling of (R)-**8** with **10** followed by hydrolysis (Scheme 6) [12]. This polymer contains sterically more bulky isopropyl groups than the hexyl groups in (R)-**9**

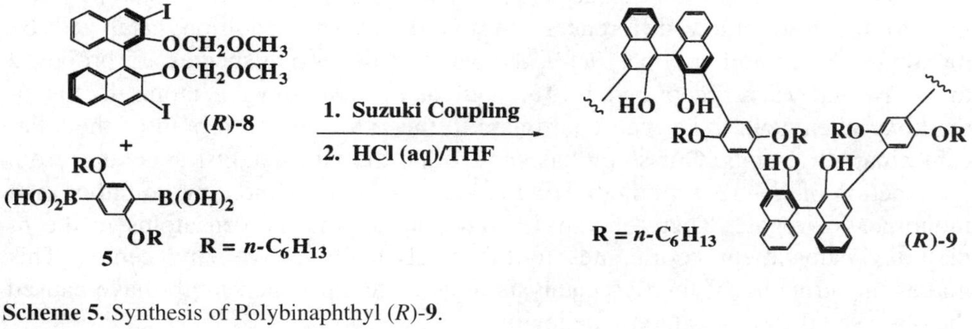

Scheme 5. Synthesis of Polybinaphthyl (R)-**9**.

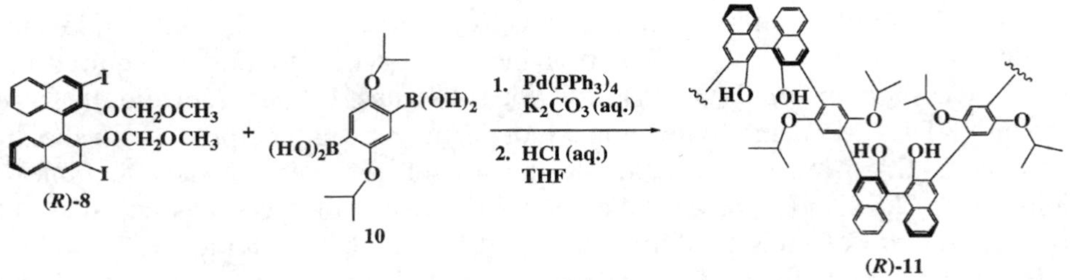

Scheme 6. Preparation of the Isopropyl Substituted Polybinaphthyl (R)-**11**.

and is also soluble in common organic solvents. The diethylzinc addition catalyzed by (R)-**11** showed slightly lower catalytic activity as well as lower enantioselectivity. This demonstrates that bulky groups in the polymer are not favorable for the reaction.

We have synthesized a monobinaphthyl ligand (R)-**13** as the model compound of polymer (R)-**9** from the Suzuki coupling of (R)-**8** with **12** followed by hydrolysis (Scheme 7) [13]. This molecule showed a very general and high enantioselectivity for the reaction of diethylzinc with a broad range of aldehydes including para-, ortho- or meta-substituted aromatic aldehydes, linear or branched aliphatic aldehydes, and aryl or alkyl substituted α,β-unsaturated aldehydes [13, 17]. It has greatly enhanced enantioselectivity over polymer (R)-**9** especially for the reaction of ortho-substituted benzaldehydes and aliphatic aldehydes.

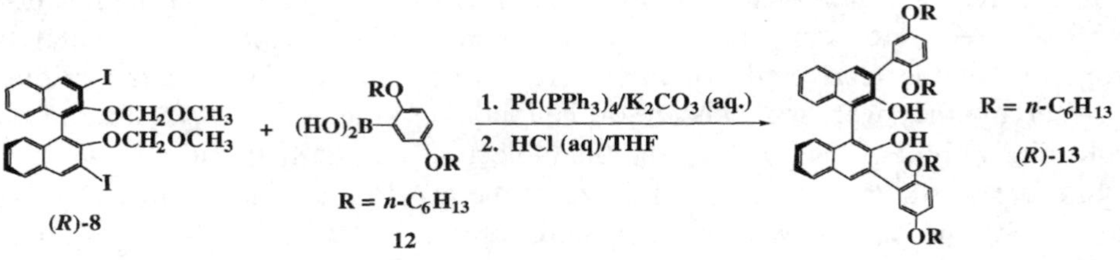

Scheme 7. Synthesis of the Model Compound (R)-**13**.

We attribute the different catalytic properties of the monomer versus the polymer to their structural differences. In the diethylzinc addition catalyzed by monomer (R)-**13** and polymer (R)-**9**, the catalytically active species are proposed to be **14** and **15** respectively. In **15**, both of the two oxygen atoms in the p-dialkoxy phenylene linker can coordinate to the zinc centers in both of the adjacent binaphthyl units. Our study has shown that the zinc catalyst generated from the reaction of (R)-**13** with diethylzinc is very likely to be monomeric rather than monomer aggregate. Therefore, in **14**, only one of the oxygen atoms in the p-dialkoxyl substituent coordinates to the catalytically active zinc center. This makes the structure of the two catalysts quite different which might have caused the observed different catalytic behavior.

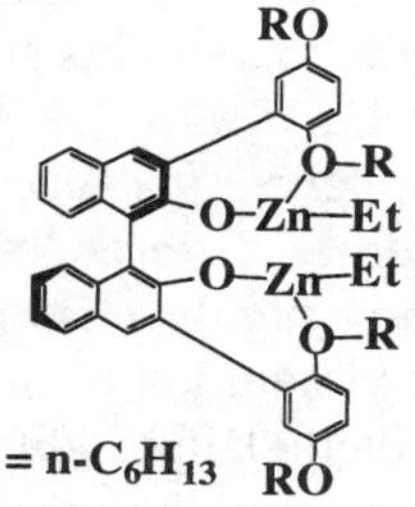

In order to preserve the excellent enantioselectivity of monomer (R)-**13** in polymer, we have synthesized polymer (R)-**17** from the Suzuki coupling of (R)-**8** with **16** followed by hydrolysis (Scheme 8) [14]. Polymer (R)-**17** contains the structural units of (R)-**13** separated by a rigid para-phenylene linker. This structure should eliminate the interference between the adjacent catalytic sites as shown in **15**. Therefore, the catalytic sites in this polymer should more closely resemble monomer (R)-**13**. Indeed, (R)-**17** exhibited very high and general enantioselectivity (91–98% ee) for the reaction of diethylzinc with a broad range of aldehydes. It is the most general polymeric catalyst yet reported for the enantioselective diethylzinc addition to aldehydes. In these reactions, (R)-**17** was recovered almost quantitatively by precipitation with excess methanol and the recovered polymer had the same enantioselectivity as the original polymer. Both monomer (R)-**13** and polymer (R)-**17** also showed high enantioselectivity for the reaction of dimethylzinc with aldehydes, though the reaction was much slower than the diethylzinc addition.

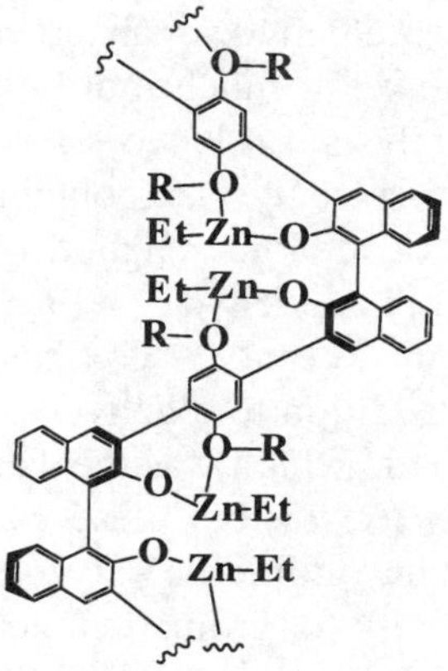

Our asymmetric alkylzinc additions to aldehydes catalyzed by the binaphthyl monomers and polymers have been extended to the diphenylzinc addition. Unlike the alkylzinc additions which are extremely slow in the absence of a catalyst, the

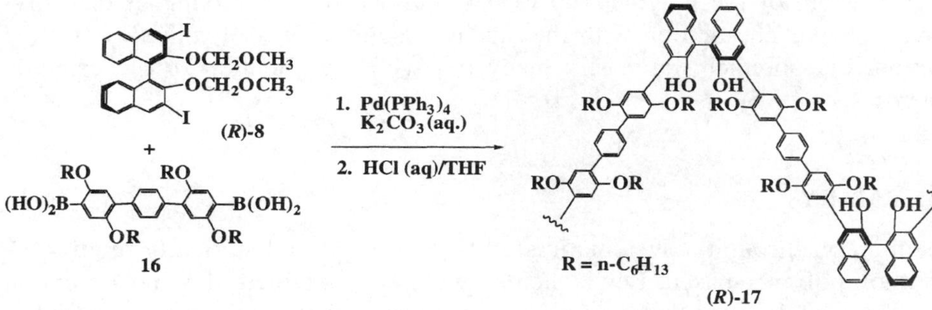

Scheme 8. Synthesis of Polybinaphthyl (R)-**17**.

diphenylzinc addition to aldehydes can proceed even without a catalyst. Such a background reaction makes it much more difficult to obtain enantioselective diphenylzinc addition catalysts. Little work has been carried out on this subject, and no highly enantioselective catalyst was obtained before [18].

We studied the use of monomer ligand (R)-13 to catalyze the reaction of diphenylzinc with propionaldehyde [19]. This process produced (S)-1-phenylpropanol with 86% ee at 0 °C. This is the first highly enantioselective catalytic diphenylzinc addition to aldehydes. The phenyl addition occurred at the re face of propionaldehyde which was the same as the diethylzinc addition to benzaldehyde catalyzed by (R)-13, but these two processes gave the opposite enantiomers. The diphenylzinc addition catalyzed by (R)-13 was also found to exhibit excellent enantioselectivity for aromatic aldehydes such as p-chlorobenzaldehyde, p-anisaldehyde and 2-naphthaldehyde (87–94% ee). In these reactions, pretreatment of the chiral ligand with diethylzinc gave a much better catalyst. The asymmetric diphenylzinc addition to aromatic aldehydes allows the enantioselective preparation of chiral diarylcarbinols that are not easily accessible by asymmetric catalysis. An example of the diphenylzinc addition is shown in Scheme 9.

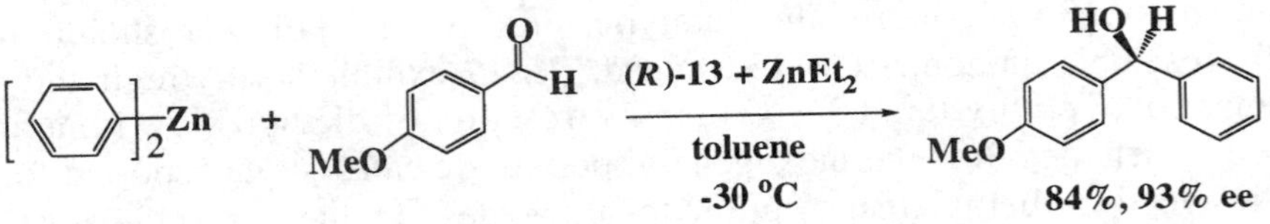

Scheme 9. An Asymmetric Diphenylzinc Addition to Aldehydes Catalyzed by (R)-13.

We have used polymer (R)-17 to catalyze the diphenylzinc addition to propionaldehyde and p-anisaldehyde [19, 20]. This polymer showed similar enantioselectivity as monomer (R)-13. However, the reaction condition had to be modified since the catalytic activity of the polymer was found to be lower than that of the monomer. Thus, the uncatalyzed diphenylzinc addition in the presence of the polymer catalyst was more competitive than in the presence of the monomer catalyst. More of the polymer (up to 0.4 equiv) and slow mixing of the polymer+ZnEt$_2$+Ph$_2$Zn solution with the aldehyde solution were required. After the reaction, the polymer was easily recovered with the retention of the catalytic properties.

3. SUMMARY

In summary, through a systematic study of the rigid and sterically regular binaphthol polymer-based Lewis acids, we have obtained the most general polymeric catalyst for the asymmetric reaction of diethylzinc with aldehydes. We have further found that the polymer catalyst is also good for the reaction of other alkyl or aryl zinc reagents. The polymer ligands used in these reactions can be easily recovered after the reaction and the recovered ligands show the same catalytic properties as the original polymers. This work not only provides highly enantioselective methods to prepare optically active secondary alcohols, but also establishes new strategies to design and synthesize polymeric chiral catalysts. The sterics and electronics of the catalytic sites in these rigid and sterically regular polymers can be systematically modified. The rigid polymer structure can also be used to preserve the catalytic properties of monomer catalysts provided that the catalytically active species are monomeric rather than monomer aggregates. These strategies may find general application in the development of enantioselective polymeric catalysts in asymmetric organic synthesis.

Acknowledgment

The work described in this article was carried out by the following postdoctoral associates and students in my laboratory: Dr. Qiao-Sheng Hu, Dr. Wei-Sheng Huang, Ms. Xiao-Fan Zheng, Ms. Dilrukshi Vitharana, Mr. Zhi-Ming Lin and Dr. Hong-Bin Yu. We are very grateful for the financial support provided by the Department of Chemistry at University of Virginia, the Department of Chemistry at North Dakota State University, the US National Science Foundation (DMR-9529805), the US Air Force Office of Scientific Research (F49620-96-1-0360), the donors of the Petroleum Research Fund—administered by the American Chemical Society, the Jeffress Memorial Trust and the US National Institute of Health (1R01GM58454).

REFERENCES

1. Santelli, M.; Pons, J.-M. *Lewis Acids and Selectivity in Organic Synthesis* CRC Press: Boca Raton, 1996.
2. Noyori, R. *Asymmetric Catalysis in Organic Synthesis* John Wiley & Sons: New York, 1994.
3. Ojima, I. Ed. *Catalytic Asymmetric Synthesis;* VCH: New York, 1993.
4. Pittman, C. U. Jr. in *Comprehensive Organometallic Chemistry, Vol. 8* (Eds.: Wilkinson, G.; Stone, F. G. A.; Abel, E. W.), Pergamon Press, Oxford, 1983, pp 553.
5. Blossey, E. C.; Ford, W. T. in *Comprehensive Polymer Science. The Synthesis, Characterisation, Reactions and Applications of Polymers, Vol. 6* (Eds.: Alien, G.; Bevington, J. C.), Pergamon Press, New York, 1989, pp 81.
6. Itsuno, S. in *Polymeric Materials Encyclopedia; Synthesis, Properties and Applications, Vol. 10* (Ed.: Salamone, J. C.), CRC Press, Boca Raton, FL, 1996, pp 8078.
7. Hu, Q.-S.; Zheng, X. -F.; Pu, L. *J. Org. Chem.* 1996, **61**, 5200.
8. Hu, Q.-S.; Vitharana, D.; Zheng, X.-F.; Wu, C.; Kwan, C. M. S.; Pu, L. *J. Org. Chem.* 1996, **61**, 8370.
9. Huang, W.-S.; Hu, Q.-S.; Zheng, X.-F.; Anderson, J.; Pu, L. *J. Am. Chem. Soc.* 1997, **119**, 4313.
10. Hu, Q.-S.; Huang, W.-S.; Vitharana, D.; Zheng, X.-F.; Pu, L. *J. Am. Chem. Soc.* 1997, **119**, 12454.
11. Yu, H.-B.; Zheng, X.-F.; Hu, Q.-S.; Pu, L. *Polymer Preprints* 1999, **40**(1), 546.
12. Lin, Z.-M.; Hu, Q.-S.; Pu, L. *Polymer Preprints* 1998, **39**(1), 235.

13. Huang, W.-S.; Hu, Q.-S.; Pu, L. *J. Org. Chem.* 1998, **63**, 1364.
14. Hu, Q.-S.; Huang, W.-S.; Pu, L. *J. Org. Chem.* 1998, **63**, 2798.
15. (a) Simonsen, K. B.; Jorgensen, K. A.; Hu, Q.-S.; Pu, L. *Chem. Commun.* 1999, 811. (b) Johannsen, M.; Jørgensen, K. A.; Zheng, X.-F.; Hu, Q.-S.; Pu, L. *J. Org. Chem.* 1999, **64**, 299.
16. Zhang, F. Y.; Yip, C. W.; Cao, R.; Chan, A. S. C. *Tetrahedron: Asymmetry* 1997, **8**, 585.
17. For reviews on asymmetric organozinc additions, see: (a) Soai, K.; Niwa, S. *Chem. Rev.* 1992, **92**, 833. (b) Noyori, R.; Kitamura, M. *Angew. Chem. Int. Ed. Engl.* 1991, **30**, 49.
18. An enantioselective diphenylzinc addition to aldehydes was found to give 57% *ee*: Dosa, P. I.; Ruble, J. C.; Fu, G. C. *J. Org. Chem.* 1997, **62**, 444.
19. Huang, W.-S.; Pu, L. *J. Org. Chem.* 1999, **64**, 4222.
20. Huang, W.-S.; Hu, Q.-S.; Pu, L. *J. Org. Chem.* 1999, **64**, 7940.

Polymer-protected metal nanocatalysts

ANDREA B. R. MAYER* and JAMES E. MARK
Department of Chemistry and the Polymer Research Center, The University of Cincinnati, Cincinnati, OH 45221-0172, USA

Abstract—Palladium and platinum colloids were generated by in-situ reductions in the presence of different polymer types. The influence of the protective polymer, both on the metal nanoparticle formation and the catalytic activities for the hydrogenation of cyclohexene as a model reaction, was qualitatively investigated.

Keywords: Metal colloids; polymers; catalysis; hydrogenation.

1. INTRODUCTION

Polymer-protected, colloidal transition metals in the nanometer-size regime are promising as novel catalyst systems, due to the significant advantages they offer relative to traditional catalyst systems [1–13]. A main benefit stems from their nanosize dimensions and the associated increase in surface area, which makes these catalyst systems highly efficient. It is also expected that novel catalytic properties such as selectivity could be associated with the small particle sizes as well [1, 14]. The presence of the protective polymer adds to the versatility of the composite materials, since it can play an active role in "tuning" the nanoparticle features and catalytic performance [1]. The polymeric matrix surrounding the nanocatalyst can be chosen to have certain properties and functions, thereby influencing the way a reactant diffuses through it in its approach to the catalytically-active sites. Various properties, such as charge, hydrophobicity, and steric hindrance of the polymer envelope are surely of significance [1]. Therefore, several types of polymers were selected and investigated for their abilities to act as protective agents for noble metal colloids, and for their influence on the catalytic activities for the hydrogenation of cyclohexene as a model reaction.

---

* To whom correspondence should be addressed. E-mail: amayer@duke.poly.edu
Present address: Department of Chemical Engineering, Chemistry, and Materials Science, Polytechnic University Brooklyn, Six Metrotech Center, Brooklyn, NY 11201, USA.

2. EXPERIMENTAL

Chemicals and reagents. The metal precursors palladium chloride ($PdCl_2$), palladium acetate [$Pd(CH_3COO)_2$; $Pd(ac)_2$], palladium trifluoroacetate [$Pd(CF_3COO)_2$; $Pd(F_3ac)_2$], palladium acetylacetonate [$Pd(CH_3COCH=C(O-)CH_3)_2$; $Pd(acac)_2$], and dihydrogen hexachloroplatinate (H_2PtCl_6), as well as the reducing agents potassium tetrahydroborate (KBH_4) were obtained from Aldrich. The homopolymers and random copolymers, cationic polyelectrolytes, and latex dispersions in water were purchased from Aldrich, Polysciences, and Monomer-Polymer & Dajac Laboratories. The block copolymers were obtained from the Polymer Source.

Colloid preparation. a) Reduction by refluxing the alcoholic solutions ("alcohol reduction method"): The platinum and palladium colloids were prepared according to the method described by Hirai *et al.* [1, 3]. The metal precursor was reduced by refluxing the alcoholic solutions (6.8×10^{-4} M, ethanol (EtOH) : water = 1 : 1 (v/v)) containing the metal precursor and the protective polymer in a mass ratio of polymer : metal = 25 : 1. For the block copolymer polystyrene-block-poly(methacrylic acid) (PS-b-PMAA) EtOH was used as a solvent; for polystyrene-block-poly(ethylene oxide) (PS-b-PEO) mixtures of tetrahydrofuran (THF) and EtOH were used. For the preparation of the PS-b-PEO samples the block copolymer was first dissolved in the desired amount of THF, and EtOH was then added dropwise, resulting in initial solvent mixtures of THF : EtOH = 1 : 9 (v/v) or 16 : 9 (v/v). The metal precursors were slowly added as solutions in EtOH (equal volumes of block copolymer and metal precursor solution were combined), thus resulting in final solvent mixtures of THF : EtOH =1 : 19 (v/v) or 1 : 2.1 (v/v).

b) Reduction by potassium tetrahydroborate: An aqueous solution containing an excess of potassium tetrahydroborate was prepared just before use and rapidly added to the stirred solutions (6.8×10^{-4} M, room temperature) which contained the metal precursor and the protective polymers in a mass ratio of polymer : metal = 25 : 1. The solvents and mass ratios were the same as described under a).

c) Reduction by stirring the solutions at room temperature: Some palladium colloids protected by the block copolymers were prepared from the palladium acetate and palladium trifluoroacetate precursors by stirring the solutions at room temperature. The concentrations, mass ratio, and solvents were the same as those described under part a).

d) Reduction by heating at 60 °C: For the palladium acetate and palladium trifluoroacetate precursors in the presence of the block copolymers some reductions were performed in a sandbath at 60 °C. The reduction conditions were the same as those described in part a).

The reduction of the metal precursors was followed by UV-vis spectroscopy from 270 to 600 nm for all samples. The spectra were recorded with a Milton Roy Spectronic 3000 Array instrument, using a 10 mm pathlength quartz cuvette. Reduction progress could be followed by the disappearance of the absorption bands for the $PdCl_2$ precursor located at about 320 nm and 425 nm, or for the 264 nm band of H_2PtCl_6. What remained was an unstructured, continuous absorption spectrum in the visible range, typical for nanosized palladium and platinum colloids [15]. All glassware was cleaned with aqua regia (for the platinum samples) or concentrated nitric acid (for the palladium samples) before use.

Characterization: The metal particle sizes and size distributions were obtained by TEM with a JEOL-100 CX II instrument (operated at an accelerating voltage of 80 kV). The samples were prepared by placing a drop of the colloidal dispersion on a formvar/carbon coated copper grid and letting the solvent evaporate at room temperature. The particle sizes were determined with a comparator, and were based on measurements of at least 150 particles.

Catalytic hydrogenations: The hydrogenations were carried out with a Parr hydrogenation apparatus (shaker type) at room temperature. For the qualitative comparison of the catalytic activities in terms of conversion of cyclohexene, cyclohexene (0.05 ml) was added to 10 ml of methanol (MeOH), and one of the metal catalysts was added as a colloidal dispersion. An amount of catalyst corresponding to 0.16 wt% platinum or 0.09 wt% palladium (with respect to cyclohexene) was added in this way, and the reaction was performed at a hydrogen pressure of 10 psi for 30 minutes. The reaction mixtures were analyzed by gas chromatography (SE-30 packed column) with a flame ionization detector, and helium as the carrier gas. Test reactions without the addition of any catalyst were performed after each evaluation reaction.

3. RESULTS AND DISCUSSION

3.1. Nonionic polymers, polyacids, and cationic polyelectrolytes

Several water-soluble nonionic polymers, polyacids, and cationic polyelectrolytes were selected, possessing varying degrees of hydrophobicity of their backbones, and different side groups. When comparing polymers possessing different hydrophobic character, larger particle sizes and size distributions were usually found for protective polymers with a restricted hydrophobic character of the backbone. For cationic polyelectrolytes small particle sizes exhibiting a narrow size distribution could be often obtained (depending on the reduction method chosen). This is due to the formation of ion pairs between negatively-charged metal precursor ions and cationic components of the polyelectrolyte. A wide variety of protective polymers are useful for the stabilization of noble metal colloids, and dispersions which were colloidally stable for months and even years could be obtained, with particles exhibiting diameters in the range of 1–15 nm.

The stable palladium and platinum samples were investigated for their catalytic activities, using the hydrogenation of cyclohexene as a model reaction. Table 1 shows results obtained for platinum and palladium catalysts prepared by the alcohol reduction method and protected by nonionic water-soluble polymers, various polyacids (anionic systems), and cationic polyelectrolytes. An increased catalytic conversion was observed for the platinum samples involving polyacids as protec-

Table 1.
Platinum and palladium nanocatalysts protected by water-soluble nonionic polymers, polyacids, and cationic polyelectrolytes

Polymer	a) Average particle size (nm) (std. dev.) b) % Cyclohexane	
	Platinum	Palladium
Nonionic polymers:		
Poly(N-vinyl-2-pyrrolidone)	a) 2.7 (1.5) b) 79.9	a) 1.5 (0.6) b) 100
Poly(N-vinyl-2-pyrrolidone-co-vinyl acetate)	a) 2.2 (0.4) b) 61.7	a) 1.5 (3.0) b) 100
Polyacids:		
Poly(methacrylic acid)	a) 1.5 (0.5) b) 100	a) 6.8 (1.8) b) 40.6
Poly(styrene sulfonic acid)	a) 2.3 (0.9) b) 80.9	a) 4.0 (1.6) b) 31.2
Poly(2-acrylamido-2-methyl-1-propane sulfonic acid)	a) ~1-4 b) 100	a) 5.1 (2.3) b) 48.8
Poly(vinyl phosphonic acid)	a) 1.6 (0.5) b) 100	a) * b) 55.5
Cationic polyelectrolytes:		
Poly(diallyldimethyl ammonium chloride)	a) 1.8 (0.7) b) 22.4	a) 5.0 (1.6) b) 2.3
Poly(methacrylamidopropyl trimethyl ammonium chloride)	a) 1.4 (0.4) b) 22.7	a) 5.0 (1.2) b) 2.4
Poly(3-chloro-2-hydroxypropyl-2-methacryloxyethyldi-methyl ammonium chloride)	a) 1.7 (0.8) b) 27.3	a) 1.5 (0.5) b) 2.5

*No TEM pictures available.

tive matrices, in comparison to the ones incorporating the nonionic polymers, and no direct influence of the platinum nanoparticle size was found. This indicates that the type of protective polymer is influential as well. The results for palladium nanoparticles prepared by the alcohol reduction method and protected by the respective polymeric matrices are also shown in Table 1. Compared to the platinum samples, the values for the catalytic conversions were higher for the palladium samples if nonionic polymers were involved. However, whereas the effect of polyacids was to increase the catalytic activities of the platinum catalysts, the values for the palladium samples incorporating the polyacids were lower. A slight size dependence may be present for the palladium particles, that is, the palladium samples protected by the polyacids showed somewhat larger particle sizes, and lower catalytic conversions. However, it is very likely that the decreased catalytic conversion was not caused exclusively by the particle size, but by the presence of the polyacids as well. A reason could be the formation of surface complexes of the acid side groups with the palladium species.

The results for the catalytic activities for several platinum and palladium samples protected by cationic polyelectrolytes are included in Table 1. The catalytic activities were found to be drastically reduced for the samples involving the cationic polyelectrolytes, even though the particle sizes were very small. This might stem from the electrostatic environment created by the polyelectrolytes around the catalytically-active platinum nanoparticles. In the case of the palladium nanocatalysts the catalytic conversions were reduced even further. In these cases an additional surface modification of the palladium nanocatalysts could be involved, for instance, by the formation of surface complexes or the polarization of the catalyst surface, in addition to the electrostatic environment around the nanoparticles. These results demonstrate clearly the strong dependence of the catalytic properties on the protective polymer matrix, which can superimpose on the expected size dependence of the catalytic activities.

3.2. Effects of polymer molecular weight

An additional factor should be the molecular weight of the protective polymer, since this affects some relevant properties such as adsorption characteristics, and its ability to stabilize the colloidal metal dispersions [16, 17]. Often, higher molecular weights of the protective matrix result in a higher protective function of the polymer, and thus improved long-time stabilization of the dispersion [16, 17]. Influences on the particle structure, due to the presence of the polymer during nanoparticle preparation, could be equally important. Also, a higher molecular weight and its associated higher protective functions could result in smaller particle sizes. However, use of a polymer above a certain molecular weight can actually promote agglomeration, due to "bridging effects" [3, 18], by interacting with several metal particles simultaneously.

The effects of the molecular weight of the protective polymer could be due to various factors, such as:

(i) the *layer thickness* of the polymer surrounding and covering the catalyst particles (this should depend to a large extent on the adsorption mechanism and conformation of the polymer onto the nanoparticles in a particular solvent),

(ii) changes in *viscosity* (which would affect the diffusion of the reactants towards the catalyst surface), and

(iii) changes in the *solubility* of the polymer in the respective solvent system (for instance, a decrease of the solubility of the protective polymer can cause stronger attachment on the metal nanoparticle, and therefore change the available surface area and reduce the catalytic activity [14, 19]).

Only a few reports are available on the influence of the molecular weight of a protective polymer in such catalytically-active metal-polymer systems. For instance, it has been shown by Hirai *et al.* [1, 20, 21] that the degree of polymerization of the protective polymer can influence the catalytic activity of metal nanocatalysts protected by poly(N-vinyl-2-pyrrolidone).

Several criteria should be fulfilled for comparisons of samples with regard to polymer molecular weight. In particular, the same preparative methods should be used for generating the polymer-protected nanoparticles, in order to exclude influences from side products during the required reductions. Finally, a solvent should be chosen which provides good solubility of the polymer over a range of molecular weights, if the adsorption effect of the polymer onto the nanoparticle surface due to decrease of solubility should be excluded as one of the parameters.

In this work palladium and platinum nanoparticles were prepared by in-situ reductions in the presence of a selection of commercially available water-soluble polymers possessing different molecular weights. Two water-soluble nonionic polymers, namely poly(N-vinyl-2-pyrrolidone) and poly(2-ethyl-2-oxazoline) were chosen. To exclude the influence of side products formed, the reduction method chosen was refluxing the alcoholic solutions, as described by Hirai *et al.* [1, 3]. Catalytic activities were qualitatively tested by the hydrogenation of cyclohexene as a model reaction. The results indicate that the molecular weight of the protective polymer is an important consideration in determining the catalytic properties of colloidal metal-polymer systems.

Nanoparticle features such as the particle size are strongly influenced by the reduction method and conditions chosen. For instance, the reflux temperature and reaction time in the alcohol reduction method are significant factors which can cause changes in nanoparticle size and shape [22, 23]. Often, larger sizes are obtained with longer reflux times. In the present investigations care was taken to obtain comparable, possibly identical, particle sizes and features by control of reflux time and temperature.

The results for the platinum samples are listed in Table 2. Very small particle diameters (below 5 nm) were obtained, and the dispersions were colloidally stable for several months. As can be seen from these results, there was hardly any influence of molecular weight on the nanoparticle features. The average particle diameters, size distributions, and particle shapes were practically identical for the same polymer types, so that differences in the catalytic activity should mainly stem from the influence of the different molecular weights of the protective polymers.

A typical feature in the alcohol reduction method for platinum nanoparticles is the coexistence of both small and separate nanoparticles, and small "nanoagglomerates" (which are made up of initially-formed smaller nanoparticles). Therefore, to further compare the samples the "agglomeration behavior" of the platinum nanoparticles had to be considered, in addition to the nanoparticle diameter, size distribution, and particle shapes. This feature has to be included in order to exclude possible effects on the catalytic conversion due to different degrees of agglomeration. A gauge for this "agglomeration behavior" is also given in Table 2, and indicates the portion of separate particles related to the total number of particles (that is, the sum of separate particles and small nanoagglomerates). A smaller value indicates a higher degree of formation of nanoagglomerates, and the value 1.0 denotes samples containing only separate particles. The platinum samples which were obtained by the same reduction methods in the presence of the same polymer type (but different molecular weights) are comparable. A typical example for platinum nanoparticles obtained by the alcohol reduction method in the presence of poly(2-ethyl-2-oxazoline) is shown in Figure 1 for weight-average

Table 2.
Platinum Nanocatalysts Protected by Polymers of Different Molecular Weight

Polymer Molecular weight	Average particle diameter (nm) (std. dev.)	Particle features (agglomeration behavior)[a]	% Cyclohexane
Alcohol reduction method:			
Poly(N-vinyl-2-pyrrolidone)			
40,000	2.5(0.9)	b; (0.64)	94
360,000	2.7(1.5)	b; (0.34)	73
Poly(2-ethyl-2-oxazoline)			
50,000	1.7(0.7)	b; (0.63)	76
200,000	1.8(0.7)	b; (0.71)	56
500,000	1.9(0.8)	b; (0.41)	44
KBH$_4$ reduction:			
Poly(2-ethyl-2-oxazoline)			
50,000	1.8(0.5)	Separate, spherical; (1.0)	72
200,000	1.7(0.8)	Separate, spherical; (1.0)	56
500,000	1.8(0.5)	Separate, spherical; (1.0)	64

[a]Agglomeration behavior = $N_S / (N_S + N_a)$; (with N_S = number of separate particles, N_a = number of small nanoagglomerates). [b]Separate, spherical particles, coexisting with small nanoagglomerates. The nanoagglomerates consist of about 3-15 initially-formed smaller nanoparticles.

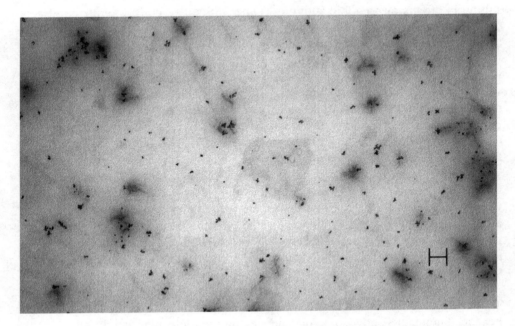

Figure 1. TEM micrograph of platinum nanoparticles obtained from H_2PtCl_6 by the alcohol reduction method in the presence of poly(2-ethyl-2-oxazoline) having the weight-average molecular weight $M_W \sim 500,000$ (bar = 33 nm).

molecular weight $M_W \sim 500{,}000$. A slight dependence of the agglomeration behavior on molecular weight could be observed for the polymers possessing the highest molecular weights. The use of a high-molecular weight polymer as protective matrix tended to result in an increased formation of nanoagglomerates. This is in agreement with findings by Hirai et al. [3] for rhodium nanoparticles protected by poly(N-vinyl-2-pyrrolidones), where lower molecular weights of the polymers resulted in a decreasing tendency for agglomeration. This was suggested as due to the more linear conformations of the lower molecular weight polymers, which could therefore cover the surface of the nanoparticles more effectively [3]. In addition, for polymers possessing very high molecular weight, as was the case investigated here, an increased bridging effect between several nanoparticles could also be responsible for the increased formation of nanoagglomerates.

Some influence of the molecular weight of the protective polymers could be detected for the platinum samples obtained by the alcohol reduction method. Lower catalytic conversions were obtained for the samples containing the higher molecular weight polymers, both for poly(N-vinyl-2-pyrrolidone) and poly(2-ethyl-2-oxazoline). Such influence could be due to decreases in polymer solubility with increasing molecular weight, and thus stronger adsorption on the nanocatalyst surface (with blocking of catalytically-active sites). This would correspond to observations made by Hirai et al. [14] where changes in the solvent caused changes in the coverage of the metal nanoparticle by the protective polymer, resulting in effects on catalytic activities. Other influences could stem from effects of the molecular weight on viscosity or adsorption mechanisms and conformations of the polymer on the nanocatalyst surface.

For comparison, some platinum samples were also prepared by the reduction with KBH_4 in the presence of the poly(2-ethyl-2-oxazolines). For these samples a value for the agglomeration behavior of 1.0 was found (that is, absence of nanoagglomerates), and the nanoparticle features among the samples were similar. Platinum nanoparticles, obtained by the reduction with KBH_4 in the presence of poly(2-ethyl-2-oxazoline) possessing a molecular weight $M_W \sim 200{,}000$ are depicted in Figure 2. Again, a lower catalytic conversion was observed for the polymer possessing the highest molecular weight ($M_W \sim 500{,}000$), compared to the one possessing the lowest, $\sim 50{,}000$. However, an even lower value for the catalytic conversion was found for the medium molecular weight of $M_W \sim 200{,}000$. In this case a less efficient control for comparable samples could be provided, due to the formation of side products during the reduction procedure. Also, the sample incorporating the polymer with $M_W \sim 200{,}000$ is the one exhibiting a somewhat broader size distribution, in comparison to the other two samples investigated.

Some palladium samples were investigated as well, with the results listed in Table 3, including the results from the hydrogenation of cyclohexene to cyclohexane. The results show some dependence on the molecular weight of the polymers. For the palladium nanoparticles protected by the poly(N-vinyl-2-pyrrolidones) of different molecular weights, very high conversions to 100% cyclohexane were

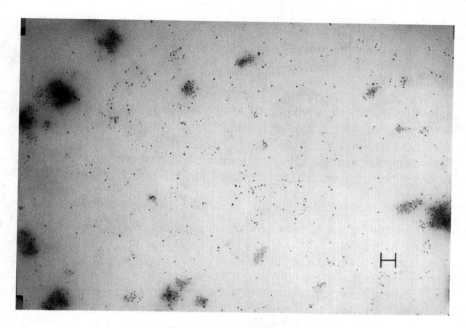

Figure 2. TEM micrograph of platinum nanoparticles obtained from H_2PtCl_6 by the reduction with KBH_4 in the presence of poly(2-ethyl-2-oxazoline) having the molecular weight $M_W \sim 200{,}000$ (bar = 38 nm).

Table 3.
Palladium Nanocatalysts Protected by Polymers of Different Molecular Weight (Preparation by the Alcohol Reduction Method)

Polymer molecular weight	Average particle diameter (nm) (std. dev.)	Particle features	% Cyclohexane
Poly(N-vinyl-2-pyrrolidone)			
40,000	1.6(0.5)	Separate, spherical	100
360,000	1.5(0.6)	Separate, spherical	100
Poly(2-ethyl-2-oxazoline)			
50,000	1.6(0.5)	Separate, spherical	92
200,000	1.6(0.7)	Separate, spherical	92
500,000	1.6(0.6)	Separate, spherical	68

obtained for both cases. However, the conversion of 100% does not allow further differentiation between the two samples, since there could be large rate differences not observed, or possibly no differences at all.

A better comparison is possible, however, for the palladium samples protected by the poly(2-ethyl-2-oxazolines). Practically no change of the catalytic conversion was found for the polymers possessing the molecular weights $M_W \sim 50{,}000$

and ~ 200,000; a catalytic conversion of about 92% was obtained for both cases. For the sample involving the polymer with the highest molecular weight, M_W ~ 500,000, however, the catalytic conversion to cyclohexane is definitely smaller, namely about 68%.

This indicates that up to a certain molecular-weight range (which is most probably different for each polymer type) the influence of the molecular weight on the catalytic activity is less pronounced. Above a certain approximate molecular weight, however, there seems to be some influence.

In any case, the results clearly demonstrate the importance of polymer molecular weight in these nanocatalyst systems.

3.3. Amphiphilic block copolymers

Another interesting class of polymers are amphiphilic block copolymers, which possess one hydrophilic and one hydrophobic block. For these investigations polystyrene-*block*-poly(methacrylic acid) (PS-b-PMAA) and polystyrene-*block*-poly(ethylene oxide) (PS-b-PEO) were selected. In polar solvent systems these block copolymers form spherical micelles possessing a hydrophobic polystyrene core and a hydrophilic corona. A variety of metal precursors and reduction methods were investigated for their influence on the catalytic activity for the hydrogenation of cyclohexene [13]. Table 4 sumarizes some representative results for palladium nanoparticles protected by PS-b-PMAA. In all cases spherical micelles were formed.

The particle size does not seem to be the only influence on the catalytic activities. The findings indicate that the more hydrophobic precursors (e.g., palladium acetate versus palladium chloride) and longer stirring times before reduction lead

Table 4.
Palladium nanocatalysts protected by PS-b-PMAA

Precursor	Reduction Method	Conversion % cyclohexane	Av. particle diameter (nm) (std. dev.)
$PdCl_2$	Refluxing	23.1	7.9 (6.3)
$PdCl_2$	KBH_4	33.5	2.1 (1.2)
$Pd(ac)_2$	Refluxing	13.8	3.4 (0.8)
$Pd(ac)_2$	KBH_4	14.6	1.3 (0.5)
$Pd(ac)_2$	RT	6.1	1.0 (0.9)
$Pd(ac)_2$	60°C	8.2	1.3 (1.2)
$Pd(F_3ac)_2$	RT	6.2	3.5 (1.5)
$Pd(F_3ac)_2$	60°C	8.3	2.7 (1.0)
$Pd(acac)_2$	Refluxing	11.8	6.3 (3.8)
$Pd(acac)_2$	Stirred/Refl.	15.8	5.4 (2.6)
$Pd(acac)_2$	KBH_4	80.7	2.0 (0.7)
$Pd(acac)_2$	Stirred/KBH_4	54.6	2.1 (0.9)

[a] Hydrogenation conditions the same as for the other reactions, but with one third of the amounts.

to lower catalytic conversions, whereas the sterically more hindered precursor palladium acetylacetonate result in increased values. This leads to the conclusion that full or partial embedding of the resulting palladium nanoparticles could be responsible for the variations in catalytic conversions. Catalyst nanoparticles which are embedded in a hydrophobic polystyrene matrix should show a higher blocking of the catalytically active sites, than particles which are embedded in a hydrophilic matrix which is less-strongly interacting with the catalyst surface. Similar results showing the same trends were also found for the palladium nanocatalysts protected by the PS-b-PEO polymer [13].

4. CONCLUSIONS

Polymer-protected noble metal nanocatalysts can lead to the design of novel catalyst systems, since they exhibit a number of advantages. The choice of the protective polymer, the preparation methods, and the metal precursors can allow the tuning of the colloidal stability and the catalytic properties of the materials. Such high versatility of these polymer-metal composite materials is intriguing, and seems promising for further developments oftailored and selective catalysts.

Acknowledgments

The authors thank Professor Randal E. Morris, Department of Cell Biology, Neurobiology and Anatomy, The University of Cincinnati Medical Center, for providing training and access for transmission electron microscopy, Professor R. Marshall Wilson for providing access to his UV-vis instrument, Professor Hans Zimmer for providing laboratory space and his hydrogenation equipment, and Professor Allan R. Pinhas, Department of Chemistry, The University of Cincinnati, for providing access to a gas chromatograph. Financial support was provided by the National Science Foundation through Grant DMR-9422223 (Polymers Program, Division of Materials Research).

REFERENCES

1. H. Hirai, N. Toshima in: *Catalysis by Metal Complexes. Tailored Metal Catalysts*, Y. Iwasawa (Ed.) D. Reidel Publishing Company, Dordrecht (1986).
2. J. S. Bradley in: *Clusters and Colloids. From Theory to Applications*, G. Schmid (Ed). VCH, Weinheim (1994).
3. H. Hirai, Y. Nakao, N. Toshima, *J. Macromol. Sci.-Chem.*, **A13**, 727 (1979).
4. N. Toshima, T. Yonezawa, K. Kushihashi, *J. Chem. Soc., Faraday Trans.*, **89**, 2537 (1993).
5. H. Hirai, H. Chawanya, N. Toshima, *Reactive Polymers*, **3**, 127 (1985).
6. M. Antonietti, E. Wenz, L. M. Bronstein, M. Seregina, *Adv. Mater.*, **7**, 1000 (1995).
7. M. Antonietti, S. Förster, J. Hartmann, S. Oestreich, *Macromolecules*, **29**, 3800 (1996).
8. S. Klingelhöfer, W. Heitz, A. Greiner, S. Oestreich, S. Förster, M. Antonietti, *J. Am. Chem. Soc.*, **119**, 10116 (1997).
9. J. F. Ciebien, R. E. Cohen, A. Duran, *Supramol. Sci.*, **5**, 31 (1998).

10. R. E. Cohen, R. T. Clay, J. F. Ciebien, B. H. Sohn in: *Polymeric Materials Encyclopedia*, Vol. 6, J. C. Salamone (Ed), p. 4143, CRC Press, Boca Raton (1996).
11. A. B. R. Mayer, J. E. Mark, *J. Polym. Sci., B: Polym. Phys.*, **35**, 1207 (1997).
12. A. B. R. Mayer, J. E. Mark, *J. Polym. Sci., A: Polym. Chem.*, **35**, 197 (1998).
13. A. B. R. Mayer, J. E. Mark, R. E. Morris, *Polym. J.*, **30**, 197 (1998).
14. H. Hirai, H. Chawanya, N. Toshima, *Makromol. Chem., Rapid Commun.*, **2**, 99 (1981).
15. J. A. Creighton, D. G. Eadon, *J. Chem. Soc., Faraday Trans.*, **87**, 3881 (1991).
16. W. Heller, T. L. Pugh, *J. Polym. Sci.*, **XLVII**, 203 (1960).
17. L. Longenberger, G. Mills, *J. Phys. Chem.*, **99**, 475 (1995).
18. R. J. Hunter in: *Foundations of Colloid Science*, Vol. 1; p. 101, Oxford University Press, NewYork (1986).
19. C.-W. Chen, M. Akashi, *J. Polym. Sci. A: Polym. Chem.*, **35**, 1329 (1997).
20. H. Hirai, H. Wakabayashi, M. Komiyama, *Chem. Lett.*, 1047 (1983).
21. H. Hirai, N. Yakura, S. Hodoshima, *Reactive & Functional Polymers*, **37**, 121 (1998).
22. S. C. Davis, K. J. Klabunde, *Chem. Rev.*, **82**, 153 (1982).
23. H. Hirai, Y. Nakao, N. Toshima, *J. Macromol. Sci. -Chem.*, **A12**, 1117 (1978).

ATRP "living"/controlled radical grafting of solid particles to create new properties

HENRIK BÖTTCHER,[1] MANFRED L. HALLENSLEBEN,[1,2,*] REINHARD JANKE,[1] MANFRED KLÜPPEL,[2] MARTIN MÜLLER,[2] STEFAN NUSS,[1] ROBERT H. SCHUSTER,[2] STEFAN TAMSEN[1] and HELLMUTH WURM[1]

[1]*Universität Hannover, Institut für Makromolekulare Chemie, Am Kleinen Felde 30, D-30167 Hannover, Germany*
[2]*Deutsches Institut für Kautschuktechnologie e.V. DIK, Eupener Straße 33, D-30519 Hannover, Germany*

Abstract—Atom transfer radical polymerisation ATRP has been applied to graft polymer chains from solid surfaces. For this purpose, new ATRP initiators have been designed which contain (i) functional sites to anchor the initiator covalently to the surface of an appropriate material, (ii) the active group which is capable to initiate the ATRP polymerisation process, and (iii) optionally a cleavable link which allows to detach the polymer grafts for polymer analysis. Surfaces have been decorated with the ATRP initiators and graft polymerisation as well as graft copolymerisation has been initiated *from* the surface. Grafting of various monomers, chain length and chain length distribution of the grafts is reported.

Keywords: ATRP; core-shell systems; grafting from; polymer films; surfaces.

1. INTRODUCTION

Physisorption or chemisorption of polymer chains onto solids may be in one case or the other the propriate solution to cover the surface with a polymer layer [1]. But for more than only a rough approach to change or even to influence the surface properties of a solid material, a procedure to bind polymer chains covalently to the solid is likely to be rather more effective. Numerous applicational reasons make a suitable polymer coating of solid materials attractive, see examples compiled in Table 1.

Grafting of preformed polymer chains onto the surface of a solid lacks good control over polymer chain density on the surface. Polymer chain conformation in solution results in steric hinderance for the chains approaching the surface and entropy reasons even more contribute to only a limited success.

For these reasons in the past more efficient techniques to graft polymer chains *from* the surface of a solid have been elaborated. The most striking advantages of such an approach are the possibility to achieve a better control over the grafted

---

*To whom correspondence should be addressed. E-mail: hallensleben@mbox.imc.uni-hannover.de

Table 1.
Tailored Solid Surface Polymer Coating of Solid Material for Different Applicational Purposes

Surface Protection – to prevent physical or chemical damage	S. Pani, Surface Coatings, J. Wiley & Sons, New York, 1986; A. Homala, M. Mate, G. B. Street, MRS Bull. **15**, 45 (1990); A. Novotny, J. D. Swalen, J. P. Raabe, Langmuir **5**, 485 (1989); J. Rühe, G. Blackman, V. J. Novotny, T. Clarke, G. B. Street, S. Kuan, J. Appl. Polym. Sci. **53**, 825 (1994)
Surface Properties – Hydrophilicity / Hydrophobicity	M. Husseman, E. E. Malmström, M. McNamara, M. Mate, D. Mecerreyes, D. G. Benoit, J. L. Hedrick, P. Mansky, E. Huang, T. P. Russell, C. J. Hawker, Macromolecules **32**, 1694 (1999)
Compatibility – of biological cells with nonbiological material	Surface and Interfacial Aspects of Biomedical Polymers, J. P. Anrade, Ed., Plenum Press, New York, 1985
Stabilisation – of dispersions or controlled flocculation	J. N. Israelachvili, Intermolecular and Surface Forces, 2nd Edit., Academic Press, London, 1992
Adhesion – between different materials	Adhesive Chemistry, L. H. Lee, Ed., Plenum Press, New York, 1985; S. C. Temin, in: Encyclopedia of Polymer Science and Engineering
Interaction – between fillers and polymer matrix	Concise Encyclopedia of Composite Materials, A. Kelly, Ed., Pergamon Press, Oxford, 1994; Interfaces in Polymer, Ceramic and Metal Matrix Composites, H. Ishida, Ed., Elsevier Science Publishers, New York, 1989
Adhesive Lubricant	J. Rühe, G. Blackman, V. J. Novotny, T. Clarke, G. B. Street, S. Kuan, J. Appl. Polym. Sci. **53**, 825 (1994)

chain density and to assure that the polymer chains assembled at the surface are bound covalently; in other words, the possibility to create a controllable and time lasting polymer coat of surfaces.

Molecular parameters which are of importance for polymer grafts on a solid surface are given in Table 2.

Rühe *et al.* have presented the most elaborated and the most convincing results on this route of polymer grafting *from* a surface of a solid. In detailed studies this group has prepared new azo initiators which contain an anchor group to bind the initiator covalently onto a surface. The Rühe group also achieved to obtain macromolecular parameters of grafts by introducing a cleavable link into the initiator to detach the grafts from the solid. Convincing results based on this concept were achieved [2, 3].

On the other hand, most of the azo initiators used to graft polymer chains *from* a solid are symmetrical in chemical architecture which means that thermal decomposition generates not only a surface bound radical but also a free radical both of which are able to initiate radical polymerisation. Therefore besides the desired grafted polymer a significant amount of ungrafted polymer is formed. Furthermore, azo initiators start polymer chain propagation in accordance with the classical

Table 2.
Polymer Grafts on Solid Surface

Questions with respect to polymer chain architecture
- chain length - controllable
- chain length distribution - controllable
- comonomer sequence in copolymers
- bloch architecture in block copolymers
- chain end functionalities

Questions with respect to covering density on the surface
- graft density - controllable
- unreacted initiating sites (*grafting from*)
---

free radical polymerisation process and therefore graft chain growth is subjected to classical free radical chain termination which allows neither control over chain length nor does it allow control over monomer sequence in copolymerisation reactions thus block copolymer grafts are not easily accessible.

In the meantime, "living"/controlled radical polymerisation techniques have been developed. In this field two different approaches have been worked out which are nominated a) the nitroxyl-mediated route (most often the nitroxyl species is chosen as 2,2,4,4-methylpiperidine N-oxide TEMPO) [4–8], and b) the so-called atom transfer radical polymerisation ATRP route [9–21], see Fig. 1. Both the mechanisms for polymer chain growth reaction include the reversible equilibrium of an inactive state in which the polymer chain end is dormant and of an active state. Activation of dormant species occurs by either a reversible temperature caused homolytic cleavage of a carbon-oxygen bond in the nitroxyl-route mechanism, or by a reversible oxidation-reduction process in the ATRP route. In the latter the alkyl halide polymer chain end interacts with a transition metal complex in its lower oxidation state. Both these new radical polymerisation techniques have allowed for a more sophisticated management of radical chain growth polymerisation as has been documented since in an increasing number of publications.

This contribution deals with a new approach to graft polymer chains *from* the surface of solid particles such as silica and nanometer scaled elastic or inelastic crosslinked microgels by the application of the "living"/controlled ATRP chain growth mechanism. For this purpose ATRP initiators have been designed which besides the ATRP initiating site have an anchor group in order to bind the initiator to the surface of a solid exhibiting reactive OH-groups, and which in addition contain a weak link which makes it possible to detach the grafted polymer chains from the solid surface in order to subject the graft to polymer analyses. A more general scheme of this initiator design is given in Fig. 2. It was our aim to have ATRP initiators at hand which can be bound to the solid surface in simple experiments from solution also in such a way as to control almost precisely the initiator density on the solid surface.

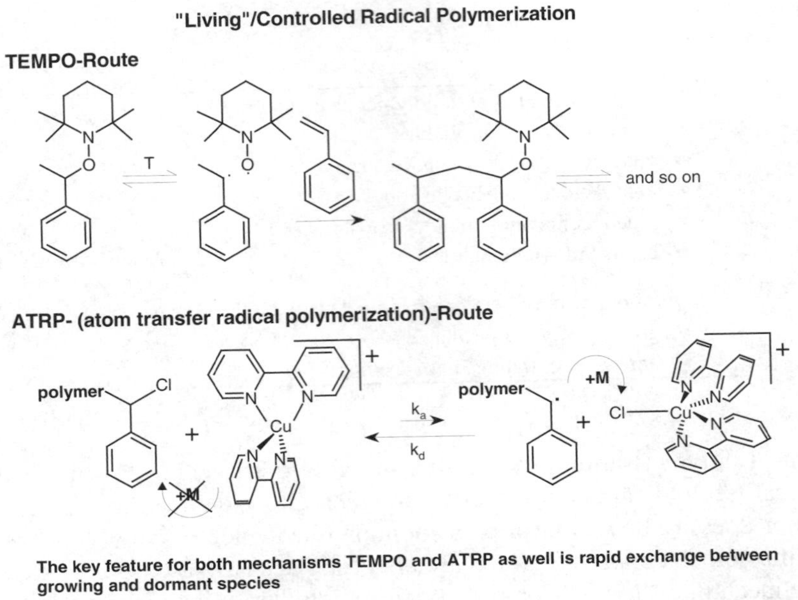

"Living"/Controlled Radical Polymerization

TEMPO-Route

ATRP- (atom transfer radical polymerization)-Route

The key feature for both mechanisms TEMPO and ATRP as well is rapid exchange between growing and dormant species

Figure 1. Presentation of the TEMPO- and the ATRP-Route for "living"/controlled radical polymerisation.

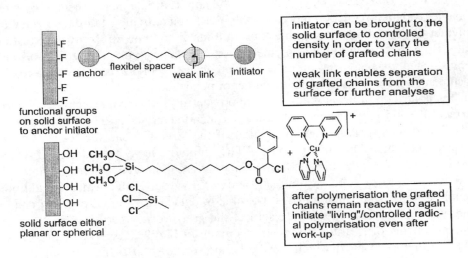

Figure 2. Schematic drawing of ATRP initiator immobilisation of the solid surface and graft polymerisation *from* the solid surface.

2. INITIATORS

2.1. To react with OH-groups

A selection of ATRP initiators also containing an anchor group to bind them to OH-group exposing surfaces is given in Fig. 3. Most of these initiators contain a weak link in form of an ester group which makes it possible to detach grafted polymer chains from the solid surface by simple hydrolysis of the ester linkage. This makes possible to investigate e.g. the influence of initiator density on the solid surface on polymer graft chain length and chain length distribution from GPC analysis and also to investigate chain architecture of the grafts by common analytical techniques. Once the reaction parameters for binding ATRP initiator to the solid surface and for carrying out graft polymerisation have been established, simple initiators such as compounds **1-3** in Fig. 3 can be used.

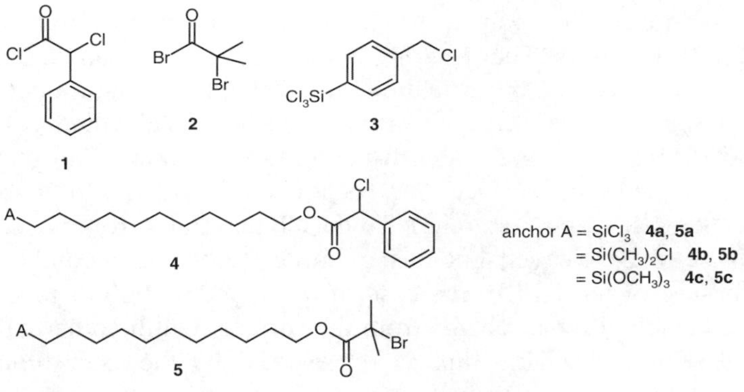

Figure 3. Selection of ATRP initiator molecules containing an anchor group to react with OH-groups on solid surfaces.

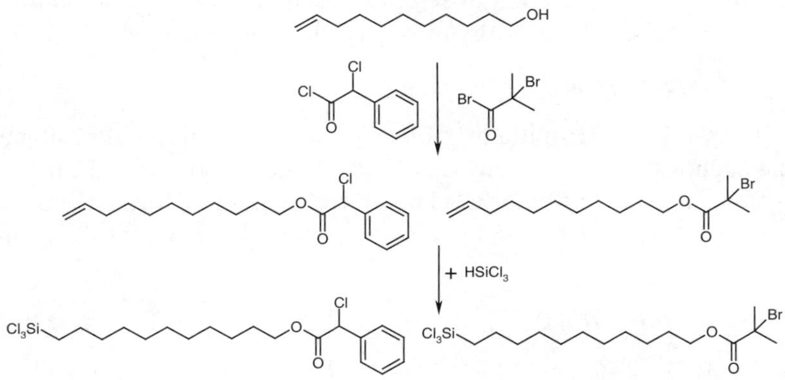

Figure 4. Easy and effective synthesis of ATRP initiator molecules comprising an anchor group and a spacer.

The spacer in the initiator molecules **4** and **5** provides the possibility to keep the ATRP initiating head group in some distance from the solid surface which may be of importance in one or the other case. A simple and general synthetic route to prepare ancor-spacer-weak link-ATRP initiator molecules is given in Fig. 4.

The molecules **4** and **5** both contain chlorosilyl or alkoxysilyl anchor groups and a spacer between initiating and anchoring sites. The reaction of chlorosilyl or alkoxysilyl derivatives with surface silanol groups of silica or silicon wafers is a well established method for surface modification of such substrates. The use of monofunctional anchor groups is perferred since the resulting bond between the ATRP initiator and the surface is well defined. Di- or trifunctional anchor groups as represented in these molecules tend to build up a network at the surface as a result of condensation reactions between individual surface active molecules, caused by remnant water in reagents and at the surface of the solid. In the case of molecules **4b** and **5b** the remnant water will cause a dimerisation of the anchor equipped ATRP initiators. The resulting compounds simply can be removed from the initiator modified silica by extraction unlike condensation products in the case of compounds **4a,c** or **5a,c**, respectively. A further disadvantage of the anchor equipped ATPR initiator **3** is the absence of a cleavable link. Graft polymerisation being accomplished, the surface may be covered with hydrophobic polymer chains as in the case of poly(styrene). To detach the grafts from the surface Si-O-Si bonds have to be cleaved but this requires strong basic conditions, i.e. aq. KOH, or the use of aq. HF. In the case of hydrophobic polymers it would take long time to detach polymer chains from the surface. On the other hand the presence of a spacered cleavable link as represented by the ester functionality in molecules **4** and **5** offers a versatile route to detach polymer chains from the surface by transesterification of hydrolysis of the ester functionality under mild conditions as described by Rühe et al. [2, 3]. The structure of the ATRP initiating site has to depend on the monomer to be polymerised. The benzyl halide structure as in **4** is useful for polymerisation of styrene and the α-halo ester structure in molecule **5** e.g. is preferred when methylmethacrylate is the monomer.

2.2. To react with glass surface

In order to install ATRP initiator molecules with an active chlorosilane or alkoxysilane anchor group to the surface of glass plates or glass beads, the glass surface has to be activated in order to increase the number of free OH-grups, e.g. by exposing the glass surface to 1 N aqueous sodium hydroxide solution for some time.

2.3. To react with Au surface

Thiol grups are well known to react with gold surface and so are disulfide groups in dialkyldisulfides. A typical ATRP initiator for these purposes is molecule which also can be prepared easily [22].

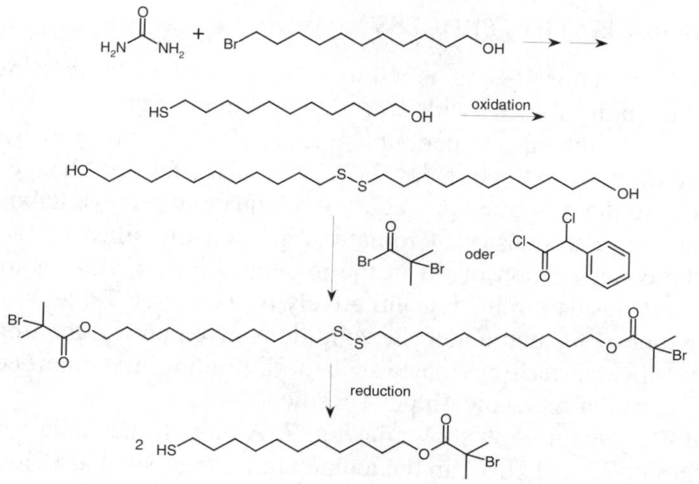

Figure. 5. Effective synthesis of ATRP initiator molecules comprising a disulfide or a thiol anchor group.

As shown in Fig. 6, a number of systems is now available for ATRP graft reactions from the surface of different solid materials independent from the physical shape of the surface.

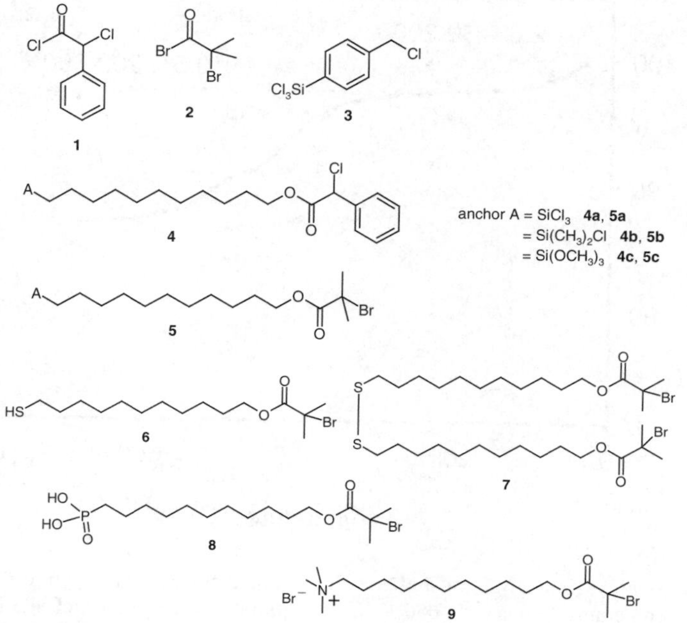

Figure 6. Selection of ATRP initiator molecules containing different anchor groups to react covalently with appropriate functionalities on the surface of a solid.

3. THE IMMOBILISATION PROCESS

As an example, initiator **4b** was bound to the silica surface by reacting pretreated and dried silica in dry toluene with **4b** for 24 h in the presence of triethylamine as a base catalyst at ambient temperature. Excess of **4b** is removed by extraction. Prolonged continuous extraction for 24 h with THF did not change the surface concentration, so the described procedure is adequate to remove unbound initiator from the surface. The density of initiating sites on the silica surface was controlled by the concentration of **4b** in the toluene solution with regard to the silica/**4b** ratio and was determined quantitatively by TGA, see Table 3.

In these experiments and under the conditions given above, the immobilisation of **4b** on the silica surface obviously tends to a limiting maximum concentration of about 0.3 mmol of molecule **4b** per 1 g silica.

A typical TGA diagram is shown in Fig. 7. Although the silica was dried in a vacuum-oven for 72 h at 50°C, in the temperature range 30–200°C loss of volatile material in a considerable quantity is detected which is ascribed to adsorbed humidity. In the subsequent temperature range 200–650°C the loss of organic material occurs whereof the density of initiating sites on the silica surface can be calculated quantitively.

From the determined amount of **4b** immobilized on the silica surface the specific area occupied by one initiator molecule **4b** can be calculated, see Table 3.

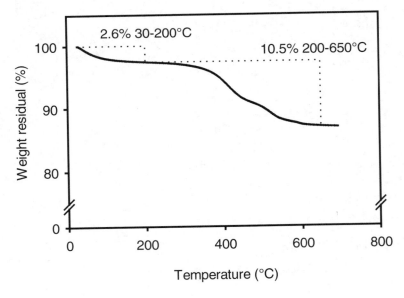

Figure 7. Thermogravitric analysis of initiator **4b** decorated silica - Temperature range 30–200°C: loss of water; temperature range 200–650°C: loss of initiator **4b**, from which the initiator density can be calculated.

Table 3.
Modification of silica with different amounts of **4b**

entry	4b / silica (mmole/g)	weight loss (%)	[4b] (mmole/g)	[4b] / m^2 (mmole/m^2)
1	0.09	1.64	0.041	2.5×10^{-4}
2	0.27	4.90	0.14	8.5×10^{-4}
3	0.51	7.93	0.23	1.4×10^{-3}
4	0.73	9.91	0.28	1.7×10^{-3}
5	1.78	10.46	0.30	1.82×10^{-3}

specific surface area of silica used was 165 m^2/g.

↓
≈ 100 Å^2 / molecule **4b**.

4. THE ATRP *GRAFTING FROM* REACTION

4.1. Grafting from *ATRP decorated silica*

A representative selection of graftings is given in Table 4. It has to be noted that even extensive extraction of the grafted silica only showed a minor quantity of 'free' that means ungrafted poly(styrene) chains to be formed under the given polymerisation conditions. The polydispersity of the detached grafts is somewhat greater than reported for ATRP styrene polymerisation in solution under the same reaction conditions; these findings are presently under investigation [24].

Once the silica is grafted and separated, dried and stored for a month or so under ambient conditions, the grafted polymer chain ends can be reactivated by the addition of CuCl and the ligand, e.b. bipyridine, and under identical reaction conditions a 'second generation' of e.g. poly(styrene) can be brought onto the surface

Table 4.
Molecular Weight of Detached Poly(styrene) (GPC)

entry	4b / silica (mmole/g)	4b / styrene (mm/mm)	conversion (%)	M$_w$	M$_n$	D	(%) 'free' polymer
1	0.041	1 : 1220	37	39.200	28.000	1.4	< 5
2	0.14	1 : 360	49	20.100	13.400	1.5	< 5
3	0.23	1 : 220	54	13.600	9.100	1.5	< 5
4	0.28	1 : 180	68	13.100	8.200	1.6	< 3
5	0.30	1 : 167	73	12.900	8.200	1.6	< 3

[styrene] = 0.05 mole, 50 mL propylenecarbonate, [**4b**] : [CuCl] : [bipy] = 1 : 1.5 : 1.5, T = 110°C, t = 24 h.

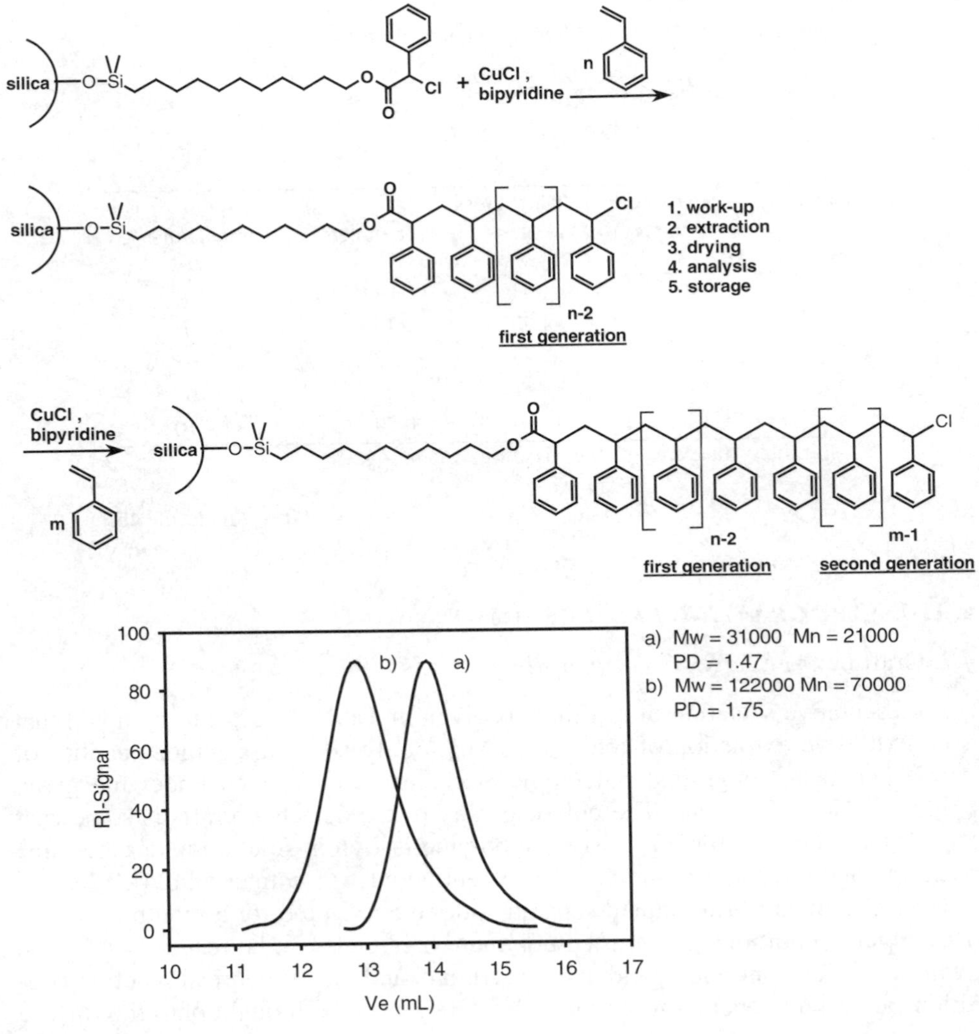

Figure 8. First and second generation of poly(styrene) grafts on silica initiated by **4b** and GPC analyses of detached first (a) and first+second (b) generation of grafts.

as is demonstrated in Fig. 8. GPC analysis of the detached grafts shows that the molecular weight distribution of the first+second generation is almost the same as for the first generation of grafts [25].

This is also the basis for the preparation of block copolymer grafts from the silica surface by applying the ATRP technique, some examples of which are given in Fig. 9. Corresponding experiments with ATRP initiator decorated microgels, which were prepared according to Fig. 10, yield a selection of polymer chain grafted microgels for which some examples are given in Fig. 11.

poly(styrene)

Mw = 7.000, 12.000, 24.000, 30.000

poly(methyl methacrylate)

Mw = 18.000, 140.000, 230.000

poly(styrene-block-p-tert.butyl styrene)

23.700 21.900

poly(styrene-block-tert.butyl acrylate)

19.700 21.000

poly(styrene-block-acrylic acid)

19.700 12.150

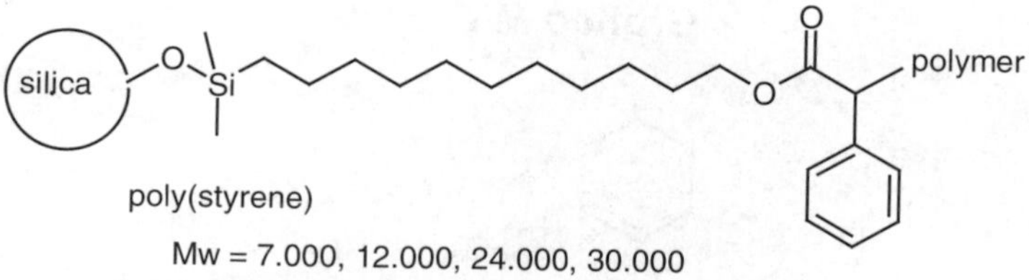

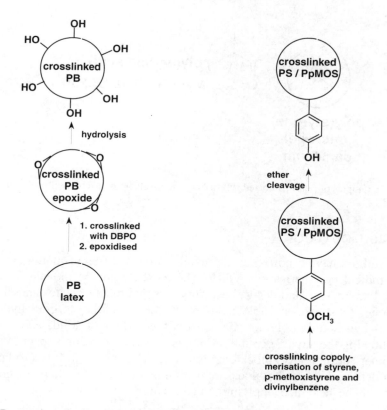

Figure 9. Silica coated with different homo and block copolymers by *grafting from.*

Figure 10. Preparation of microgels with different T_gs and OH-groups at the surface.

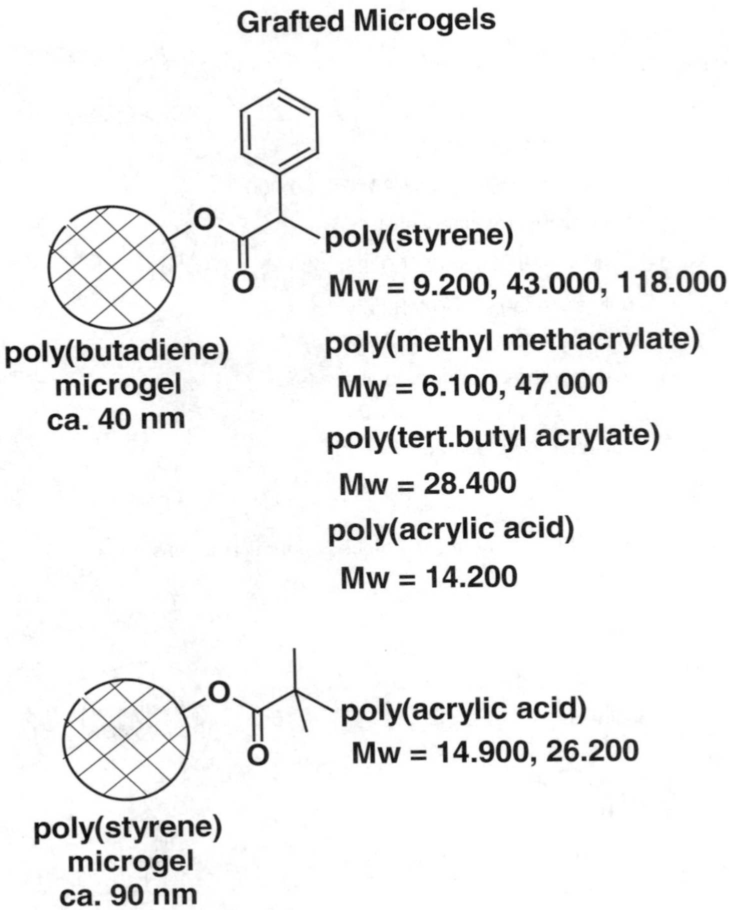

Grafted Microgels

poly(butadiene) microgel ca. 40 nm

poly(styrene)
Mw = 9.200, 43.000, 118.000

poly(methyl methacrylate)
Mw = 6.100, 47.000

poly(tert.butyl acrylate)
Mw = 28.400

poly(acrylic acid)
Mw = 14.200

poly(styrene) microgel ca. 90 nm

poly(acrylic acid)
Mw = 14.900, 26.200

Figure 11. Examples of polymer coated soft or hard microgels with different polymer chemistry.

5. GENERAL OUTLOOK

Several authors have reported on the availability of different monomers to atom transfer radical polymerisation ATRP. The most prominent molecules so far investigated are styrene and derivatives, acrylates and methacrylates and also vinylpyrrolidone. As shown in Table 5 this opens the possibility to graft hydrophobic as well as hydrophilic polymer chains *from* the surface of solids and solid particles. Following the advice given by Matyjaszewski *et al.* with respect to the order of monomer addition [23], block copolymers can be obtained. It has to be investigated in detail what are the block architectures and how the shape of core-shell-systems can be designed applying ATRP. These investigations are in progress [24].

Table 5.
ATRP Copolymerization; Monomers and Comonomers Reported in Literature

Monomers and Comonomers	Reactivities	Literature
(structures: styrene derivative, methyl acrylate, n-butyl acrylate, methacrylate, hydroxyethyl acrylate, acrylonitrile, vinyl acrylate, acrylamide, N-vinylpyrrolidone, N-(2-hydroxypropyl)acrylamide)	According to prominent investigations, ATRP is likely to preceed via a free radical mechanism. This makes it possible to apply free radical polymerisation copolymerisation parameters.	D. M. Haddleton, M. C. Crossman, K. H. Hunt, C. Topping, C. Waterson, K. G. Suddaby, *Macromolecules* **30**, 3992, (1997) A. Kajiwara, K. Matyjaszewski, *Macromol. Rapid Commun.* **19**, 319, (1998) T. E. Patten, K. Matyjaszewski, *Adv. Mater.* **10**, 901, (1998)

Acknowledgment

The authors are endebted to Fonds der Chemischen Industrie and to the Government of Lower Saxonie for financial support. Discussions with H. Menzel are appreciated. Mr. G. Koerner, DIK, contributed to the DSC and TGA measurements. Silica (Ultrasil 3370) with a specific surface area of 165 m^2/g (BET) was donated by Degussa AG, Dr. B. Freund.

REFERENCES

1. G. J. Fleer, M. A. Cohen Stuart, J. M. H. M. Scheutjens, T. Cosgrove, B. Vincent, Polymers at Interfaces, Chapman & Hall, London 1993.
2. O. Prucker, J. Rühe, Macromolecules **31**, 592 (1998).
3. O. Prucker, J. Rühe, Macromolecules **31**, 602 (1998).
4. M. K. Georges, R. P. N. Veregin, P. M. Kazmaier, G. K. Hamer, Macromolecules **26**, 2987 (1993).
5. M. K. Georges, R. P. N. Veregin, G. K. Hamer, P. M. Kazmaier, Macromol. Symp. **88**, 89 (1994).
6. C. J. Hawker, J. Am. Chem. Soc. **116**, 11314 (1994).
7. C. J. Hawker, G. G. Barclay, A. Orellana, J. Dao, W. Davenport, Macromolecules **29**, 5245 (1996).
8. C. J. Hawker, Acc. Chem. Res. **30**, 373 (1997).
9. J. S. Wang, K. Matyjaszewski, J. Am. Chem. Soc. **117**, 5614 (1995).
10. J. S. Wang, K. Matyjaszewski, Macromolecules **28**, 7901 (1995).
11. V. Percec, B. Barboiu, Macromolecules **28**, 7970 (1995).
12. T. E. Patten, J. Xia, T. Abernathy, K. Matyjaszewski, Science **272**, 866 (1996).

13. V. Percec, B. Barboiu, A. Neumann, J. C. Ronda, M. Zhao, Macromolecules **29**, 3665 (1996).
14. D. Haddleton, C. B. Jasieczek, M. J. Hannon, A. J. Schooter, Macromolecules **30**, 2190 (1997).
15. C. Granel, P. Dubois, R. Jerome, P. Teyssie, Macromolecules **29**, 8576 (1996).
16. H. Uegaki, Y. Kotani, M. Kamigaito, M. Sawamoto, Macromolecules **30**, 2249 (1997).
17. M. Kato, M. Kamigaito, M. Sawamoto, T. Higashimura, Macromulecules **28**, 1721 (1995).
18. T. Ando, M. Kato, M. Kamigaito, M. Sawamoto, Macromolecules **29**, 1070 (1996).
19. T. Ando, M. Kamigaito, M. Sawamoto, Macromolecules **30**, 4507 (1997).
20. K. Matyjaszewski, M. Wie, J. Xia, N. E. McDermott, Macromolecules **30**, 8161 (1997).
21. T. E. Patten, K. Matyjaszewski, Adv. Mater. **10**, 901 (1998).
22. S. Nuß, Diploma Thesis, Universität Hannover 1999.
23. K. Matyjaszewski, D. A. Shipp, J.-L. Wang, T. Grimaud, T. E. Patten, Macromolecules **31**, 6836 (1998).
24. H. Böttcher, M. L. Hallensleben, S. Nuß, H. Wurm, under preparation.
25. H. Böttcher, M. L. Hallensleben, S. Nuß, H. Wurm, Polym. Bull. **44**, 223 (2000).

Functionalization of polymers prepared by "living" free radical polymerization

E. BEYOU,* P. CHAUMONT, C. DEVAUX, N. JARROUX and N. ZYDOWICZ

Laboratoire d'Etudes des Matières Plastiques et des Biomatériaux, UMR CNRS n°5627, Université Claude Bernard-Lyon 1, 43 Bd du 11 Novembre 1918, F-69622 Villeurbanne Cedex, France

Abstract—Nitroxide- and bromo-terminated polystyrenes resulting from "living" free radical polymerization processes have been "functionalized" by combination reaction with thiuram disulfides compounds i.e. tetraethyl thiuram disulfide and N,N'-diethyl-N,N'-bis{2-(trimethylsilyloxy)ethyl} thiuram disulfide, or tetraphenylethane based derivatives i.e. 1,1,2,2-tetraphenyl-1,2-bis(trimethylsilyloxy)ethane. The substitution of the terminal moiety by dithiocarbamate or diphenylmethyl groups have been characterized by ^{1}H NMR, ^{13}C NMR, IR spectroscopy, MALDI-TOF mass spectrometry, ESI mass spectrometry and LSI mass spectrometry.

Keywords: Living free radical polymerization; TEMPO; ATRP; functionalization; thiuram disulfide compounds; tetraphenylethane based-derivatives.

1. INTRODUCTION

In a similar manner to anionic "living" polymerization, ATRP and nitroxide-mediated "living" radical polymerization [1] of olefins lead to the formation of polymers with narrow polydispersities, and allow a perfect control of the molecular mass of the polymer obtained. In those radical processes, the living character is respectively a consequence of the reversible formation of radicals from alkyl halides accompanied by reduction/oxidation of a copper halide catalyst or of the fast reversible combination reaction of a growing polymer radical with a stable nitroxide radical to form adducts; which are "dormant" species (Scheme 1).

Heating these polymers to temperatures above 100°C leads to a homolytic cleavage of the alkoxyamine-styrene bond or of the Halogen-carbon bond in the case of ATRP, therefore it was interesting to study a free radical process allowing the quantitative substitution of the terminal group by another chemical species carrying some functional group. In the attempt to generate monochelics as in anionic "living" polymerization processes (for example, the first step of Schulz and Milkovich's method [2] for the synthesis of polystyrene macromonomers), the elimination of alkoxyamine- or halogen-group was investigated. Pionteck *et al.*

---

*To whom correspondence should be addressed. E-mail: beyou@matplast.univ-lyon1.fr

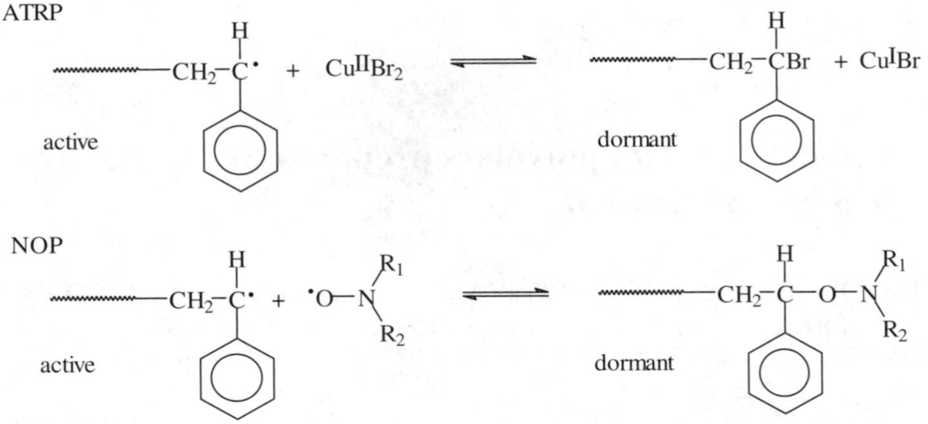

ATRP

Scheme 1. Equilibrium between active and dormant species in ATRP and in nitroxide-mediated "living" polymerization.

[3] tried different reagents such as lithium aluminium hydride, ionol and m-chloroperbenzoic acid to obtain an inert end group. For our research, we have tested compounds containing S-S or C-C bond with low dissociation energy. A speculative scheme for such a substitution is the addition to the alkoxyamine- or halogen-terminated polymers of an organic species leading to an irreversible combination with the polymer radical coming from the thermal homolysis of the polymer chain end, the temperature being high enough to dissociate the alkoxyamine group or the halogen. If the combination reaction of the new group is irreversible, such an addition might result in a quantitative substitution of the nitroxide moiety or of the halogen.

Thiuram disulfides are known to behave as iniferters, i.e. are able to act in radical polymerization media as initiators, transfer agents and termination agents [4]. The termination reaction is the consequence of the formation of dithiocarbamate radicals in the reaction medium because of either the thermal homolysis of the S-S bond of the thiuram disulfide, or the transfer reaction of alkyl radicals onto the S-S bond (Scheme 2). Thus, a nitroxide terminated polymer heated in the presence of a thiuram disulfide might lead to the substitution of the nitroxide moiety by a dithiocarbamate group (Scheme 3).

Endo *et al.* have found some "living" nature in the thermal polymerization of styrene in the presence of disulfides [5] attributed to the dissociation of the C-S bond. For example, the dissociation energy is 217 kJ/mol in the case of $C_6H_5CH_2$-SCH_3. This value may be compared to the calculated dissociation energy of alkoxyamine groups [6], ranging from 60 to 160 kJ/mol. In a recent study, Fukuda *et al.* [7] have experimentally determined an apparent value of 124 kJ/mol for the dissociation energy of nitroxide-terminated polystyrene. From these values, the substitution can be expected to be *quasi* quantitative.

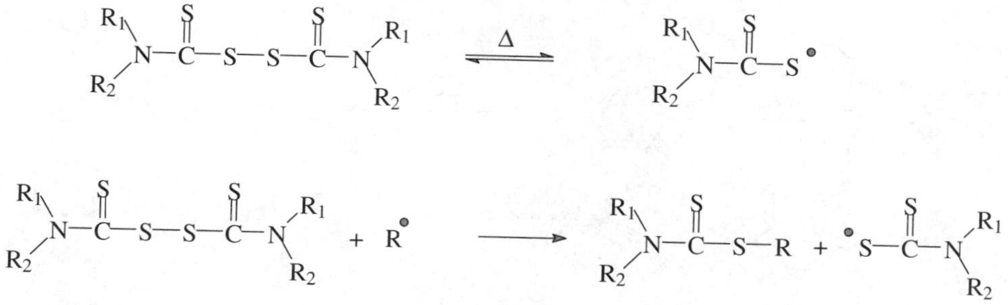

Scheme 2. Formation of dithiocarbamate radical with thiuram disulfides.

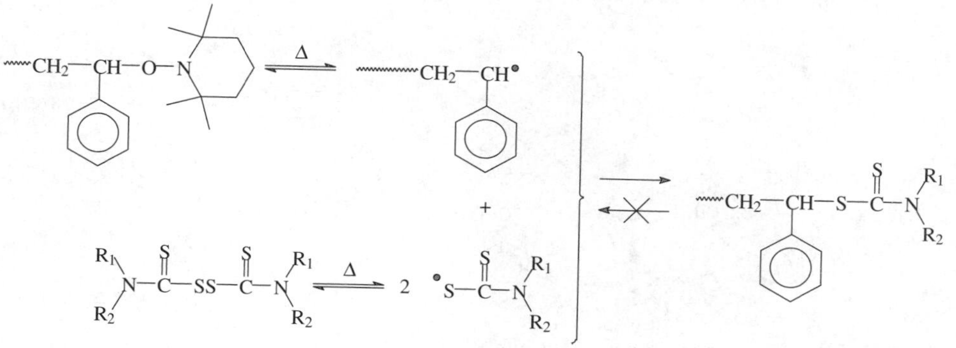

Scheme 3. Substitution of nitroxide moieties by dithiocarbamate groups.

We have also tested the substitution of the nitroxyde moiety or of the halogen, in the case of ATRP, with other compounds that generate radicals upon decomposition which are referred to as "initers": the tetraphenylethane based derivative [8] (Schemes 4 and 5). Bledzki et al. [9, 10] have used them in the polymerization of styrene and methyl methacrylate but, in the case of styrene, the formed oligomers were not able to initiate once more the polymerization because the end groups formed by initiator radicals and styrene units were rather stable at normal polymerization temperatures (60-120°C). It's one of the reasons why the elimination of nitroxyde group or halogen in the presence of diphenylmethyl radicals seems to be possible. This paper deals with the study of such systems.

The samples obtained were characterized by Size Exclusion Chromatography (SEC), Matrix-Assisted Laser Desorption Ionization Time-of-Flight Mass Spectrometry (MALDI-TOF MS), Electrospray Ionization Mass Spectrometry (ESI MS), Liquid Secondary Ions Mass Spectrometry (LSI MS), ^{1}H NMR, ^{13}C NMR, and IR spectroscopy.

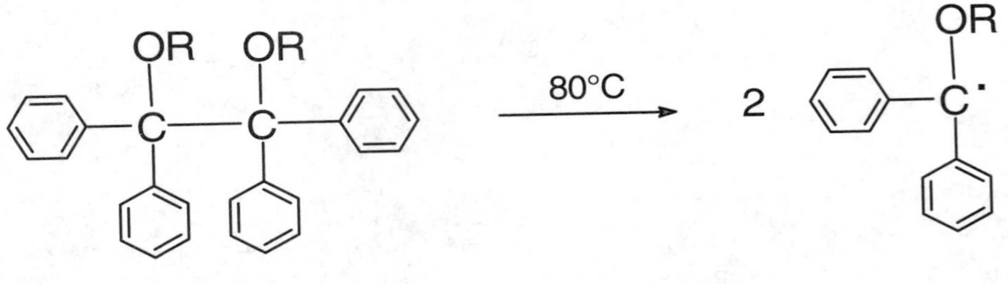

Scheme 4. Homolysis of tetraphenylethane based derivatives.

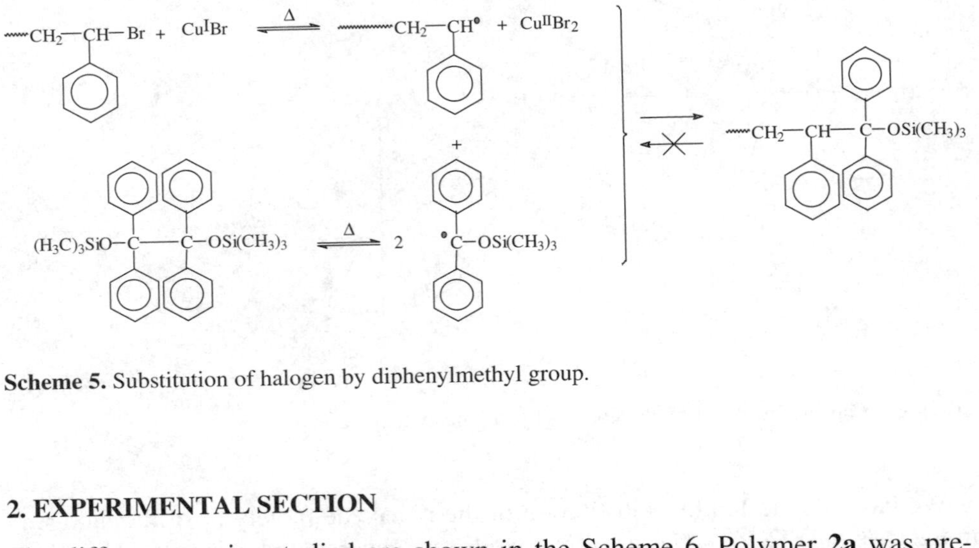

Scheme 5. Substitution of halogen by diphenylmethyl group.

2. EXPERIMENTAL SECTION

The different species studied are shown in the Scheme 6. Polymer **2a** was prepared by the Hawker's method [11], i.e. the synthesis of an alkoxyamine initiator (**1a**), the 1-benzoyl-2-phenyl(2',2',6',6'-tetramethyl-1'-piperidinyloxy)ethane, followed by the polymerization of styrene [12] in the presence of **1a**. According to this procedure, polystyrene sample **2a** was prepared. The synthesis of polymer **2b** was conducted as described by Matyjaszewski [1d, 1e].

The thiuram disulfide compounds used were (**3a**) tetraethyl thiuram disulfide, and (**3b**) N,N'-diethyl-N,N'-bis{2-(trimethylsilyloxy)ethyl} thiuram disulfide. The tetraphenylethane based derivative was (**3c**) 1,1,2,2-tetraphenyl-1,2-bis(trimethylsilyloxy)ethane.

Materials. Commercially available styrene and benzoyl peroxide (BPO, Aldrich) were purified by standards methods. Tetraethyl thiuram disulfide (**3a**), TEMPO and trimethylsilylchloride were purchased from Aldrich and used as received. Silica gel for flash chromatography was Merck Kieselgel 60 (70-230 mesh).

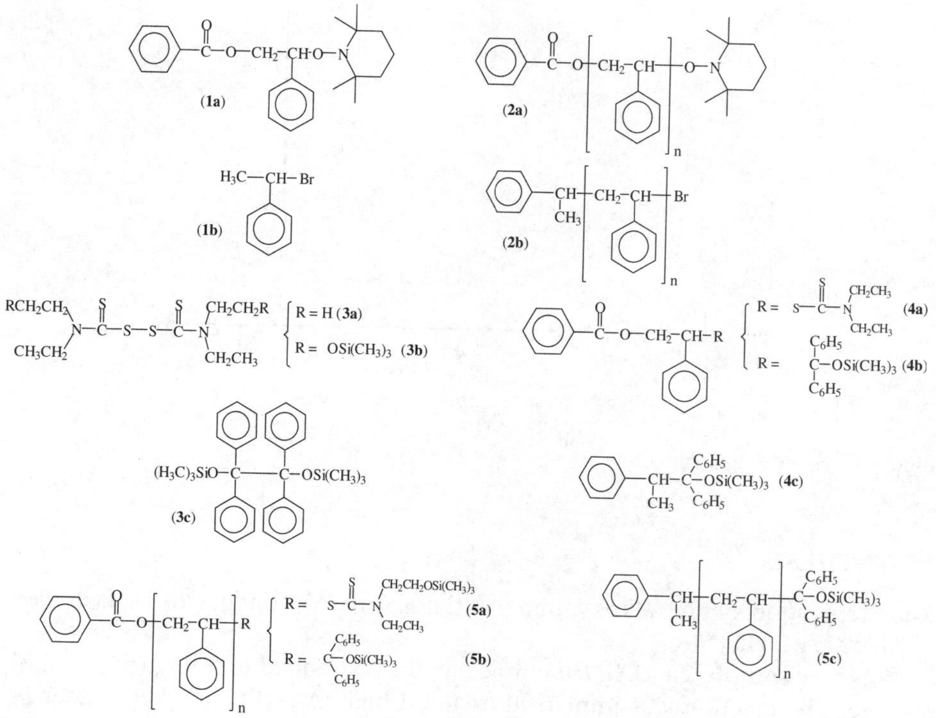

Scheme 6. Studied compounds.

Instruments. Infrared spectra were recorded on a Perkin-Elmer spectrophotometer. ^{1}H NMR spectra were recorded in solution with a Bruker AM 250 (250 MHz) spectrometer. ^{13}C NMR spectra were recorded at 62.9 MHz on a Bruker AM 250 spectrometer, using the carbon signal of the solvent as the internal standard. SEC characterizations were carried out on a Waters chromatograph connected to a multi angle LS detector (Wyatt). THF was used as the solvent. MALDI-TOF-MS experiments were conducted using a VOYAGER ELIT from Perseptive Biosystem equipped with a N_2 laser (λ = 337nm). The matrix was made of 1,8,9-anthracenetriol containing silver trifluoroacetate. The weight concentration of the polymer was 0.1%. The spectra were recorded in linear and reflectron modes over the range 800-7000. ESI-MS: polymer samples were investigated using a VG Platform from Micromass, equipped with a monoquad as the analyzer. The spectra were acquired over the range 400-3000 at different cone voltages (100, 140 and 180 V). The solvent was methanol containing $2 \cdot 10^{-4}$ mol.L^{-1} of NaI or dichloromethane/methanol/formic acid mixture. LSI MS: The apparatus used was a ZAB 2 SEQ from Micromass. Ionization was obtained with Cs$^+$ LSI. Polymer samples were dissolved in dichloromethane and the solvent was nitrobenzylalcohol. Mass spectra were acquired over the range m/z 360-3600.

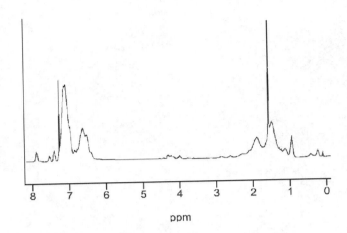

Figure 1. ^{1}H NMR spectrum of **2a**.

3. SYNTHESIS

1a. Preparation of the alkoxyamine initiator **1a** was conducted as described by Hawker et al. [11].

Polymer sample 2a. 0.2g (0.5 mmol) of **1a** was added to 5.95g (57.2 mmol) of styrene. The reaction was immersed in an oil bath at 130°C and kept under nitrogen for 2 hours. (^{1}H NMR: Figure 1).

Polymer sample 2b. 0.33g (2.3 mmol) of CuBr, 0.21g (1.15 mmol) of **1b** and 1.01g (6.9 mmol) of bipyridine was added to 12g (115 mmol) of styrene. The reaction was immersed in an oil bath at 80°C and kept under nitrogen for 16 hours.

3b. To a solution of 3.3g (30.48mmol) of trimethylsilylchloride in 5ml of THF was added 3.1g (30.48mmol) of triethylamine. The solution was stirred for 10mn. Then, 5g (15.24mmol) of N,N'bis(2-hydroxyethyl)-N,N'diethyl thiuram disulfide [12] in 10ml of THF was added dropwise. The reaction was allowed to proceed overnight. After filtration, the solvent was evaporated and the residue washed with water, dried over MgSO$_4$ and purified by flash chromatography (1:1 heptane/dichloromethane) to afford a yellow oil (m = 4.5g, 62%)

^{1}H NMR (CDCl$_3$) δ 0.09 (s, 18H, CH_3), 1.24 (t, 3H, CH_3), 1.45 (t, 3H, CH_3), 3.92-4.14 (m, 12H, CH_2)

3c. Tetraphenylethane based derivative was prepared according to Calas' method [14].

4. STUDY OF THE SUBSTITUTION

Reaction between 1a and 3a (synthesis of 4a). To a solution of **3a** [199mg (0.67mmol)] in 10ml of xylene, was added dropwise 330mg (0.87mmol) of **1a** in 20ml of xylene. The reaction was allowed to proceed for 5h at 140°C. After cooling, the solution was evaporated to dryness and purified by flash chromatography eluting with: 1 heptane/ dichloromethane to give **4a** as a yellow oil (130mg, 40%).

4a: ^{1}H NMR (CDCl$_3$) δ 1.27 (t, 6H, CH_3), 3.68-3.85 (m, 2H, CH_2), 3.90-4.15 (m, 2H, CH_2) 4.81 (d, 2H, CH_2), 5.72 (t, 1H, CH), 7.20-7.65 (m, 8H, ArH), 7.90-8.15 (m, 2H, ArH).

^{13}C NMR (CDCl$_3$) δ 11.58, 12.65, 46.84, 49.82, 54.12, 66.51, 126.78, 127.89, 128.34, 128.67, 128.95, 129.46, 139.99, 138.02, 166.19, 193.23.

Anal. Calcd. for C$_{18}$H$_{23}$NSO$_2$: C, 68.1; H, 7.3; O, 10.1; N, 4.4; S, 10.1 Found: C, 67.5; H, 6.9; N, 4.1; O, 9.8; S, 10.3.

Reaction between 2a and 3b (synthesis of polymer 5a). To a solution of 21.1mg (0.07mmol) of **3b** in 10ml of xylene, was added 400mg (0.14mmol) of **2a** in 20ml of xylene. The reaction was immersed in an oil bath for 20h under nitrogen atmosphere at 140°C. After cooling, the solution was evaporated to dryness and the polymer **5a** was purified by precipitation (twice) in methanol and dried overnight under vacuum (m = 199mg, 0.066mmol, 44%).

5a: ^{1}H (CDCl$_3$) δ 0.10 (s, 9H, CH_3), 1.20-2.85 (m, 65H, CH_3, CH_2, CH), 3.75-4.45 (m, 8H, CH_2N, CH_2O), 6.13-7.54 (m, 108H, ArH), 7.85-7.92 (m, 2H, ArH) (cf. Figure 2).

^{13}C NMR (CDCl$_3$) δ 1.10, 29.78, 40.54-46.07 (m), 59.60, 70.50, 125.75-129.55 (m), 130.92, 132.89, 145.39, 166.35, 195.31.

Reaction between 2a and 3c (synthesis of polymer 5b). To a solution of 600mg (1.17mmol) of **3c** in 10ml of xylene, was added 353mg (0.117mmol) of **2a** in 20ml of xylene. The reaction was immersed in an oil bath for 20h under nitrogen atmosphere at 140°C. After cooling, the solution was evaporated to dryness and the polymer **5b** was purified by precipitation (twice) in methanol and dried overnight under vacuum (m = 310mg, 0.101mmol, 86%).

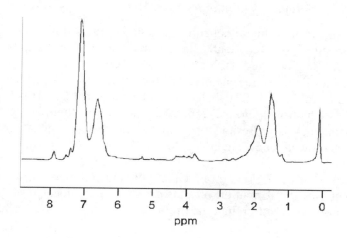

Figure 2. ^{1}H NMR spectrum of **5a**.

5b: ¹H (CDCl₃) δ -0.35 (d, 9H, C*H*₃), 1.18-2.75 (m, 78H, C*H*₃, C*H*₂, C*H*), 3.75-4.45 (m, 2H, C*H*₂O), 6.10-7.55 (m, 143H, Ar*H*), 7.90-8.05 (m, 2H, Ar*H*).

Reaction between 1b and 3c (synthesis of 4c). To a solution of **3c** [1g (1.96mmol)] in 10ml of toluene, was added 363mg (1.96mmol) of **1b**, 562mg (5.92mmol) of copper bromide and 1.38g (5.88mmol) of bipyridine. The reaction mixture was degassed and was allowed to proceed for 16h at 80°C. After cooling, the solution was evaporated to dryness and purified by flash chromatography eluting with 1 heptane/ dichloromethane to give **4c** as a yellow oil (473mg, 66%).

4c: ¹H NMR (CDCl₃) δ 0.05 (s, 9H, C*H*₃), 1.55 (d, 3H, C*H*₃), 4.25 (q, 1H, C*H*), 6.95-7.65 (m, 15H, Ar*H*)

¹³C NMR (CDCl₃) δ 2.38, 18.51, 47.79, 49.82, 85.09, 126.24-130.78, 142.95, 143.74, 145.34.

Reaction between 2b and 3c (synthesis of polymer 5c). To a solution of **3c** [28.4mg (0.11mmol)] in 10ml of toluene, was added 400mg (0.11mmol) of **2b**, 319mg (0.22mmol) of copper bromide and 26.05mg (0.11mmol) of bipyridine. The reaction mixture was degassed and was allowed to proceed for 114h at 80°C. After cooling, the solution was evaporated to dryness and the polymer **5c** was purified by precipitation (twice) in methanol and dried overnight under vacuum (m = 346mg, 0.094mmol, 86%).

5c: ¹H (CDCl₃) δ -0.30_-0.15 (m, 9H, C*H*₃), 0.85-2.55 (m, 64H, C*H*₃, C*H*₂, C*H*), 6.35-7.45 (m, 115H, Ar*H*).

¹³C NMR (CDCl₃) δ 1.65, 1.70, 1.85, 40.45, 42.75-47.50 (m), 118.95, 123.25-130.35 (m), 145.15-147.50

²⁹Si NMR (CDCl₃) δ 9.30, 9.55-9.65 (m), 9.80

5. RESULTS AND DISCUSSION

5.1. Study of the substitution on initiators **1a** *&* **1b** *(synthesis of* **4a-c***)*

The ¹H NMR spectrum of the model compound **4a** is reported in the experimental section. This spectrum showed clearly the disappearance of the signals located at 0.76, 1.00, 1.27 (singlet), and 1.37 ppm respectively, attributed to the methyl groups attached to the nitroxide moiety, and the formation of signals at 3.73 and 4.0 ppm attributed to the CH₃-C$\underline{H}$₂-N hydrogen atoms, and 1.27 (triplet) attributed to the C$\underline{H}$₃-CH₂-N hydrogen atoms, respectively, i.e. all of them being carried by the dithiocarbamate moiety. Besides, the comparison between the ¹³C NMR spectra of **1a** and **4a** showed a new signal located at 193 ppm, which was attributed to the $\underline{C}$=S carbon.

The comparison of the IR spectra of **1a** and **4a** showed many differences. The most significant was the formation of a strong sharp peak located at 1418 cm⁻¹. Such a signal, also observed in the case of **3a**, was assigned to the C=S stretching band.

The model compound **4b** were not isolated because many products are observed by thin layer chromatography and we did not manage to get it.

The reaction between **1b** and **3c** gave **4c** whose structure was confirmed by ^{1}H NMR spectrum: a strong signal at δ 0.10 ppm, which could be unambiguously attributed to the 9 hydrogen atoms of the trimethylsilyl group located on the diphenylmethyl group, was observed. Besides, the signal at 5.3 ppm attributed to the C$\underline{H}$Br (product **1b**) disappeared and a new signal at 4.2 ppm was observed (C$\underline{H}$-C(C$_6$H$_5$)$_2$OSiMe$_3$).

5.2. Study of the substitution on polymers **2a** & **2b** (synthesis of **5a-c**)

*5.2.1. SEC analysis of polymers **5a-c***. Due to the slight solubility of the polymer samples into methanol, the yield of the reactions between nitroxide terminated polystyrene and thiuram disulfides was 40% in terms of weight percentage of polymer recovered. In fact, the low molar mass part of such a sample is more soluble in methanol than the high part. This behavior explains the slight increase of both the number and the weight average of molar masses of **5a-c** compared to **2a** and **2b** (cf. Table 1 and 2). However, such precipitation could not be avoided because of the necessity to remove all traces of thiuram disulfide, tetraphenylethane based derivatives and/or nitroxide species in the recovered polymer.

Table 1.
Characterization of nitroxide- and bromo-terminated polymers **2a** and **2b**

Run	[Styrene]/[1]	Monomer Conversion ρ	Dpn [a] Theoretical [c]	Dpn SEC	PI [b] SEC	Dpn ^{1}H NMR [d]
2a	109	0.18	20	27	1.1	25
2b	109	0.16	17	25	1.1	24

[a] Dpn: average number of the degree of polymerization.
[b] PI: polydispersity.
[c] calculated as ρ x [Styrene]/[1].
[d] calculated from the ratio of the number of hydrogen atoms carried by both BPO-C$\underline{H}_2$- and –C$\underline{H}$(C$_6$H$_5$)-O-N= fragments to C$_6\underline{H}_5$ (BPO included).

Table 2.
Characterization of dithiocarbamate- and diphenylmethyl-terminated polymers (**5a-c**)

Run	Polymer recovered Weight %	Dpn [a] SEC	PI [b] SEC	Dpn ^{1}H NMR	Dpn MALDI-TOF MS
5a	31	29	1.1	29 [c]	26 [d]
5b	80	30	1,1	–	–
5c	80	26	1,2	–	–

[a] Dpn: average number of the degree of polymerization.
[b] PI: polydispersity.
[c] Calculated from the ratio (C$\underline{H}_3$)$_3$Si to C$_6\underline{H}_5$ (BPO included).
[d] Calculated from the series A in the MALDI-TOF spectrum.

The SEC trace showed only one peak. In particular, the possible formation of polymer species showing a molar mass being twice the molar masses of **2a** or **2b**, as the consequence of some hypothetical recombination between the polymer radicals, could be totally excluded.

5.2.2. ^1H, ^{13}C & ^{29}Si NMR analysis of polymers 5a-c. As in the spectrum of **4a**, the ^1H NMR diagram of **5a** did not show any signal that could be attributed to alkoxyamine groups. Besides, the signal at 0.1 ppm which could be attributed to the 9 hydrogen atoms of the trimethylsilyloxy group located on the dithiocarbamate moiety allowed, with the ratio between this signal and the signals of the hydrogen atoms located on the polymer chain, the determination of the molar mass of this polymer which was in good agreement with the SEC characterization (cf. Table 2). The ^{13}C NMR spectra of **5a** also showed a signal located at 195 ppm, being attributed to the C=S carbon.

For the polymer **5b**, the ^1H NMR spectrum was more surprising because there are two large signals for the nine hydrogen atoms of the trimethylsilyloxy group around –0.35 ppm and when we observed the diagram after heating the sample at 130°C, the two signals are closer which indicate that those protons are located on the same atom: we assumed the presence of diastereoisomers to explain those signals. If we evaluate the proportion of substitution with the ratio between the trimethylsilyloxy group and the protons located at 6.3-7.6 ppm, the result is 62%. The analysis of the signals at 1.2-2.3 ppm confirms that the substitution is not quantitative because we observed a singlet at 1.6 ppm which is characteristic of C$\underline{H}_3$-CH$_2$-N hydrogen atoms (thiocarbamate group).

This result led us to testing a substitution reaction on the polymer **2b** in the same experimental conditions to give **5c**. The ^1H NMR spectrum gave a similar result: two large signals for hydrogen atoms located on the trimethylsilyloxy group (Figure 7) in spite of the result obtained with model compound **4c** whose ^1H NMR spectrum contains only one signal. The evaluated proportion of substitution in **5c** was 65% with the use of a large excess of the tetraphenylethane based derivatives. The ^{29}Si NMR spectrum indicated three large signals which were correlated with the two signals observed in the ^1H NMR spectrum around –0.35 ppm (Figure 8): there is no trace of the 1,1,2,2-tetraphenyl-1,2-bis(trimethyl-silyloxy)ethane.

5.3. Discussion of NMR and IR characterizations

In all cases NMR and IR characterizations are in good agreement with the expected substitution even if the reaction is not quantitative for **5b** and **5c**. However, the NMR or IR signals resulting from the dithiocarbamate- or diphenylmethyl-fragment contained in the polymer **5a**, **5b** and **5c** could also be attributed to some traces of **3b** or **3c** in the recovered polymer, despite the careful precipitation of these polymer samples. For this reason, the presence of the dithiocarbamate- or of the diphenylmethyl-fragment at the polymer chain end must be proved by mass spectrometry.

5.4. Mass Spectrometry

In these studies, benzoyloxy moiety is noted BPO, nitroxide moiety is noted TEMPO and N-2-(trimethylsilyloxy)ethyl-N-ethyl-dithiocarbamate moiety is noted NTSE, ethylbenzene moiety is noted EB and trimethylsilyloxydiphenyl-methyl moiety is noted DPMS.

5.4.1. MALDI-TOF MS.

MALDI-TOF of the three products **2a** and **5b** were investigated in both reflectron and linear modes. The reflectron mode providing a higher flight path allows a better resolution. The spectra were surprisingly the same, offering the same distributions, although it has been proved by ^{1}H NMR that they had different structures. The weakness of the bounding between the last styrene residue of the polymer chain and the further heteroatom (O or S) is well known, thus **2a** and **5a** are probably subjected to fragmentation during ionization, providing the cleavage and the loss of a fragment corresponding to the polymer chain end. This behavior has already been observed by Haddelton et al. [15] in the case of nitroxide terminated polystyrene, i.e. **2a**.

Therefore, the MS spectra of the polymers appear similar, independent of the nature of the chemical group located at the polymer chain end. Thus, MALDI-TOF MS does not seem to be adapted to the characterization of such products, in the experimental conditions chosen. Nevertheless, three principal series have been assigned in both linear and reflectron modes, for Dp ranging from 9 to 50, as $[BPO-(sty)_n-CH=CHC_6H_5 + Ag]^+$, series A; $[BPO-(sty)_n-BPO + H]^+$, series B; $[BPO-(sty)_n-BPO + Na]^+$, series C, and $[(sty)_{n+1}-CH=CHC_6H_5 + Ag]^+$, series D, respectively. The series D corresponds to the next homologue of series A but subjected to a fragmentation with a loss of 122 Da, i.e. a BPO residue. The series C is the less important series, the alkali metal ion coming probably from the glassware, and thus, its presence is variable (it is not observed in the MS spectrum of **2a**). The series A has been used to calculate the molar mass number average of the sample **5a** (cf. Table 2 and Figure 3). The molar mass average determined by MALDI is slightly lower than the values obtained by both SEC and NMR analyses. As discussed by Davis et al. [16], the comparison between SEC and MALDI data shows that MALDI generally underestimates the high molar mass tail in a polymer distribution. The origin of these differences have various origins. In the present case, the difference may be attributed to the ionization efficiency and/or the detector response at higher mass.

5.4.2. ESI MS.

ESI-MS of the samples **2a** and **5a** gives different results to those obtained by MALDI-TOF MS. The spectra of the polymers are quite different, suggesting that ESI-MS seems to be more adapted to the study of such samples than MALDI-TOF MS. As it is known from the literature, various parameters such as solvent, nature and concentration of added cations and cone voltage strongly influence the ionization and the desorption efficiency from the droplets [17]. The experimental conditions must be adapted to each sample. In the present

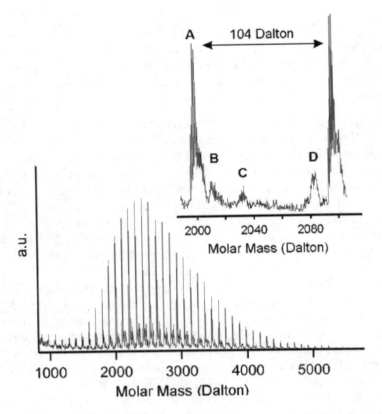

Figure 3. MALDI-TOF spectrum of polymer **5a**.

study, two solvents were tried: (a) (dichloromethane/methanol 50/50 mixture with 1% formic acid) and (b) methanol with the presence of NaI ($2 \cdot 10^{-4}$ M).

2a. In a first attempt, **2a** was studied in methanol with Na^+ addition. The predominant species observed are $[BPO-(sty)_n-TEMPO]^+$, i.e. corresponding to M^+, and not MNa^+ or MH^+ as expected. The observation of M^+ is quite surprising, because of the presence of cations in the solvent, i.e. usually leading to the formation of the cationized species [15], unless the formation of this species depends on the nature of the cations [18]. Unlike some results reported by Haddelton et al. [15], the series $[BPO-(sty)_n-CH=CHC_6H_5 + cation]^+$ is not observed. Thus, methanol seems not adapted to the analysis of **2a**.

With the dichloromethane/methanol/formic acid solvent mixture, the expected $[BPO-(sty)_n-TEMPO, H]^+$ is observed (Table 3 and Figure 4), series E, with a Dp ranging from 4 to 20 for a cone voltage of 120V. Another minor series (series F) is observed and corresponds to $[(sty)_n-TEMPO, H]^+$, that corresponds to the loss of the BPO residue (122 Da). The intensity of these series slightly increases with increasing cone voltage, indicating that the series F is the product of fragmenta-

Table 3.
Comparison of Experimental versus Theoretical Mass for ESI-MS spectrum of **2a** in dichloromethane/methanol/formic acid solvent mixture

Peak	Origin	Formula	Theor mass (Da)	Exptl mass (Da)
E	BPO-(sty)$_{11}$-TEMPO, H$^+$	C$_{104}$H$_{111}$O$_3$N, H	1422.8	1422.8
F	(sty)$_{13}$-TEMPO, H$^+$	C$_{113}$H$_{122}$ON, H	1510.0	1509.3

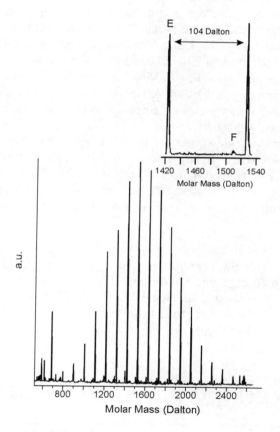

Figure 4. ESI-MS spectrum of polymer **2a**.

tion [19]. As already described in the literature [17], slight modifications of the relative abundance of the various oligomers with exit capillary voltage is observed. Varying this voltage from 80 to 120V resulted in an increase of the higher oligomers. Various hypotheses can be made to explain these changes, such as the neutralization by loss of cations though collisions in the interface which occurs preferentially for low-mass compounds [17].

5a. Unfortunately, the ESI-MS of **5a** cannot prove the quantitative substitution of the nitroxide moiety. For example, the species [BPO-(sty)$_{14}$-NTSE + Na]$^+$, whose theoretical mass is expected at 1836.9 Da, is observed at 1838.1 Da (Table 4). However, this value also corresponds to the theoretical molecular mass of **2a** with a DP equal to 15, i.e. M$^+$ in the experimental conditions. Thus, due to the bad resolution of this experiment, no conclusion about the structure of **5a** is obtained by ESI-MS, even if the quantitative substitution has been proved by ^{1}H NMR.

For the two compounds **2a** and **5a** studied in methanol with Na$^+$ addition, at low m/z, two other series with mass distributions of 150 Da and 82 Da peculiarly abundant as cone voltage is decreased, are attributed to the presence of traces of some contaminant during the sample preparation, because they do not appear on the spectrum of **2a** in the dichloromethane/methanol/formic acid solvent mixture.

5b. The ESI-MS spectrum of **5b** indicates that the predominant species are BPO-(sty)$_n$-DPMS, i.e. corresponding to MNa$^+$ as expected. A minor series is also observed, i.e. corresponding to MK$^+$ or MNa$^+$ with the loss of the trimethylsilyloxy fragment. Another minor series is attributed to the presence of traces of the nitroxide-terminated polystyrene.

5c. The predominant species observed in the diagram ESI-MS (Table 7, Figure 9) are EB-(sty)$_n$-DPMS, i.e. corresponding to MNa$^+$. Likewise **5b**, minors series are observed, i.e. corresponding to MK$^+$ or MNa$^+$ with the loss of the trimethylsilyloxy fragment. There is no trace of the bromo-terminated polystyrene but this seems to be normal because there is no ionization of this polymer when we study it by ESI-MS.

5.4.3. LSI MS. **2a.** In this case, the protonated species [BPO-(sty)$_n$-TEMPO + H]$^+$ were observed, with the presence of minor series, confirming the correct structure of the polymer (cf. Table 5 and Figure 5).

5a. In this case, the protonated species [BPO-(sty)$_n$-NTSE + H]$^+$ was observed. This result confirms the expected structure of the polymer, with the presence of minor series (Table 5-6, and Figure 5-6).

Table 4.
Comparison of Experimental versus Theoretical Mass for ESI-MS spectrum of **5a** in methanol with addition of NaI (2·10^{-4} M)

Peak	Origin	Formula	Theor mass (Da)	Exptl mass (Da)
G	BPO-(sty)$_{14}$-NTSE, Na$^+$	C$_{127}$H$_{135}$O$_3$NS$_2$Si, Na	1836.9	1838.1
H	BPO-(sty)$_{14}$-NTSE, Na$^+$ Loss of CH$_2$CH$_3$	C$_{125}$H$_{130}$O$_3$NS$_2$Si, Na	1807.9	1809.3
I	BPO-(sty)$_{14}$-NTSE, Na$^+$ Loss of CH$_3$	C$_{126}$H$_{132}$O$_3$NS$_2$Si, Na	1821.9	1824.0
	BPO-(sty)$_{15}$-TEMPO$^+$	C$_{136}$H$_{143}$O$_3$N	1838.1	Observed ?

Table 5.
Comparison of Experimental versus Theoretical Mass for LSI-MS spectrum of **2a**

Peak	Origin	Formula	Theor mass (Da)	Exptl mass (Da)
L	BPO-(sty)$_{18}$-TEMPO, H$^+$	C$_{160}$H$_{167}$O$_3$N, H	2151.3	2152.6

Table 6.
Comparison of Experimental versus Theoretical Mass for LSI-MS spectrum of **5a**

Peak	Origin	Formula	Theor mass (Da)	Exptl mass (Da)
Q	BPO-(sty)$_{19}$-NTSE, H$^+$	C$_{167}$H$_{175}$O$_3$NS$_2$Si, H	2335.3	2337.0
R	BPO-(sty)$_{20}$-NTSE, H$^+$ Loss of OSi(CH$_3$)$_3$	C$_{172}$H$_{174}$O$_2$NS$_2$, H	2350.3	2353.5
S	?			2367.9
T	?			2384.0
U	BPO-(sty)$_{20}$-NTSE, H$^+$ Loss of CH$_3$	C$_{174}$H$_{180}$O$_3$NS$_2$Si, H	2424.3	2425.9
(L)	BPO-(sty)$_{20}$-TEMPO, H$^+$	C$_{176}$H$_{183}$O$_3$N, H	2359.4	Not observed

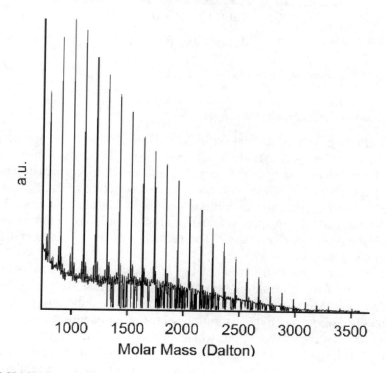

Figure 5. LSI-MS spectrum of polymer **2a** (series L).

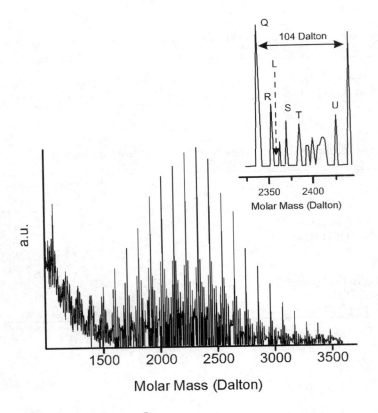

Figure 6. LSI-MS spectrum of polymer **5a**.

5.5. Discussion of MS characterizations

The particular behavior of **2a** and **5a** in MALDI-TOF has to be noted. This behavior seems to be attributed to the weakness of the C-O bonds, respectively C-S, between the polymer and both the nitroxide and the dithiocarbamate chain end. ESI MS spectra are compatible with the expected structure of **2a**, **5b**, **5c** and the influence of the experimental conditions on the ionization efficiency has been shown with **5a**. LSI MS appears to be adapted to the study of the structure of **5a**, although some fragmentations occur. The pathways of these fragmentations have not yet been explained.

However, the expected structures of the polymer samples, i.e. resulting from the substitution of the nitroxide group by a dithiocarbamate fragment have been clearly identified. Moreover, the spectrum of **5a** did not show any peaks that can be attributed to **2a**; i.e. such a substitution seems to be quantitative (cf. Figure 5-6). As expected in the introduction, whatever the rate of dissociation of the C-S bond between the polymer backbone and the dithiocarbamate moiety, it is low enough compared to the dissociation rate of the C-O bond of the alkoxyamine group to be neglected.

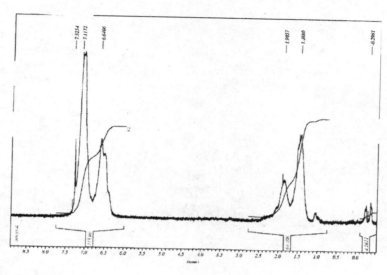

Figure 7. ^{1}H NMR spectrum of **5c**.

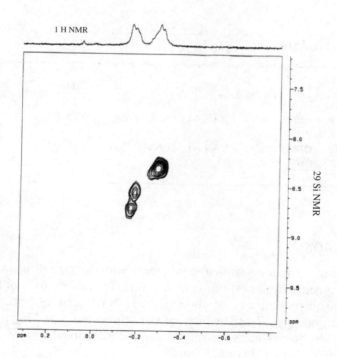

Figure 8. COSY ^{1}H/^{29}Si spectrum of **5c**.

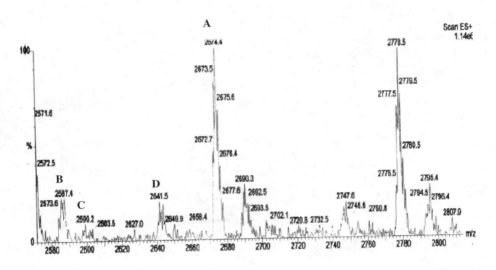

Figure 9. ESI-MS spectrum of polymer **5c**.

Table 7.
Comparison of Experimental versus Theoretical Mass for ESI-MS spectrum of **5c** in methanol with addition of NaI ($2 \cdot 10^{-4}$ M)

Peak	Origin	Formula	Theor mass (Da)	Exptl mass (Da)
A	EB-(sty)$_{22}$-DPMS, Na$^+$	C$_{127}$H$_{135}$O$_3$NS$_2$Si, Na	2672.6	2672.7
B	EB-(sty)$_{22}$-DPMS, Na$^+$ Loss of OSi(CH$_3$)$_3$	C$_{125}$H$_{130}$O$_3$NS$_2$Si, Na	2583.5	2584.5
C	EB-(sty)$_{22}$-DPMS, Na$^+$ Loss of Si(CH$_3$)$_3$?	C$_{126}$H$_{132}$O$_3$NS$_2$Si, Na	2599.5	2598.2
D	?	?	–	2640.5

6. CONCLUSION

The reaction of nitroxide terminated polymers with thiuram disulfide compounds leads to the quantitative substitution of the nitroxide moiety by a dithiocarbamate group. This substitution can be shown by ^{1}H NMR and IR spectroscopy, and clearly identified by LSI MS, which appears to be the best MS method for the characterization of polymers bearing chain end functional groups linked to the polymer chain by both C-O or C-S bonds.

The substitution reaction with tetraphenylethane based derivatives is not quantitative: we obtained about 60% of substitution. We hope to increase this proportion by increasing the temperature or by using other ligands for the copper complex to get a better solubility. The two large signals observed in the ^{1}H NMR spectrum and corresponding to the nine hydrogen atoms of the trimethylsilyloxy group were attributed to the presence of diastereoisomers.

We also suppose that the substitution reaction is more efficient in the case of thiuram disulfides than in the case of tetraphenylethane-based derivatives because thiuram disulfides can act as transfer agent or as termination agent so they would have two possibilities to remove the chain end group of polystyrene.

Other chemical species (halogenated tetraphenylethane derivatives) with the same reaction pattern, but leading to the formation of other interesting functional polymers, are under investigation.

Acknowledgements

The authors thank ESSILOR for industrial support and F. Delolme (Service Central d'Analyse, CNRS, Solaize, LSI MS), I. Zanella-Cléon (Service Central d'Analyse, CNRS, Solaize, ESI MS), J. C. Blais (LCSOP, Université Pierre et Marie Curie, Paris, MALDI-TOF MS), and J. M. Lucas (LEMPB, Université de Lyon 1, Villeurbanne, SEC) for characterizations.

REFERENCES

1. (a) D. H. Solomon, E. Rizzardo, P. Cacioli, U.S. Patent 4,581,429. (b) E. Rizzardo, *Chem Aust.*, **54**, 32, (1987), (c), M. K. Georges; R. P. N. Veregin; P. M. Kazmaier, G. K. Hammer, *Trends Polym. Sci.*, **2**, 66, (1994). (d) D. Greszta, D. Mardare, K. Matyjaszewski, *Macromolecules*, **27**, 638, (1994). (e) K. Matyjaszewski, S. Gaynor, D. Greszta, D. Mardare, T. Shigemoto, *J. Phys. Org. Chem.*, **8**, 306, (1995).
2. (a) G. O. Schulz, R. Milkovich, *J. Appl. Polym. Sci.*, **27**, 4773, (1982). (b) G. O. Schulz, R. Milkovich, *J. Polym. Sci., Polym. Chem. Ed.*, **22**, 1633, (1984).
3. H. Malz, H. Komber, D. Voigt, J. Pionteck, *Macromol. Chem. Phys.*, **199**, 583 (1998).
4. (a) T. E. Ferington, A. V. Tobolsky, *J. Am. Chem. Soc.*, **77**, 4510, (1955). (b) T. E. Ferington, A. V. Tobolsky, *J. Am. Chem. Soc.*, **80**, 3215, (1958). (c) E. Staudner, J. Beniska, *Eur. Pol. J.* Suppl.
5. K. Endo, K. Murata, T. Otsu, *Macromolecules*, **25**, 5554, (1992).
6. G. Moad, E. Rizzardo, *Macromolecules*, **28**, 8722, (1995).
7. T. Fukuda, A. Goto, K. Ohno, Y. Tsujii, *Controlled Radical Polymerization*, K. Matyjaszewski, Ed.; ACS Symp. Series, **685**, 190, (1998).
8. R. Guerrero, P. Chaumont, J. E. Herz, G. J. Beinert, *Eur. Polym. J.*, 28, 1263, (1992).
9. A. Bledzki, D. Braun, K. Titzschkau, *Makromol. Chem.*, **184**, 745, (1983).
10. A. Bledzki, D. Braun, *Makromol. Chem.*, **187**, 2599, (1986).
11. C. J. Hawker, G. G. Barclay, A. Orenella; J. Dao, W. Devonport, *Macromolecules*, **29**, 5245, (1996).
12. C. J. Hawker, *J. Am. Chem. Soc.*, **116**, 11185, (1994).
13. Nair, C. P. R.; Clouet, G.; Chaumont, P. *J. Polym. Sci.*, **27**, 1795, (1989).
14. R. Calas, N. Duffaut, C. Biran, P. Bourgeois, F. Pisciotti, J. Dunoguès, *C. R. Acad. Sci., Paris*, **267**, 322, (1968).
15. Jascieczek, C. B.; Haddelton, D. M.; Shooter, A. J.; Buzy, A.; Jennings, K. R.; Gallagher, R. T. *Polym. Prepr.*, **37**, 845, (1996).
16. Zammit, M. D.; Davis, T. P.; Haddleton, D. M.; Suddaby, K. G. *Macromolecules*, **30**, 1915, (1997).

17. Guittard, J.; Tessier, M.; Blais, J. C.; Bolbach, G.; Rozes, L.; Marechal, E.; Tabet, J. C. *J. Mass Spectrom.*, **31**, 1409, (1996).
18. Jascieczek, C. B.; Buzy, A.; Haddelton, D. M.; Jennings, K. R. *Rapid Commun. Mass Spectrom.* **10**, 509, (1996).
19. Loo, J. D.; Usdeth, H. R.; Smith, R. D. *Rapid Commun. Mass Spectrom.*, **2**, 207, (1988).

Functional beaded polymers *via* metathesis polymerization: Concepts and applications

MICHAEL R. BUCHMEISER,* FRANK SINNER and M. MUPA

Institute of Analytical Chemistry and Radiochemistry, University of Innsbruck, Innrain 52a, A-6020 Innsbruck, Austria

Abstract—Ring-opening metathesis polymerization (ROMP) as well as alkyne metathesis polymerization were combined with traditional polymerization techniques such as precipitation polymerization or surface grafting and used for the synthesis of functionalized polymer beads. By this approach, a large variety of functionalized supports as well as polymer-immobilized catalysts have been prepared and subsequently used for applications in separation science such as chiral high-performance liquid-chromatography (chiral HPLC), high-performance cation- and anion-chromatography (HPIC), and the selective extraction of lanthanides. Finally, they have been used in heterogeneous catalysis, e. g. for palladium-mediated coupling reactions (Heck-type reactions, aminations).

Keywords: Ring-opening metathesis polymerization (ROMP); metathesis polymerization; norbornenes; alkynes; functional polymer beads; HPLC; catalysis.

1. INTRODUCTION

Due to the growing demand for well-defined, surface-functionalized high performance materials in heterogeneous catalysis and separation sciences, the development of such supports is still an intensively investigated area. Despite the significant progress that has been made so far, there still remain many problems that need to be solved. Supports in a broad sense are usually based on surface-modified silica or organic polymers such as certain acrylates or polystyrene-divinylbenzene (PS-DVB). Derivatization reactions of silica-based materials are easy to perform and fairly easy to control. Nevertheless, even highly sophisticated end-capping procedures were not capable of overcoming the pH - instability from which these materials still suffer. The resulting limitations in working pH and temperature entail several problems, significantly limiting their range of applicability, in particular the use of acidic or basic substrates. While PS-DVB-based materials overcome the problem of hydrolysis, a major problem encountered with these carriers lies in poorly controllable surface-derivatizations. Consequently, the nature of the "working" functionality is often rather based on assumption than on

---

* To whom correspondence should be addressed. E-mail: michael.r.buchmeiser@uibk.ac.at

real analysis. In order to avoid any derivatization reactions or polymer transformations, the *polymerization* of *functional monomers* seemed therefore highly favorable. Polymerization was seeked to be carried out in a living manner [1, 2] in order to obtain access to block copolymers and to control molecular weight. The final goal was the preparation of well-defined, chemically and mechanically stable polymers suitable for use in separation sciences (HPLC, HPIC, etc.) as well as in heterogeneous catalysis. In this contribution, the design and synthesis of such materials as well as some applications will be summarized.

Among the few techniques suitable for the living polymerization of functional polymers, ring-opening-metathesis polymerization (ROMP) using well defined, high-oxidation state metathesis catalysts based on molybdenum, tungsten [3] and ruthenium [4–9] represented the most attractive one. Two major advantages have to be stated, which demonstrate the high versatility of this polymerization technique [10]. On one hand, ROMP allows the living polymerization of (almost any) functional monomer based on cyclic olefins. This permits the stoichiometric design of block-copolymers including cross-linked polymers. On the other hand, the formation of a certain backbone structure with regard to tacticity and *cis-trans*-configuration may be predetermined by the choice of a certain initiator [11–13]. These characteristics and the fact, that neither the sterical nor the conformational situation of any monomer are changed in course of the polymerization, lead to a high reproducibility in the synthesis of of these tailor-made materials. Consequently, this facilitates correlations between polymer structure, capacity and the resulting properties. Recently, we described the use of ROMP for the preparation of beaded, functionalized polymers [14, 15] for applications in SPE of organic compounds [16–19] as well as for the selective extraction of lanthanides [20, 21] and transition metals [22, 23]. In the following, some recent results in the preparation of functionalized beaded polymers will be summarized.

2. RESULTS AND DISCUSSION

2.1. Materials prepared by spin-coating of silica with ROMP-based polymers

2.1.1. Preparation of HPIC-materials [24]. For cation chromatography, monomers bearing either sulfonic or carboxylic acid groups are of general interest. In this context, 7-oxanorborn-2-ene-5,6-dicarboxylic acid (ONDCA) represents an interesting molecule for various reasons. In case this monomer is polymerized *via* ROMP-techniques, the resulting polymer backbones consist of a vinylen-spaced poly-tetrahydrofurane (poly-THF) with each unit bearing two *vic, cis*-configured carboxylic acids (Scheme 1). These vinylen groups in the linear polymer may further be used for subsequent cross-linking reactions. The oxygen in the resulting five-membered ring significantly elevates the hydrophilicity of this ligand compared to the parent poly(norborn-2-ene-5,6-dicarboxylic acid). The *cis*-configuration of the two carboxylic acid groups, which would be lost e. g. by

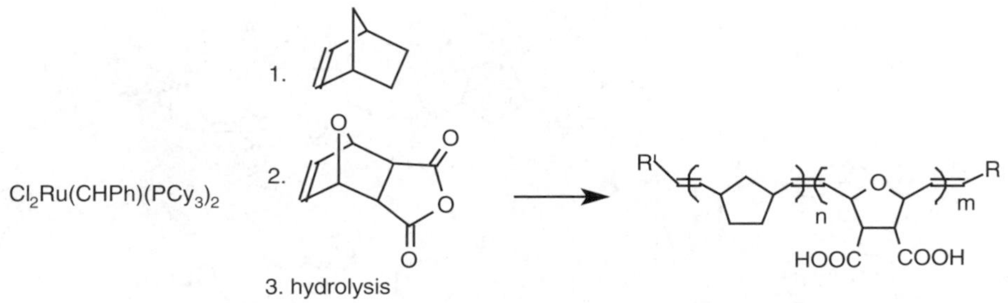

Scheme 1. Synthesis of NBE-ONDCA-copolymers *via* ROMP (NBE = norbornene).

radical polymerization of maleic anhydride [25], may be used for complexation (*vide infra*). Homopolymers of 7-oxanorborn-2-ene-5,6-dicarboxylic anhydride (ONDCA) and norborn-2-ene (NBE), repectively, as well as copolymers of ONDCA and NBE with well-defined block sizes and molecular weights have been prepared by ROMP using Grubbs-type initiators of the type $Cl_2Ru(CHPh)(PCy_3)_2$, Cy = cyclohexyl. Despite the fact that initiation of ONDCA does not proceed stoichiometrically with this type of initiator, a living polymerization within the limits of the classical definitions (no chain termination, no chain transfer) [1, 2] may be performed. In case polymerizations were started with NBE, well defined block-copolymers with regard to block-size and molecular weight, one block consisting of poly(ONDCA) and another of poly-NBE, may be prepared. The linear polymers and copolymers prepared by this approach were spin-coated onto various vinyl-silanized silica materials and subsequently cross-linked employing azobis-*i*-butyronitrile (AIBN) to yield a poly(succinic acid)-derivatized support (Figure 1). Generally, speaking, coating leads in combination with the use of well-defined polymers and copolymers to more reproducible materials, as the amount of polymer and consequently the capacity (expressed in mmol of functional group/g) of the material is simply determined by weight. Different amounts of prepolymer were deposited at the surface and finally cross-linked using thermally initiated radical polymerization. Besides of a higher solubility, the general advantage of ONDCA-NBE-copolymers lies in the entirely different chemical properties of the two polymer blocks. While the poly-NBE block is highly hydrophobic, the poly-ONDCA-block is highly hydrophilic. In course of the coating the polymer starts to precipitate onto the surface. A self-assembly of the block-copolymer is believed to occur, where major amounts of the apolar NBE-part get deposited onto the apolar surface vinylsilyl groups and major amounts of the polar, THF-solubilized poly-ONDCA-part form the new support surface. The general quality of the spin-coating method in terms of perfect surface coverage was confirmed applying a standard Engelhardt test [26–28] to a poly-NBE coated column. The optimum coating thickness with regard to the accessibility of the carboxylic acid groups was determined to be in the range 50–120 µg/m^2. Finally, stability criteria in terms of pH-stability were investigated by

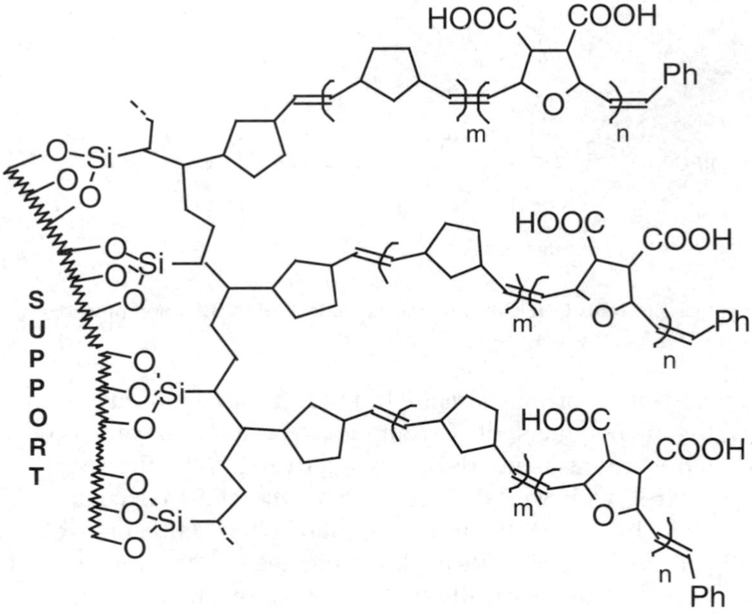

Figure 1. Structure of thermally cross-linked, ONDCA-*co*-NBE spin-coated silica.

depositing different amounts of polymer (10–120 mg/g) onto the surface of various silica materials. In course of these investigations, ONDCA-NBE-copolymer layers were found to be extremely inert to acids and bases. Under acidic conditions, they may be treated repeatedly with even 15% nitric without any loss of polymer coating. In basic media (aqueous sodium hydroxide), the stationary phases were operated up to a pH of 12 for several days without any change in elution time, order or even peak-shape and half-widths. Surprisingly, homopolymers based on poly-ONDCA did not show the desired pH-stability, despite the fact, that the polymer layer was subsequently cross-linked with the surface vinyl groups. This might be explained by the highly hydrophilic character of poly-(ONDCA) polymers, where both H^+ as well as OH^- groups may reach the silica surface due to diffusion.

Spin-coated silica based on NBE-ONDCA copolymers were found suitable for the separation of isomeric phenols, anilines, lutidines, flavones and hydroxyquinolines. An example of a fast baseline separation of isomeric hydroxyquinolines (compounds 1-4) within 6 minutes is shown in Figure 2.

2.1.2. Extraction of metal ions. Generally, the complexation capabilities of dicarboxylic acids such as oxalic or succinic acid for lanthanides are known for long [29]. Consequently, the extraction properties of the sorbents prepared as described above were for lanthanides were investigated. As expected, extraction

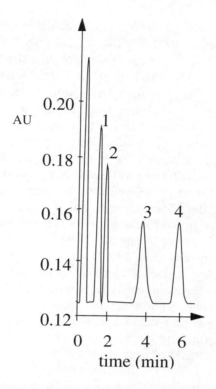

Figure 2. Separation of 2-hydroxyquinoline (1), 4-hydroxyquinoline (2), 6-hydroxyquinoline (3), 8-hydroxyquinoline (4). For conditions refer to lit [24].

efficiencies of this type of resin for lanthanides were found to be almost quantitative (>97 %) in a concentration range of 20 ng/l up to 250 μg/ml with relative standard deviations (RSD) ≤ 1 % at a sorption pH the range of 4.5–5.5. As a consequence of the high chemical stability and high extraction efficiencies, these coated materials represent attractive sorbents for the rapid and quantitative extraction or screening of inactive and radioactive lanthanides from complex sources such as rocks or atomic power plant waste [30].

2.2. Materials prepared by precipitation polymerization [22, 23]

For industrial purposes, the demand for highly active and stable heterogeneous catalytic systems is high [31–34]. Only few reports exist on the synthesis of heterogeneous palladium catalysts e. g. based on palladium-loaded porous glass [35] or palladium colloids [36, 38]. While phosphine-based ligands possess excellent binding properties for palladium (II), they are easily transformed into the corresponding phosphine oxides. This reaction results in a slow but permanent release of palladium (catalyst bleeding) which significantly aggravates its industrial use. Standard procedures for the preparation of polymer-supported ligands usually entail the surface modification of commercially available polymer supports, e. g. PS-DVB or chloromethylated PS-DVB (Merrifield polymer). Nevertheless, this synthetic route is characterized by some drawbacks. In order to achieve maximum derivatization, porous materials with high surface areas have to be chosen. As a major part of the specific surface area (σ) results from internal pores, large amounts of the desired ligand are located at the interior of the particle. This leads to a diffusion-controlled reaction during catalysis, which usually reduces the overall reaction rate constant [31]. Another critical point lies in the usually employed divergent synthetic approach for surface derivatizations. The synthetic protocol often consists of at least two to three heterogeneous steps, and each one may not be accomplished in a quantitative way. In contrast to homogeneous reactions, the resulting "byproducts" are not removed. This leads to a situation, where a significant amount of the initial functionality is not transformed into the desired ligand. Poor definitions in terms of chemical structure and problems of catalyst poisoning are often the consequence.

Using precipitation polymerization techniques, ROMP may be used for the preparation of new heterogeneous catalysts. Thus, ROMP of norborn-2-ene-5-(N,N-di(pyrid-2-yl)carbamide followed by cross-linking with 1,4,4a,5,8,8a-hexahydro-1,4,5,8-*exo-endo*-dimethanonaphtalene [23] and subsequent loading of the resulting resin with palladium(II) chloride results in the formation of a heterogeneous Pd (II) catalyst [39]. The synthesis of the support and loading of the resin is summarized in Scheme 2. The resulting material represents a typical small particle catalyst [31]. The carrier particles have a mean diameter of 20–40 µm, possess only macropores (> 1000 Å) and are therefore characterized by a low specific surface area (4–6 m^2/g) both in the dry as well as in the wetted state. Consequently, swelling is a minor problem. As the palladium-loaded chelating groups are located at the surface of the particle [22, 23], they are easily accessible and diffusion plays a minor role. As a consequence of the high selectivity of the ligand for palladium, no substitution by other metal ions which might be present in technical grade chemicals occurs. The high affinity of the ligand for palladium also leads to a high temperature stability of up to 150°C and to an extraordinary pH-stability of the resulting complex. Thus, palladium may not be removed from the ligand within a pH range of 0–12.

This heterogeneous catalyst turned out to be highly active in the vinylation of aryl halides (Heck-type couplings) with turn-over numbers (TONs) of up to 210000. Even higher TON's (up to 350000) may be achieved in the arylation of alkynes. Moderate yields (65 %) and TONs ($\leq$ 4000) may additionally be achieved in the amination of aryl bromides. In general iodo- and bromoarenes have to be used, nevertheless, chloro-compounds may be activated by addition of tetrabutylammonium bromide (TBAB), which is know to promote Heck-type couplings [40–43]. The elevated overall coupling rate constant and the high stability of this catalytic system may give access to a permanent system suitable for industrial purposes.

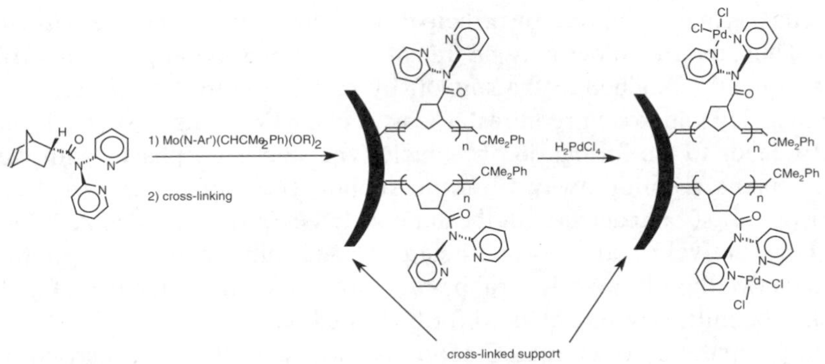

Scheme 2. Preparation of heterogeneous Pd-catalyst *via* ROMP.

2.3. Preparation of functionalized polymer supports via ROMP-grafting

2.3.1. Materials for chiral HPLC [44]. Materials suitable for chiral LC have to fulfill the standard criteria for HPLC such as mechanical stability. Taking all requirements and restrictions that are entailed with a surface functionalization into consideration, a straight-forward yet broadly applicable synthetic route may be accomplished by two steps: (i) a simple surface derivatization by copolymerizable anchoring groups and (ii) the use of a well-defined polymer chemistry such as ROMP for the attachment of the actual working functionalities. This attachment must be performed in a way that the final derivatization is less or not dependent on the actual amount of anchoring groups, yet the entire synthetic approach provides a high chemical stability of the final material, which is especially of importance in the case of silica-based materials. In other words, the anchoring groups that serve as a linker between the support and the actual functional groups may not be broken mechanically or chemically, e.g. by low or high pH-values. To achieve this goal even in the case of silica-based materials, surface-grafting has to be carried out in a way that the final graft-copolymers form impervious layers which prevent any contact of the support with the mobile phase. Consequently, we elaborated a new grafting chemistry based on ROMP for the reproducible surface derivatization of monosized polymer supports [45].

Suitable anchoring groups for the preparation of a ROMP-graft-copolymer are surface-attached norborn-2-ene-5-yl-groups. These may easily be introduced in the case of silica materials using trichloro-norborn-2-ene-5-ylsilane. Subsequent "endcapping" with a mixture of chlorotrimethylsilane and dichlorodimethylsilane leads to a sufficient derivatization of a major part of the surface silanol groups. In the case of PS-DVB-based materials, more sophisticated methods have to be applied. Bromomethylations using trioxane, tin tetrabromide and trimethylbromosilane [46] and, alternatively, conversion of the chloromethyl groups into to the corresponding bromomethyl groups *via* halogen exchange [47, 48] were performed. In a final step, the bromomethylated PS-DVB resins were converted into the norborn-2-ene-5-ylmethylethers. In principal, two different approaches may be performed. Thus, the monomer may be transformed into a living polymer *via* ROMP and subsequently attached to the support by reaction with the surface norborn-2-ene groups. This approach requires at least a class-IV living system [2] and consequently leads to the formation of tentacle type stationary phases with the linear polymer chains pointing away from the support (Method A). Alternatively, the initiator may first be reacted with the support to become heterogenized. Monomer that is consecutively added to this heterogenized initiator becomes grafted onto the surface (Method B). While Grubbs-type initiators require the use of method B, Schrock-type initiators may be used for both methods.

For the preparation of chiral LC-supports surface grafting was carried out with a series of chiral monomers either based on L-amino acids or β-cyclodextrin according to method B (Scheme 3). Some evidence for the assumption, that an impervious copolymer layer was formed during the grafting procedure is provided by the high pH-stability (pH 2 - 10) which was found for the grafted silica supports. Chiral separations of proglumide on a poly-NBE-β-cyclodextrin grafted silica remained constant even after excessive treatment with mobile phases adjusted to pH values from 2 - 10. Any hydrolysis within this pH range may strictly be ruled out as chiral separations are known to be drastically affected by the presence of any free silanol groups (unspecific retention). The chromatographic data that may be deduced from these separations were compared with those obtained with a commercially available CSP (Cyclobond I™ 2000) and revealed superior or at least comparable results in terms of selectivity and resolution. In addition, separation times, which represent a crucial point in chiral separations, were found to be dramatically reduced using ROMP-based columns. Thus, separation of proglumide was achieved on a poly-NBE-β-cyclodextrin-grafted Nucleosil 300-5 in less than 2 minutes. An illustration is given in Figure 3. Such highly efficient separations underline in combination with the enhanced selectivities the utility of the entire concept.

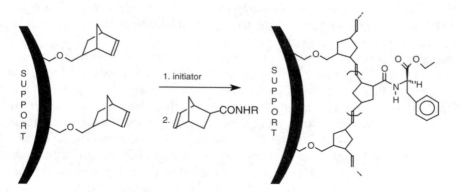

Scheme 3. Surface-derivatization of norbornene-modified polymer supports *via* ROMP-grafting.

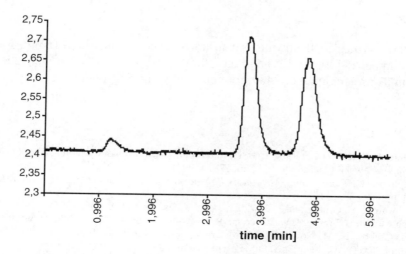

Figure 3. Separation of proglumide on poly(NBE-β-CD)-grafted silica.

2.3.2. Poly(octamethylmetallocenyl)-based materials for HPIC. Until now, the polymerization of 1-alkynes [49–53] and in particular of metallocenylalkynes [54–57] has been investigated intensely and is now well understood. Based on these studies, poly(ethynylferricinium)-based anionic exchangers have been prepared following the grafting concept described above. Thus, metathesis polymerization of 4-ethynyl-1-(octamethylferrocenylethenyl)benzene using a Schrock-type catalyst and subsequent grafting of the living polymer onto a NBE-derivatized silica support yields the desired octamethylferrocene-grafted stationary phase (Figure 4). Oxidation is easily performed in acetonitrile, resulting in a metallocene-based anion-exchanger that may be used effectively for the separation of oligonucleotides.

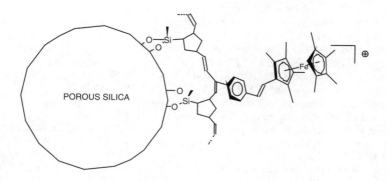

Figure 4. Structure of octamethylferricinium-grafted silica.

Acknowledgement

The author wishes to thank the "Austrian Science Foundation" (FWF, Vienna, project number P-12963 GEN) and the "Jubiläumsfonds der Österreichischen Nationalbank", project number 7489, for generous financial support.

REFERENCES

1. S. Penczek, P. Kubisa and R. Szymanski, Makromol. Chem. Rapid. Commun., **12**, 77 (1991).
2. K. Matyjaszewski, Macromolecules, **26**, 1787 (1993).
3. R. R. Schrock, *Ring-Opening Metathesis Polymerization*; D. J. Brunelle, Ed.; Hanser: Munich, 1993, pp 129.
4. P. Schwab, R. H. Grubbs and J. W. Ziller, J. Am. Chem. Soc., **118**, 100 (1996).
5. P. Schwab, M. B. France, J. W. Ziller and R. H. Grubbs, Angew. Chem., **107**, 2179 (1995).
6. R. H. Grubbs, B. M. Novak, D. M. McGrath, A. Benedicto, M. France and S. T. Nguyen, Polym. Prepr. (Am. Chem. Soc., Div. Polym. Chem.), **33**, 1225 (1992).
7. S. T. Nguyen and R. H. Grubbs, J. Am. Chem. Soc., **115**, 9858 (1993).
8. M. B. France, R. H. Grubbs, D. V. McGrath and R. A. Paciello, Macromolecules, **26**, 4742 (1993).
9. R. H. Grubbs, J. M. S. Pure Appl. Chem., **A31**, 1829 (1994).
10. M. R. Buchmeiser, Chem. Rev., **100**, 1565 (2000).
11. R. R. Schrock, Acc. Chem. Res., **23**, 158 (1990).
12. R. R. Schrock, J.-K. Lee, R. O'Dell and J. H. Oskam, Macromolecules, **28**, 5933 (1995).
13. R. R. Schrock, Polyhedron, **14**, 3177 (1995).
14. M. R. Buchmeiser, N. Atzl and G. K. Bonn, Int. Pat. Appl., AT404 099 (181296), PCT /AT97/00278.
15. M. R. Buchmeiser, N. Atzl and G. K. Bonn, J. Am. Chem. Soc., **119**, 9166 (1997).
16. D. Ambrose, J. S. Fritz, M. R. Buchmeiser, N. Atzl and G. K. Bonn, J. Chromatogr. A, **786**, 259 (1997).
17. G. Seeber, M. R. Buchmeiser, G. K. Bonn and T. Bertsch, J. Chromatogr. A, **809**, 121 (1998).
18. K. Eder, M. R. Buchmeiser and G. K. Bonn, J. Chromatogr. A, **810**, 43 (1998).
19. M. R. Buchmeiser and G. K. Bonn, Am. Lab., **11**, 16 (1998).
20. M. R. Buchmeiser, R. Tessadri, G. Seeber and G. K. Bonn, Anal. Chem., **70**, 2130 (1998).
21. M. R. Buchmeiser and R. Tessadri, Austrian Pat. Appl., A 1132/97 (020797).
22. M. R. Buchmeiser, F. Sinner, R. Tessadri and G. K. Bonn, Austrian Pat. Appl., AT 405 056 (010497).
23. F. Sinner, M. R. Buchmeiser, R. Tessadri, M. Mupa, K. Wurst and G. K. Bonn, J. Am. Chem. Soc., **120**, 2790 (1998).
24. M. R. Buchmeiser, M. Mupa, G. Seeber and G. K. Bonn, Chem. Mater., **11**, 1533 (1999).
25. M. Rätzsch, Progr. Polym. Sci., **13**, 277 (1988).
26. R. Grüner, F. Schwan and H. Engelhardt, LaborPraxis, **9**, 24 (1998).
27. H. Engelhardt, M. Arangio and T. Lobert, LC-GC-Int., **12**, 803 (1997).
28. H. Engelhardt, H. Löw and W. Götzinger, J. Chromatogr. A, **554**, 371 (1991).
29. K. Jung and H. Specker, Fresenius Z. Anal. Chem., **289**, 48 (1978).
30. G. Seeber, P. Brunner, M. R. Buchmeiser and G. K. Bonn, J. Chromatogr. A, **848**, 193 (1999).
31. R. J. Wijngaarden, A. Kronberg and K. R. Westerterp, Industrial Catalysis, Wiley-VCH, Weinheim (1998).
32. B. Cornils, W. A. Herrmann and R. W. Eckl, J. Molec. Catal. A: Chemical, **116**, 27 (1997).
33. W. A. Herrmann and C. Köcher, Angew. Chem., **109**, 2257 (1997).
34. W. A. Herrmann and B. Cornils, Homogeneous Catalysis-Quo vadis?, Applied Homogeneous Catalysis with Organometallic Compounds, VCH, Weinheim (1996).

35. J. Li, A. W.-H. Mau and C. R. Strauss, Chem. Commun., 1275 (1997).
36. M. Beller, H. Fischer, K. Kühlein, C.-P. Reisinger and W. A. Herrmann, J. Organomet. Chem., **520**, 257 (1996).
37. S. Klingelhöfer, W. Heitz, A. Greiner, S. Oesterreich, S. Förster and M. Antonietti, J. Am. Chem. Soc., **119**, 10116 (1997).
38. S. Bräse, D. Enders, J. Köbberling and F. Avemaria, Angew. Chem., **110**, 3614 (1998).
39. M. R. Buchmeiser, Austrian Patent Appl., A 344/99 (020399) (020399).
40. L. Lavenot, C. Gozzi, K. Ilg, I. Orlova, V. Penalva and M. Lemaire, J. Organomet. Chem., **567**, 49 (1998).
41. W. A. Herrmann, C. Broßmer, C.-P. Reisinger, T. H. Riermeier, K. Öfele and M. Beller, Chem. Eur. J., **3**, 1357 (1997).
42. C. Gozzi, L. Lavenot, K. Ilg, V. Penalva and M. Lemaire, Tetrahedron Lett., **38**, 8867 (1997).
43. G. T. Crisp and M. G. Gebauer, Tetrahedron, **52**, 12465 (1996).
44. M. R. Buchmeiser, J. Chromatogr. A, in press (2000).
45. M. R. Buchmeiser, F. Sinner, M. Mupa and K. Wurst, Macromolecules, **33**, 32 (2000).
46. S. Itsuno, K. Uchikoshi and K. Ito, J. Am. Chem. Soc., **112**, 8187 (1990).
47. J. H. Batler and K. P. Spina, Synth. Comm., 14, (1984).
48. K. B. Yoon and J. K. Kochi, J. Chem. Soc. Chem. Commun, 1013 (1987).
49. R. R. Schrock, S. Luo, N. Zanetti and H. H. Fox, Organometallics, **13**, 3396 (1994).
50. R. R. Schrock, S. Luo, J. C. Lee Jr., N. C. Zanetti and W. M. Davis, J. Am. Chem. Soc., **118**, 3883 (1996).
51. S. Koltzenburg, E. Eder, F. Stelzer and O. Nuyken, Macromolecules, **32**, 21 (1999).
52. F. J. Schattenmann, R. R. Schrock and W. M. Davis, J. Am. Chem. Soc., **118**, 3295 (1996).
53. H. H. Fox, M. O. Wolf, R. O'Dell, B. L. Lin, R. R. Schrock and M. S. Wrighton, J. Am. Chem. Soc., **116**, 2827 (1994).
54. M. Buchmeiser and R. R. Schrock, Macromolecules, **28**, 6642 (1995).
55. M. Buchmeiser, Macromolecules, **30**, 2274 (1997).
56. M. R. Buchmeiser, N. Schuler, N. Kaltenhauser, K.-H. Ongania, I. Lagoja, K. Wurst and H. Schottenberger, Macromolecules, **31**, 3175 (1998).
57. M. R. Buchmeiser, N. Schuler, H. Schottenberger, I. Kohl, A. Hallbrucker and K. Wurst, Designed Monomers & Polymers, in press (2000).

Electron-beam initiated cationic polymerization

JAMES V. CRIVELLO,[1,*] THOMAS C. WALTON[2] and RANJIT MALIK[1]

[1]*Department of Chemistry, Rensselaer Polytechnic Institute, Troy, NY 12180*
[2]*Aeroplas Corp., International, 265 B Proctor Hill Road, Hollis, NH 03049*

Abstract—During the course of this investigation, it was demonstrated that epoxy monomers can be efficiently polymerized using both e-beam and γ-ray irradiation. The key to this chemistry is the use of onium salts as initiators to facilitate cure. Studies have shown that diaryliodonium salts are more efficient initiators than triarylsulfonium salts for this purpose. Silicon-containing epoxides are especially reactive monomers and polymerize very rapidly when irradiated with ionizing radiation in the presence of onium salt initiators. A mechanism involving the reduction of the onium salt initiators by free radicals generated by radiolysis of the monomer has been proposed. The use of this new methodology for the rapid, efficient e-beam cure of high performance carbon fiber reinforced composites is described. The potential use of this new methodology for the rapid fabrication of light-weight, damage-tolerant, thermally and oxidatively resistant composites is detailed.

1. INTRODUCTION

In this laboratory, we have been exploring the UV and thermally induced cationic polymerizations of a wide variety of monomers including epoxides [1, 2]. Our work is based on the key discovery that certain onium salts such as diaryliodonium **I** and triarylsulfonium salts **II** efficiently initiate the cationic polymerization of virtually all known types of cationically polymerizable monomers when irradiated with UV light.

$$\text{Ph—I}^+\text{—Ph} \quad MtX_n^- \qquad \text{Ph}_3\text{S}^+ \quad MtX_n^-$$

$$\text{I} \qquad\qquad \text{II}$$

Scheme 1 shows an abbreviated generalized mechanism of an onium salt photoinitiated cationic polymerization illustrated with triarylsulfonium salt photoinitiators [3–6]. A similar scheme can be written for diaryliodonium salt photoinduced polymerizations.

---

*To whom correspondence should be addressed. E-mail: crivej@rpi.edu

$$Ar_3S^+ \ X^- \xrightarrow{h\nu} [Ar_3S^+ \ X^-]^* \longrightarrow \begin{Bmatrix} Ar_2S^{+\bullet}X^- + Ar\bullet \\ Ar_2S + Ar^+ X^- \end{Bmatrix} \quad \text{eq. (1)}$$

$$\begin{Bmatrix} Ar_2S^{+\bullet}X^- + Ar\bullet \\ Ar_2S + Ar^+ X^- \end{Bmatrix} \xrightarrow{RH} HX + \text{Products} \quad \text{eq. (2)}$$

$$HX + nM \longrightarrow H(M)_{n-1}M^+ \ X^- \quad \text{eq. (3)}$$

Scheme 1.

Interaction of a triarylsulfonium salt with light affords the excited sulfonium salt which decays with either the homolytic or heterolytic rupture of a carbon-sulfur bond (eq. 1). The resulting highly reactive species (cations, cation-radicals and free radicals) decay by a number of pathways through reaction with the solvent or monomer (RH) (eq. 2) to generate a variety of products and the acid, HX. If the HX generated is a strong Brønsted acid, protonation of the monomer, M, (eq. 3) occurs rapidly to initiate cationic polymerization. Photoinitiated cationic polymerizations may be applied to a wide variety of types of cationically polymerizable monomers including: epoxides, vinyl ethers, oxetanes, oxazolines, styrene, and many others.

Cationic photopolymerizations have found many commercial uses due to the rapidity of these processes, their low energy requirements and, particularly, because such polymerizations can be carried out in bulk, eliminating the need for solvents. Cationic photopolymerizations are ideally suited for such thin film applications as coatings, printing inks, adhesives, and many others. Although the uses of cationic crosslinking photopolymerizations are continuing to expand, there are certain limitations of this chemistry which restrict its use in still other applications. For example, highly pigmented, optically opaque, or very thick substrates cannot be effectively polymerized using light induced polymerizations. Thin films deposited on complex shapes with interior surfaces that cannot be reached by light are also not effectively polymerized using this technique.

To address these latter applications, forms of radiation with a greater penetrating power (i.e. higher energy) must be used. Electron-beam (e-beam) and γ-ray radiation are currently being employed for this purpose. The interaction of organic substrates with either e-beam or γ-ray radiation produces free radicals by primary bond cleavage reactions. These radicals can be used to initiate the polymerization of multifunctional acrylate of methacrylate monomers to generate crosslinked polymers. While the latter monomers are effectively polymerized using either e-beam or γ-ray radiation, the polymers generated typically possess rather low thermal, oxidative and mechanical performance characteristics. Accordingly, alternative polymerization schemes are being investigated.

The direct, e-beam or γ-ray induced "free cationic" (i.e. without initiators) ring-opening polymerization of epoxide and other heterocyclic monomers has been previously described and proceeds vigorously at room temperature [7–10]. However, these polymerizations require monomers which are rigorously purified and the reactions must be carried out under scrupulously oxygen and water-free conditions. Furthermore, high e-beam or γ-ray doses are required to produce even small amounts of polymer. For these reasons, ionizing radiation induced cationic polymerizations have not attracted much attention from either the scientific or industrial communities [11].

It occurred to us that onium salt photoinitiators might be responsive to ionizing radiation. In particular, the heavy mass iodine atom in diaryliodonium salts has a fairly high cross section to ionizing radiation and radiolysis would be expected to result in the fragmentation of these compounds in a manner similar to the photolysis pathway shown in Scheme 1. Accordingly, a program was begun to determine: 1) if onium salts were responsive to ionizing radiation, 2) whether practical polymerizations could be carried out, 3) to determine what radiation dose was required and, finally, 4) whether useful polymer properties can be obtained.

2. EXPERIMENTAL

2.1. Materials

Silicone-epoxy monomers **VI-VIII** were prepared as previously described [12, 13] from the corresponding silane and 4-vinylcyclohexene oxide (3-vinyl-7-oxabicyclo[4.1.0]heptane) using the hydrosilation reaction. The resins were characterized by various techniques. Infrared spectra were obtained on a Buck Model 500 Infrared Spectrometer. ^{1}H NMR spectra were recorded on a Varian 500 MHz Nuclear Magnetic Resonance Spectrometer. Elemental analyses were performed by Quantitative Analysis, Bound Brook, NJ. Other epoxy resins, Tactix 123, Epon resins (Epon 862; bisphenol-A diglycidyl ether and Epon 825; bisphenol-F diglycidyl ether) and 3,4-epoxycyclohexylmethyl 3',4'-epoxycyclohexanecarboxylate (UVR 6105) were obtained respectively, from the Dow Chemical, Shell Chemical and the Union Carbide corporations and used without further purification. The onium salt initiators (4-n-octyloxyphenyl)phenyliodonium hexafluoroantimonate (**IOC-8**) [14] and S,S-diphenyl-S(4-thiophenoxyphenyl)sulfonium hexafluoroantimonate (**SS**) [15], were prepared as previously reported. AG193P plain weave 3K (AS-4) carbon fiber fabric was supplied by the Hercules Corporation.

2.2. Preliminary photochemical reactivity evaluations

Tack-free photopolymerization rates were determined in air using a Model QC 1202 UV Processor obtained from the RPC Equipment Company, Plainfield, IL. This apparatus is equipped with two 12 in (30.5 cm) medium pressure mercury arc lamps mounted perpendicular to the direction of travel of the conveyor. The lamps were operated together or independently at 380 V and 9.8 ± 0.8 amps. The lamps were also operated at either high (300 W), medium (300 W), or low (120 W) power levels. The conveyor speed was varied between 10 and 500 ft/min (3 and 15 m/min). Samples of the silicone-epoxy resins obtained by dissolving 0.5 mol % **IOC-8** were coated as 25 μm liquid films onto glass plates, passed through the curing chamber and immediately tested for tack on exiting the chamber. The minimum speed determined at a given light intensity required to produce a tack-free film was recorded as the tack-free speed.

2.3. Low intensity e-beam irradiation studies

Liquid samples to be irradiated were coated onto 2 mil poly(ethylene terephthalate) foils to give 75 μm films. An Energy Sciences Inc. Electrocurtain Model CB-150 electron beam irradiator operating at 165 KeV and equipped with a 15 cm linear cathode was used to irradiate the samples. The wet film samples were attached to a continuous web and passed through the beam. Experiments were run under nitrogen and air at a constant web speed. The dose was varied by changing the amperage applied to the filament. Samples were tested for tackiness immediately on exiting the irradiation chamber. This qualitative test was performed by placing a thumb under moderate pressure on the cured film and twisting. If the film did not deform under these conditions, it was classified as tack-free.

2.4. High intensity e-beam irradiation studies

High intensity e-beam irradiation studies were carried out using the Atomic Energy Canada Limited (AECL) Research 1-10/1 Electron Linear Accelerator in Pinawa, Manitoba, Canada. A 10 MeV (electron energy) pulsed electron beam was configured with a vertical horn to scan over a 50 cm wide path onto a variable speed conveyor belt. All irradiations were conducted behind a heavily shielded concrete and lead barrier. The electron beam was pulsed at 13-19 Hz with a pulse width of 4 μs. The beam is scanned at a repeat rate of 2-7 Hz with a circular spot size at the surface of the conveyor of approx. 10 cm. The dose rate delivered at the conveyor surface is 130 Gy/second (50 cm scan). This e-beam source was employed for the fabrication of bars for mechanical tests and for the cure of carbon fiber reinforced composites.

2.5. γ-ray irradiation studies

Similarly, γ-ray irradiations were conducted at AECL in a ^{60}Co gamma cell at a dose rate of 9.8 Mrad/min. Samples (3 g) were placed in small screw cap vials and sealed. To the outside wall of the vials were attached a thermocouple connected to a recorder for direct temperature readout. The samples were ballasted against a cured sample to eliminate the thermal effects due to radiation induced heating. The entire apparatus was insulated with polyurethane foam and the samples irradiated in the ^{60}Co gamma cell for periods up to 80 minutes. The progress of the polymerizations was followed by continuous monitoring of the temperature of the sample with time using an XY plotter.

2.6. Fabrication of Test Specimens

Graphite fiber composites were fabricated in a 0°, 90° fiber orientation by standard vacuum bag techniques and by resin transfer molding methods. Vacuum assisted resin transfer molding (VARTM) was carried out using a 12 in x 12 in x 0.25 in aluminum mold with a removable top plate for easy removal of the cured composite. The mold was prepared by coating with Frekote Sealer B-15 (Dexter Hysol Corp.) and baking for 1h at 200°F. This was followed by the spray application of a dry lubricant (Dexter Hysol Number 1 Release agent). Four layers (8 plys) of carbon fiber fabric was laid in the mold and the mold sealed. A vacuum was placed on the mold and the liquid resin containing the initiator was drawn into the mold from a reservoir. Final consolidation was carried out after closing off the resin flow exit and reducing the pressure to 40 psi at the resin entry port for at least ten minutes. The resin charged VTRM preform panels were stored precatalyzed for up to one and a half days before irradiating. The filled molds and laid-up laminates were irradiated by placing them horizontally on a conveyor and passing them through the 10 MeV electron beam at such a rate as to produce a total dose of 7.5 Mrad. All e-beam irradiations were conducted behind concrete shielded maze walls inside the accelerator room.

Vacuum bag lay-up was carried out employing the usual techniques and using a nylon-66 bagging film to form the lay-up radiation window. Pro-Wrap which is a type of thick polypropylene bagging film was also used during one set of e-beam irradiation conditions.

2.7. Mechanical Testing

Measurements of the flexural and tensile strengths and moduli were determined in accordance with ASTM D790 and ASTM D638 procedures respectively. In-plane shear strength (ASTM D3846) and dynamic mechanical properties (ASTM D4065) were measured in air on either a Rheometrics System IV instrument (NASA, LaRC) or a Rheometrics RSA dual cantilever instrument (AECL) in air at a heating rate of 10°C/min.

3. RESULTS AND DISCUSSION

3.1. Initial photochemical and e-beam studies

Initial screening work in this laboratory was directed towards the use of diaryliodonium and triarylsulfonium tetrafluoroborate photoinitiators as rate enhancing additives for the e-beam and γ-ray induced polymerization of epoxide monomers. E-beam irradiations of the three representative monomers and oligomers shown below were carried out in the presence of the above mentioned onium salts [16].

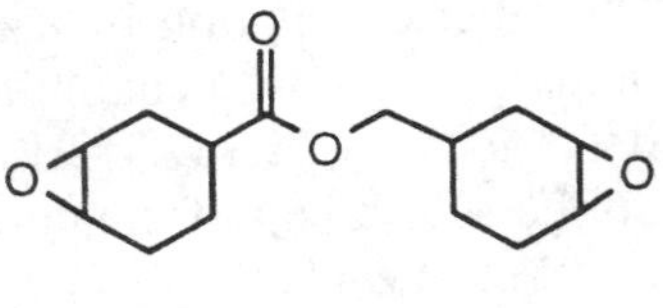

III

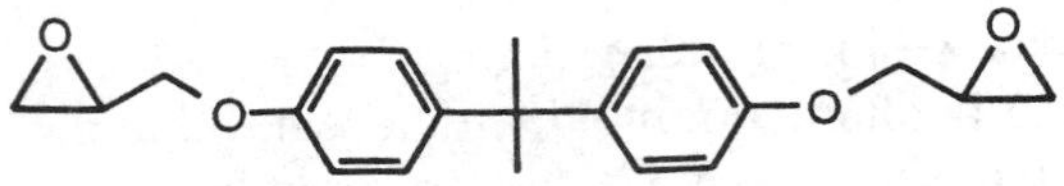

IV

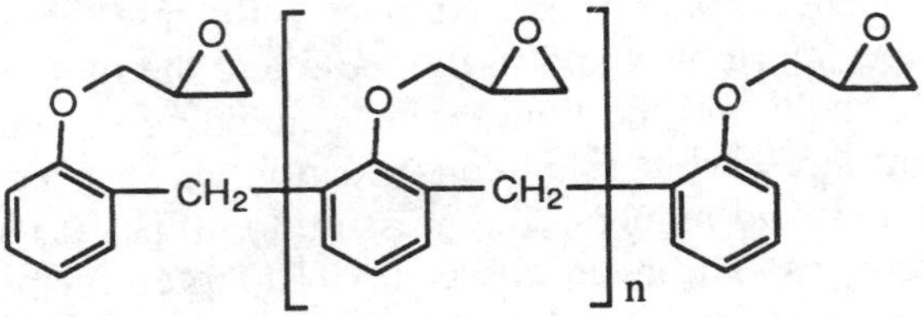

V

The polymerizations of these monomers did take place, however, the doses required were too high i.e >150 Mrad (1 Mrad = 10 KGy = 1 x 10^8 ergs/g absorbed energy) to be useful for practical applications. More recently, Russian workers [17, 18] have employed sulfonium, diazonium, ferrocenium and iodonium fluoroborate salts as e-beam cure adjuvants, while Davidson and Wilkinson [19] have used the hexafluorophosphate salts of the same onium salt compounds. Although it was clear from the data generated by these three groups that the use of onium salt promoters was effective in enhancing the rate of epoxide polymerization and lowering the required dose, dose levels remained much too high to be of practical use. Not only are such doses impractical, but in many cases either the substrate matrix resin or the fiber reinforcement suffer significant radiation damage under these conditions. A reduction in the required dose by at least a factor of 10 was required.

To improve the poor dose response of the above epoxy systems, two major reaction parameters can be modified. First, the reactivity of the epoxy resin can be maximized. Second, the structure and concentration of the onium salt photoinitiator can be adjusted. A study conducted in this laboratory [20], of the reactivity of epoxy compounds in photoinitiated cationic polymerization showed that the range of reactivity of these monomers to be very large and to be dependent on such factors as ring strain, steric factors as well as the presence or absence of other functional groups in the molecule. The most reactive monomers were those bearing epoxycyclohexyl groups. Particularly reactive were a class of monomers we have called silicone-epoxy resins [21, 22]. The structures of two of these novel monomers are shown below.

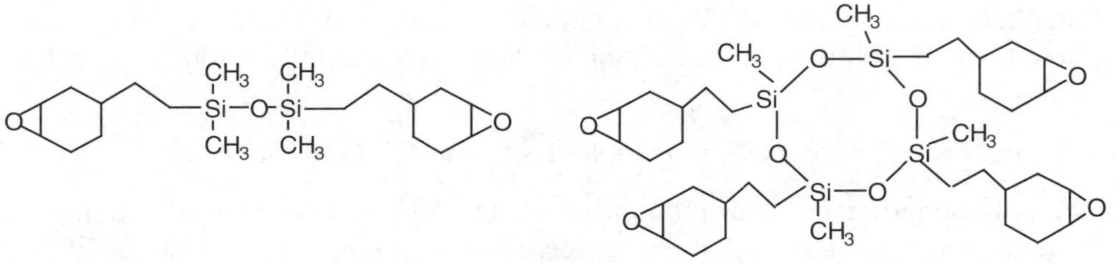

VI   **VII**

Monomers **VI** and **VII** contain siloxane linkages to which are attached highly reactive epoxycyclohexyl groups. The monomers may be prepared by simple hydrosilation of 4-vinylcyclohexene oxide with the appropriate silicon-hydride compound in the presence of a noble metal catalyst containing platinum or rhodium. Shown below in equation 4 is the synthesis of monomer **VI**.

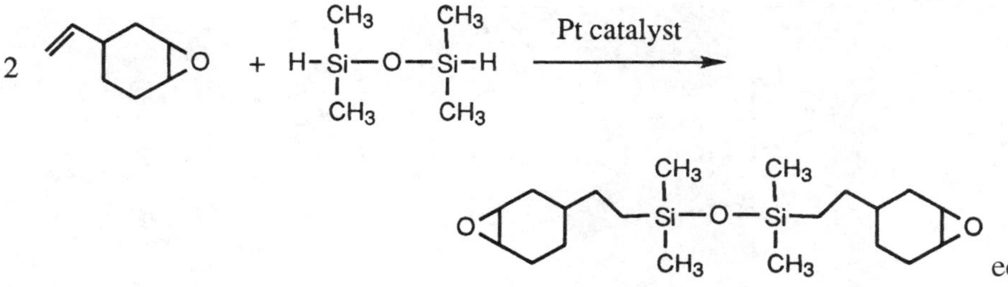

eq. (4)

The above synthesis can be conducted at low temperatures in the presence or absence of a solvent and high yields of the desired monomers are obtained.

Similarly, a wide variety of multifunctional silicon-hydrogen compounds can be combined with 4-vinylcyclohexene oxide to make cyclic, branched or even polymeric epoxy resins containing the highly reactive epoxycyclohexane group. Such condensation reactions are typically rapid and virtually quantitative. All the silicone-epoxy monomers are colorless fluids. The monomers have no UV absorption and are, therefore, stable to long term weathering. Additional attractive features of these monomers are that they are non-toxic, have a very low order of skin and eye irritation and they do not outgas on UV, e-beam or thermal cure. Monitoring the UV photoinitiated cationic polymerization of these monomers by several methods including differential scanning photocalorimetry, real-time infrared spectroscopy and direct tack-free times has confirmed that these monomers possess extraordinarily high rates of polymerization. An example of a comparison of the tack-free cure times of monomers **VI** and **VII** with a commercially available "high reactivity" biscycloaliphatic epoxy resin, 3.4-epoxycyclohexylmethyl 3'.4'-epoxycyclohexanecarboxylate (**III**), is shown in Table 1. As may be noted in Table 1, the reactivities of monomers **VI** and **VII** are at least one to two orders of magnitude greater than **III**. These silicon-containing epoxide monomers undergo photoinitiated cationic polymerization at rates comparable to multifunctional acrylates.

3.2. Thin Film E-beam Polymerization of Silicone-Epoxy Monomers

The polymers derived from monomers **VI** and **VII** by UV induced cationic polymerization have glass transition temperatures (T_gs) as high as 180-200 °C and thermal degradation temperatures exceeding 300-350 °C in air [23]. Some preliminary mechanical tests have also shown that these materials possess flexural

Table 1.
Tack-free cure rates for the UV cure of epoxy siloxane monomers

Monomer	PI Conc. mol% (wt %)	Tack-Free Cure Rate (ft/min)		
		2 Lamps (300 W)	1 Lamp (300 W)	1 Lamp (120 W)
(siloxane-epoxy monomer VI)	0.5* (0.84)	>500	>500	300
	0.25* (0.42)	>500	>500	250
	0.10* (0.17)	>500	>500	150
	0.05* (0.08)	>500	400	50
(siloxane-epoxy monomer VII)	0.5* (0.44)	>500	>500	250
	0.25* (0.22)	>500	>500	150
	0.10* (0.09)	>500	450	100
	0.05* (0.04)	400	150	40
	0.05‡ (0.04)	500	>500	150
(bis-cycloaliphatic epoxy III)	0.5* (1.3)	40	30	N.R.⁺

*(4-octyloxyphenyl)phenyliodonium SbF_6^-, ‡(4-thiophenoxyphenyl)diphenylsulfonium SbF_6^-, ⁺No Reaction.

moduli measured in a three-point bend mode of 1.7-3.0 GPa. Based on the above photopolymerization results, a study of the behavior of monomers **VI** and **VII** in the presence of various onium salt photoinitiators under e-beam irradiation conditions was carried out [24–26]. Shown in Tables 2 and 3 are the results of these studies conducted with the aid of an Energy Sciences, Inc. Model CB-150 Electrocurtin Electron Beam Irradiator operating at 165 KeV. Films 75 µm in thickness of the monomer containing the indicated photoinitiator were drawn onto poly(ethylene terephthalate) foils and irradiated at room temperature under an atmosphere of nitrogen. The samples were immediately tested after irradiation for their state of polymerization (cure) as indicated by tackiness. In Tables 2 and 3 (+) indicates the film was tack-free, while (±) indicates that it was slightly tacky, denoting partial polymerization, and (-) means a wet film and no polymerization was detected. Table 2 shows that efficient crosslinking polymerization of monomer **VI** takes place in the presence of 0.25-1.0 mol% diaryliodonium and triarylsulfonium salt photoinitiators bearing the SbF_6^- anion at doses from 2-3 Mrad. Transparent, colorless films were obtained which were insoluble in all solvents due to crosslinking. In the absence of the photoinitiators, polymerization does not take place even on prolonged irradiation at high doses (>10 Mrad). Photoinitiators

Table 2.

The Cationic E-Beam Polymerization[*] of

Initiator	Initiator Conc. (mole %)	Dose (Mrad)	Result[#]
None	–	3	-
$Ph_2I^+\ SbF_6^-$	1.0	2	+
$Ph_2I^+\ SbF_6^-$	0.75	2	+
$Ph_2I^+\ SbF_6^-$	0.50	2	+
$Ph_2I^+\ SbF_6^-$	0.50	3	+
$Ph_2I^+\ SbF_6^-$	0.25	2	+
$Ph_2I^+\ PF_6^-$	0.50	3	-
$(4\text{-}C_8H_{17}OPh)Ph_2I^+\ SbF_6^-$	0.5	2	+
$Ph_3S^+\ AsF_6^-$	0.5	3	±
$Ph_3S^+\ PF_6^-$	0.5	3	-
$(4\text{-}PhSPh)Ph_2S^+\ SbF_6^-$	0.5	3	+
$(4\text{-}PhSPh)Ph_2S^+\ PF_6^-$	0.5	3	-

[*]Cured as 75 µm films on poly(ethylene terephthalate) foils under nitrogen at ambient temperature (~25°C). [#](+) Indicates the film was tack-free while (±) means that slightly tacky denoting partial cure and (-) indicates no cure.

Table 3.
The Cationic E-Beam Polymerization of Monomers **VII**, **VIII** and **IX**[*]

Monomer	Initiator[#]	Dose (Mrad)	Result[•]	T_g
VII	none	3	−	
VII	$Ph_2I^+ \, SbF_6^-$	2	+	190
VII	$Ph_3S^+ \, SbF_6^-$	2	+	
VII	$Ph_3S^+ \, SbF_6^-$	3	+	
VII	$(4\text{-}C_8H_{17}OPh)Ph_2I^+ \, SbF_6^-$	2	+	
VII	$(4\text{-}PhSPh)Ph_2S^+ \, SbF_6^-$	2	+	
VIII	$Ph_2I^+ \, SbF_6^-$	2	+	176
IX	$Ph_2I^+ \, SbF_6^-$	3	+	185

[*]Cured as 75 μm films on poly(ethylene terephthalate) foils under nitrogen at ambient temperature (~25°C). [#]In all cases 0.5 mole % photoinitiator was used. [•](+) Indicates the film was tack-free while (±) means that slightly tacky denoting partial cure and (−) indicates no cure.

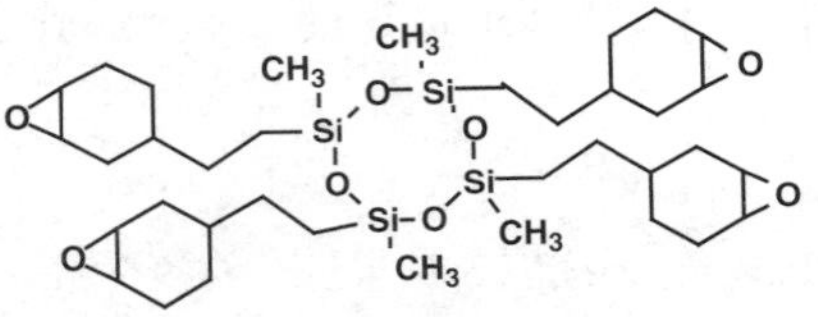

VII

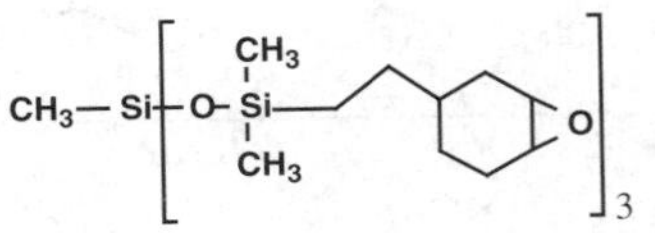

VIII 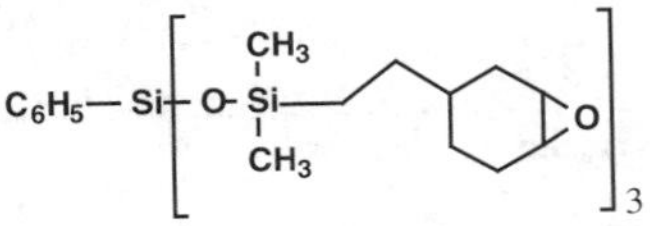   **IX**

bearing the SbF_6^- anion were by far the most active while those with the AsF_6^- anion were less so. Polymerization could not be achieved with either diaryliodonium or triarylsulfonium salts bearing the PF_6^- anion at doses at or below 3 Mrad. Table 3 shows similar data for tri- (**VIII**, **IX**) and cyclic tetrafunctional (**VII**) silicone-epoxy monomers. Again, efficient polymerization takes place with these monomers at does between 2-3 Mrad. Also included in Table 3 are representative T_g data measured for the e-beam cured resins. It may be noted that high T_gs (176-190°C) were obtained for these materials. The conversions have been shown by differential scanning calorimetry to be essentially quantitative. We were very encouraged by these results which indicate that the e-beam sensitivity of these silicone epoxy systems is nearly the same that observed for comparable acrylate and methacrylate systems (3-4 Mrad) [27, 28].

While preliminary results using very low intensity irradiation sources were encouraging, studies at higher radiation intensities in thicker samples of the silicone epoxy monomers had not been undertaken nor had an effective and simple means of monitoring these polymerizations been developed. For this reason, a study of the effects of resin structure on the polymerization rate was conducted using high energy ^{60}Co γ–ray irradiation in a gamma cell. The effects of γ–rays with respect to their ability to induce polymerization are very similar to e-beam irradiation and both involve bond breaking reactions with resulting free radical formation. The advantage of using γ-ray irradiation for this study not only consists of the intrinsic highly penetrating nature of this radiation which makes possible the cure of thick samples, but also the low dose rate which permitted us to conveniently follow the course of the cationic polymerizations. These reactions were carried out in small vials to which thermocouples were attached. The progress of the exothermic polymerizations could be directly followed by monitoring the increase of the temperature of the samples as a function of time.

The reactivities of various commercially available epoxide monomers were compared with that of monomer **VI** and this data is given in Figure 1. Each polymerization was conducted with bulk monomer containing 0.5 mole% of (4-octyloxypheny)phenyliodonium hexafluoroantimonate (**IOC-8**).

IOC-8   **SS**

Inspection of Figure 1 shows that a threshold γ-ray radiation dose of 0.8-1.0 Mrad is required to initiate polymerization in monomer **VI**, whereas typical aromatic glycidyl ether resins (bisphenol-A diglycidyl ether and bisphenol-F diglycidyl ether) required much higher threshold doses of 4.0-5.5 Mrad to induce polymerization. It is interesting to note that in this experiment, the polymerization of the bisphenol-A glycidyl ether-based epoxy resins are about as reactive as the biscycloaliphatic epoxy resin, **III**. It should be also pointed out that the above samples had a mass of 3 g and a maximum thickness of 1 cm. E-beam induced cationic ring-opening polymerizations of epoxides are exothermic as noted in Figure 1 and are markedly autoaccelerated by the rise in temperature during the reaction. For this reason, differences in the monomer reactivities observed here may be somewhat different than those obtained in previous thin film photo- or e-beam induced polymerization studies.

In Figure 2, is shown a comparison of the results of a ^{60}Co γ-ray irrradiation study of the polymerization of monomer **VI** in which **IOC-8** and (S-4-

thiophenoxyphenyl)diphenylsulfonium hexafluroantimonate (**SS**) were used as photoinitiators. Both of these photoinitiators bear the SbF_6^- anion. From the results given in Figure 2, it can be concluded that there is an additional effect on the rate of the reaction which is related to the type of onium salt cation which is employed and that the diaryliodonium salt, **IOC-8** is considerably more effective as an accelerator than **SS**.

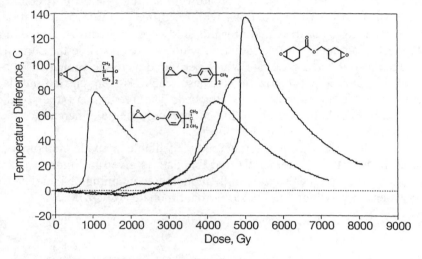

Figure 1. Effect of epoxide structure on the ^{60}Co γ-ray induced cationic polymerization using 0.5 mol% IOC-8. (γ-ray dose rate = 9.8 Mrad/min).

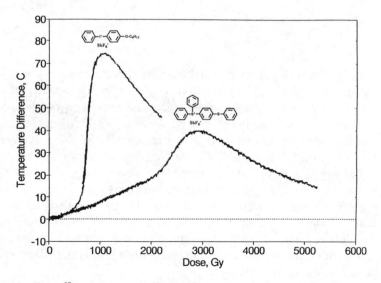

Figure 2. Study of the ^{60}Co γ-ray induced cationic polymerization of monomer **VI** using 0.5 mol% of onium salts IOC-8 and SS. (γ-ray dose rate = 9.8 Mrad/min).

3.3. Mechanism of Electron-Beam Induced Cationic Polymerizations

Shown in Scheme 2 is the mechanism which we propose for the electron-beam induced cationic polymerization of epoxides.

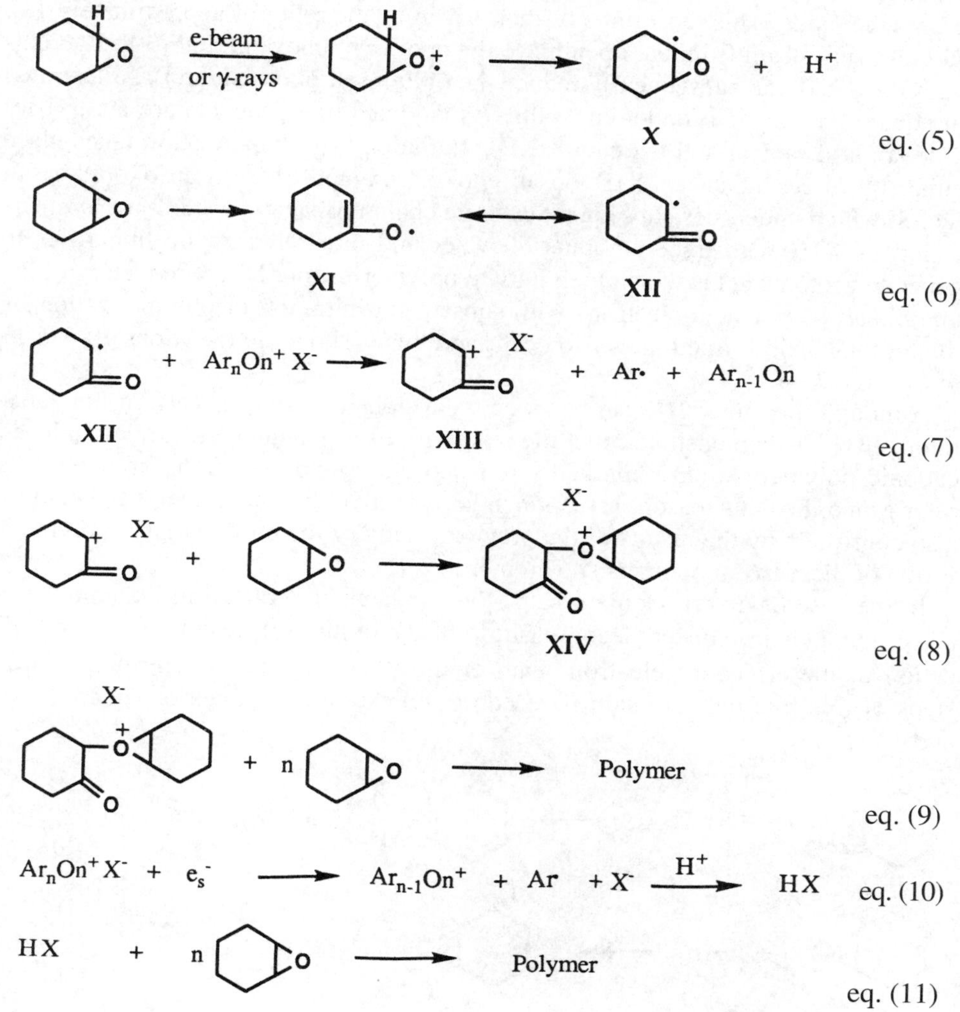

eq. (5)

eq. (6)

eq. (7)

eq. (8)

eq. (9)

$Ar_nOn^+ X^- + e_s^- \longrightarrow Ar_{n-1}On^+ + Ar\cdot + X^- \xrightarrow{H^+} HX$ eq. (10)

$HX + n \text{(cyclohexene oxide)} \longrightarrow$ Polymer eq. (11)

Scheme 2.

In Scheme 2, for the sake of clarity, the basic polymerizable epoxide unit of silicone-epoxide monomers such as **VI** is represented by cyclohexene oxide. E-beam or γ-ray irradiation of the monomer (eq. 5) initially produces free radical species by first, ionization of the epoxide then, loss of the tertiary proton on the carbon adjacent to the epoxide oxygen. Next, the cleavage of a carbon-oxygen bond (eq. 6) is facilitated by the facile rearrangement of the initially formed carbon-centered radical **X** to the resonance stabilized radicals **XI** and **XII**. This reaction has considerable literature precedent and an analogous mechanism has been proposed by Gritter and Wallace [29] for the free radical induced ring opening of propylene oxide. This reaction is further driven by the relief of approximately 112 kJ mol^{-1} ring strain from the opening of the epoxycyclohexane ring [30]. Radicals **X**, **XI** and **XII** can subsequently reduce the onium salt photoinitiator as illustrated in equation 7 for diaryliodonium salts. As depicted, the products are an aryl iodide, an aryl radical and the cation **XIII**. Initiation of polymerization takes place either by direct attack by **XIII** on an epoxide oxygen to form an oxonium salt (eq. 8) which undergoes subsequent cationic chain propagation (eq. 9) or by interaction of **XIII** with trace amounts of water and other hydroxylic impurities to generate protonic acids which then initiate polymerization. Direct reduction of the onium salt by solvated electrons with subsequent initiation of polymerization by the protonic acid formed may also occur and this is shown in equations 10 and 11 of Scheme 2.

Although monomer **III** also bears epoxycyclohexyl groups, work in this laboratory [31] has demonstrated that the reactivity of this monomer in photoinduced cationic polymerization is markedly reduced due to side reactions involving the ester group. For this reason, **III** is much less reactive than monomer **VI**. This was also confirmed by the results of the previous study of the photoinitiated polymerization of these two monomers shown in Table 1.

In the case of diaryliodonium salts, the mechanism depicted in Scheme 2 can constitute a chain induced decomposition of the onium salt resulting the amplification of the effects of electron-beam or the γ-ray irradiation. The propagation steps of this free-radical chain induced amplification are shown in equations 12 and 13.

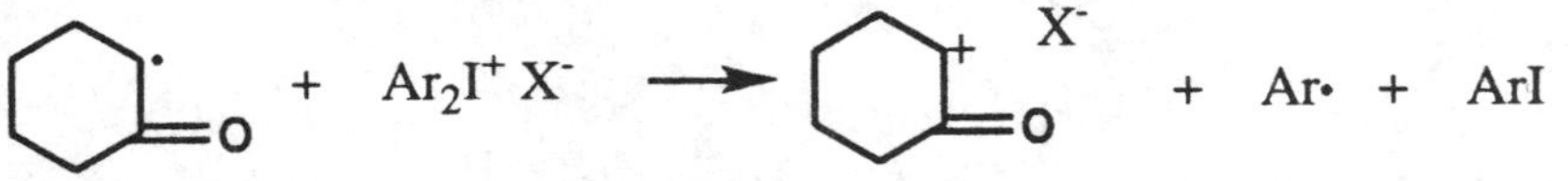

$$\text{(cyclohexyl radical)=O} + Ar_2I^+ X^- \longrightarrow \text{(cyclohexyl cation)=O} \; X^- + Ar\cdot + ArI \quad \text{eq. (12)}$$

$$\text{(cyclohexyl)-O} + Ar\cdot \longrightarrow \text{(cyclohexyl radical)-O} + ArH \quad \text{eq. (13)}$$

There is considerable literature precedent for many portions of the above reaction mechanism. In 1978, Ledwith and Yagci reported the induced decomposition of diaryliodonium salts by either photo- or thermally generated free radicals [32, 33]. As noted in equations 14 and 15 of Scheme 3, this process involves a redox reaction in which the diaryliodonium salt is reduced while the dialkoxyphenylmethyl radical is oxidized. The diaryliodine free radical generated in equation 15 undergoes irreversible decomposition to form an aryl radical and an aryliodide (eq. 16).

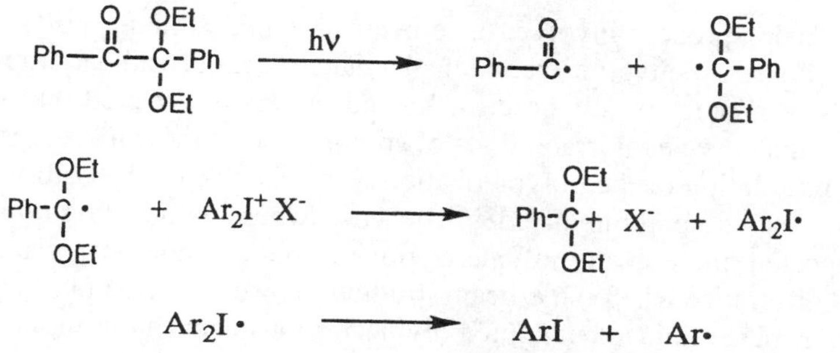

$$\text{Ph}-\underset{\text{O}}{\overset{\text{O}}{\text{C}}}-\underset{\text{OEt}}{\overset{\text{OEt}}{\text{C}}}-\text{Ph} \xrightarrow{h\nu} \text{Ph}-\overset{\text{O}}{\text{C}}\cdot + \cdot\underset{\text{OEt}}{\overset{\text{OEt}}{\text{C}}}-\text{Ph} \quad \text{eq. (14)}$$

$$\text{Ph}-\underset{\text{OEt}}{\overset{\text{OEt}}{\text{C}}}\cdot + \text{Ar}_2\text{I}^+\text{X}^- \longrightarrow \text{Ph}-\underset{\text{OEt}}{\overset{\text{OEt}}{\text{C}}}+ \text{X}^- + \text{Ar}_2\text{I}\cdot \quad \text{eq. (15)}$$

$$\text{Ar}_2\text{I}\cdot \longrightarrow \text{ArI} + \text{Ar}\cdot \quad \text{eq. (16)}$$

Scheme 3.

Similarly, Bi and Neckers [34] have reported that certain photoexcited dyes can abstract protons from N,N-dimethylaniline to produce radicals which on oxidation by a diaryliodonium salt generate cations that can initiate the cationic ring-opening polymerization of epoxides. Lastly, recent work in this laboratory has uncovered strong evidence for the free radical induced decomposition of diaryliodonium salts and we have used this principle to design a series of very highly reactive epoxy monomers [35, 36].

Since a key process in the mechanism of an e-beam initiated cationic polymerization involves the reduction of an onium salt by carbon centered free radicals **X**, **XI** or **XII**, the free energy of the process can be described by equation 17 in which E_r^{ox} and E_o^{red} are respectively the oxidation potential of the radical and the reduction potential of the onium salt [37].

$$\Delta G = E_r^{ox} - E_o^{red} \quad \text{eq. (17)}$$

For this reaction to occur, the reduction potential of the radical should be lower than that for the onium salt. Further, the magnitude of the value of ΔG should be at least -30 to -40 kJmole^{-1} or higher for the electron transfer to occur with any appreciable efficiency. We estimate the oxidation potential of radicals **X**, **XI** or **XII** to be of the order of -120-130 kJ mole^{-1}. Since the reduction potential of **IOC8** is -14 kJ mole^{-1} and that of **SS** is approximately -89 kJ mole^{-1}, diaryliodonium salt photoinitiators are more reactive than triarylsulfonium salts in e-beam induced cationic polymerizations because they are more easily reduced by free radicals generated by e-beam irradiation of the epoxide substrate. Further, due to their higher reduction potentials, triarylsulfonium salts do not undergo the free-radical induced chain decomposition shown in equations 12 and 13 while the corresponding diaryliodonium salts do. Ledwith [38] obtained a quantum yield of 3 for the photolysis of certain diaryliodonium salts in cyclohexene oxide. This is further suggestive of the free radical induced catalytic cycle shown in equations 12 and 13.

Since the major process involved in e-beam initiated cationic polymerization has been proposed to involve free-radical chain reaction, radical traps should markedly slow or interrupt this process. Indeed, it was discovered that when the onium salt mediated e-beam irradiations of silicone epoxy monomers were carried out in air, a retardation of the polymerization rate was observed. Inhibition of polymerization was more pronounced in the case where thin film samples were used. As expected, the e-beam polymerizations of bulk or composite samples exhibited little difference when the e-beam irradiations were carried out in air or under nitrogen. E-beam induced cationic polymerizations carried out in the presence of the radical trap, nitrobenzene, were also retarded. In accord with the mechanism of cationic epoxide ring-opening polymerizations, all e-beam and γ-ray polymerizations were inhibited by amine or other basic substances.

3.4. Fabrication, e-beam cure and mechanical testing of carbon fiber reinforced composites

The rapid and efficient e-beam and γ-ray initiated cationic polymerization of silicone-epoxide monomers in the presence of onium salt photoinitiators is ideal for many of the applications which were mentioned at the outset of this article. Herein we will specifically discuss the application of this technology to high performance composites.

At the present time, one of the main impediments to the use of composites for both general as well as high performance applications is their high cost. The high cost of composites is directly related to the current methods which are employed in composite fabrication. These include: complex hand lay-up and preforming processes, long solvent devolatilization and long, complex thermal curing cycles involving high temperatures. The high cost of tooling is also a major contributor to the overall cost. There is also a high reject rate due to imperfections introduced during the complex fabrication process. Clearly, there is a need for simpler, alternative techniques for the fabrication and cure of high performance structural composites.

The use of e-beam radiation to carry out rapid, pollution free, low energy crosslinking polymerizations (curing) make it a highly attractive technique for replacing the complex, time-consuming thermal cures for composite fabrication [39]. "E-beam curing" as it is called has been applied chiefly to multifunctional monomers polymerizable by free radical mechanisms, i.e. monomers such as unsaturated polyesters, and multifunctional epoxy acrylates and methacrylates. Although these monomers have been employed for the fabrication of composites [40–45], the thermal and mechanical properties of such composites are notably deficient. In this laboratory, we have been exploring the use of cationic e-beam curing for the fabrication of fiber reinforced silicone-epoxy composites [46]. To accomplish this task, a source high energy electrons with high penetrating power was required. We have employed the 10 MeV e-beam accelerator, operated by the Atomic Energy Commission of Canada Limited, located in Pinawa, Manatoba,

Canada. 10 MeV electrons are sufficient to penetrate and cure heavily fiber reinforced composites of several inches thickness. Light metals such as aluminum are virtually transparent to the e-beam irradiation and were used to fabricate sample molds.

Initial e-beam polymerization studies were carried out on pure bulk monomers. The properties of the resulting polymers were determined using dynamic mechanical testing (DMA) techniques in a torsion mode. Identical modulus profiles were obtained for polymers derived from silicone-epoxy monomer **VII** containing 0.125 to 2.0 mol% **IOC-8** or **SS** at 8.0 Mrad which is much higher than the threshold dose (Figure 2) for both onium salt initiators. Typical DMA curves for **VII** polymerized in the presence of 0.5 mol% **IOC-8** are shown in Figure 3. The room temperature dynamic modulus of e-beam cured polymers obtained for **VII** was 1.2 GPa and the modulus showed little change as the sample was heated to 320°C. The Tan δ peak that is observed at about 80 °C in Figure 3 may not be the actual T_g and could be indicative of a sub-T_g transition due to localized deformations within the crosslinked matrix. In any case, the value of the Tan δ at the peak is about 0.05 which is characteristic of a very highly crosslinked polymer.

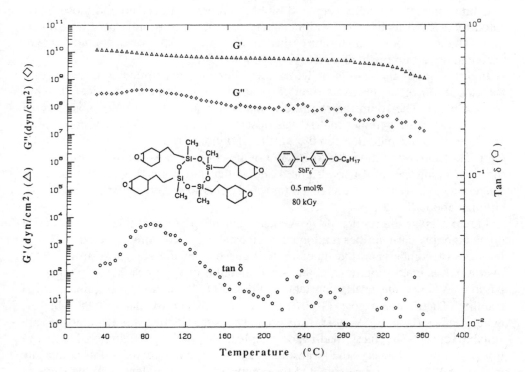

Figure 3. Mechanical properties of e-beam cured monomer **VIII**. Polymerization carried out using 0.5 mol% IOC-8 at a dose of 8.0 Mrad.

Employing a variety of different monomers, eight-ply high modulus graphite fiber reinforced composites were fabricated in a 0°,90° orientation using conventional prepreg and lay-up (vacuum bag) methods and by vacuum assisted resin transfer molding (VRTM) techniques and cured using 10 MeV e-beam irradiation. VRTM samples were fabricated using a specially designed aluminum mold described in the experimental portion of this paper. After some preliminary investigations, it was determined that an optimum dose required to cure the resin transfer molded composites inside the aluminum mold was 7-8 Mrad. Table 4 lists the epoxy monomers and the fabrication parameters which were employed and gives the results of the mechanical testing which were performed on the resulting composites.

These studies definitively demonstrate that graphite fiber reinforced high performance epoxy composites may be readily fabricated and efficiently cured using e-beam irradiation. Mechanical properties were measured on the composites directly after irradiation without a thermal postcure. All mechanical properties reported in Table 4 are a statistical average of seven samples. A similar composite fabricated with carbon fibers, but using a bisphenol-A diglycidyl ether resin (Tactix 123) and thermally cured with an aromatic amine hardener (Dow Hardener H41) according to the manufacturer's recommendations served as a control with which to compare the mechanical properties of the e-beam cured composites.

Composites were readily prepared by both prepreg and lay-up and those fabricated by VRTM techniques exhibited very low porosity by microscopic evaluation. The best room temperature mechanical properties were obtained using prepreg and lay-up methods employing monomer **VII**.

Figure 4 gives the results of DMA measurements performed in a three point flex mode on an eight-ply carbon fiber reinforced composite fabricated using monomer **VI**. The room temperature dynamic modulus for this composite is 16 GPa. A T_g for the polymer matrix was observed at about 160 °C. The room temperature dynamic modulus for the unfilled polymer from monomer **VI** was 0.7 GPa. The increase in modulus to 16 GPa in the composite is due to the stiffness of the reinforcing carbon fibers. Table 4 shows that the static flexural modulus for this composite was 29 GPa. The composite begins to undergo thermal decomposition at about 280 °C.

Figure 5 gives the results of DMA measurements performed in a torsional mode on an eight ply carbon fiber reinforced composite prepared from monomer **VII**. For the sake of comparison, the modulus (G') curve B of the control resin has been overlaid. It is interesting to note, that shear modulus of the e-beam cured composite based on **VII** remains nearly constant well past 300 °C indicating very high thermal stability. The damping peak (Tan δ) has a value of approximately 0.08 which is again characteristic of a very stiff material. This suggests that these composites are suitable for such structural, load-bearing applications as I-beams. As noted earlier in this paper, there is a very small transition seen in the DMA curve for this composite at about 80 °C. The high stiffness of the composite is also evident from the value of the static flexural modulus; 42 GPa (6.16×10^6 psi).

Table 4.
Composite Test Panel[a] Fabrication Parameters and Static Mechanical Property Test Results

Composite Resin Matrix	Initiator[b]	Fabrication Method	Dose[c] (KGy)	Flexural		Tensile		In-plane Shear Strength
				Strength psi(std-dev)	Modulus psi(std-dev)	Strength psi(std-dev)	Modulus psi(std-dev)	psi(std-dev)
Control Tactix 123/H31	—	RTM	1h @ 80°C 2h @ 150°C[e]	103175 (±2635)	6.22×10^6 (±2.6×10^7)	66053 (±4600)	9.15×10^6 (±2.4×10^6)	4962 (±168)
VI	SS	RTM	177.5	29859 (±5331)	4.50×10^6 (±3.8×10^7)	50405 (±1338)	6.26×10^6 (±1.06×10^6)	2430 (±217)
VI	IOC8	RTM[d]	76.5	49038 (±4067)	4.27×10^6 (±3.3×10^7)	—	—	—
VII	SS	Hot melt prepreg	81.6	52404 (±4425)	6.16×10^6 (±3.9×10^7)	72215 (±2114)	9.84×10^6 (±2.7×10^6)	1789 (±424)
VII[d]	SS	Hot melt prepreg	81.6	49241 (±4257)	5.88×10^6 (±4.7×10^7)	63350 (±2912)	9.39×10^6 (±2.5×10^6)	1848 (±339)

[a]Fiber volume for all test panels was approximately 50%; all test panels were 8 plies except Ib which was 20 plies. [b]Initiator conc. 1.6%; SS used as 50% solution in propylene carbonate. [c]Cured using AECL 1-10/1 1kW E-Beam accelerator at 10 MeV. [d]Monomer VII heated for 8 hr at 110°C to increase viscosity. [e]Thermal cure as recommended by supplier.

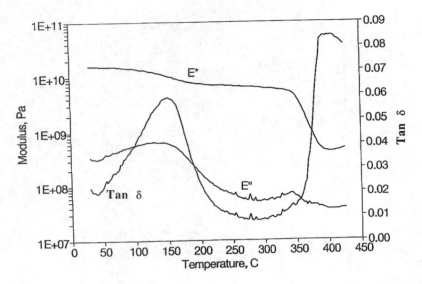

Figure 4. DMA Study of an 8 ply carbon fiber-reincorced composite prepared from monomer **VI** using VRTM.

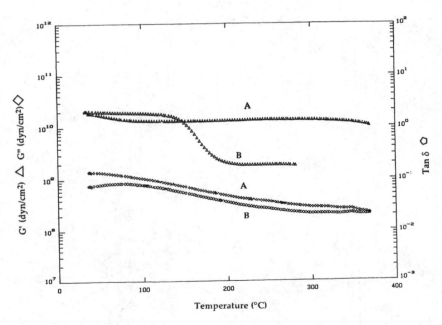

Figure 5. DMA Study of an 8 ply carbon fiber-reincorced composite prepared from tetrafunctional monomer **VII** using VRTM.

4. CONCLUSIONS

E-beam induced cationic ring-opening polymerizations of epoxides proceed efficiently in the presence of onium salt photoinitiators bearing the SbF_6^- anion. Particularly reactive are silicone-epoxy monomers containing epoxycyclohexane rings. These monomers require low doses and produce crosslinked polymers with high T_g values. Confirmation of the feasibility of using low dose e-beam radiation to cure fiber reinforced epoxy-functional silicone monomers to rapidly and efficiently fabricate carbon fiber reinforced composites has been demonstrated. Initial studies have shown that these unoptimized epoxy-based composites display better mechanical and thermal properties than those obtained by the e-beam curing of vinyl monomers. Using such epoxy monomers, fully formed vacuum bag or VRTM fabricated composites can be e-beam cured in a time scale of several seconds. We anticipate that many additional applications will be found for e-beam initiated cationic polymerizations.

Acknowledgment

The authors would like to acknowledge financial support from the National Aeronautics and Space Administration many helpful discussions from John W. Connell of NASA Langley.

REFERENCES

1. J. V. Crivello, *Ring-Opening Polymerization*, D. J. Brunelle, editor, Hanser Pub., Munich, 1993, p. 157.
2. J. V. Crivello and J. L. Lee, *Polym. J.*, **17**(1), 73 (1985).
3. J. V. Crivello, *Makromol Chem, Macromol Symp*, **13/14**, 145 (1988).
4. R. J. DeVoe, M. R. V. Sahyun, N. Serpone, D. K. Sharma, *Can J Chem*., **65**, 2342 (1987).
5. J. L. Dektar and N. P Hacker, *J Org Chem.*, **55**, 639 (1990).
6. J. L. Dektar and N. P Hacker, *J Org Chem.*, **56**, 1838 (1991).
7. *Applied Radiation Chemistry: Radiation Processing*, By R. J. Wood and A. K. Pikaev, pp. 272–340, Wiley-Interscience, New York, 1994.
8. *Radiation Processing of Polymers*, edited by A. Singh and J. Silverman, pp. 187–203, Carl Hanser Verlag, New York, 1992.
9. S. Penczek, J. Wieteszka and P. Kubisa, *Makromol. Chem.*, **97**, 225 (1966).
10. S. Okamura and K. Hayashi, *Makromol. Chem.*, **47**, 230 (1961).
11. L. W. Dickson and A. Singh, *Radiat. Phys. Chem.*, **31**(4–6), 587 (1988).
12. J. V. Crivello, J. L. Lee, *J. Polym. Sci. Polym. Chem. Ed.*, **28**, 479 (1990).
13. J. V. Crivello, J. L. Lee, ACS Symp. Ser., 417, edited by C. E. Hoyle and J. F. Kinstle, 1990, p. 398.
14. J. V. Crivello and J. L. Lee, *J. Polym. Sci. Part A: Polym. Chem.*, **27**, 3951 (1989).
15. S. R. Akhtar, J. V. Crivello and J. L. Lee, *J. Organic Chemistry*, **55**, 4222, (1990).
16. Unpublished work from this laboratory.
17. A. A. Kozlov, V. N. Doroshenko and A. P. Meleshevich, *Vysokomol. Soyed.*, **A25**, 1505 (1983).
18. A. P. Maleshevich, A. A. Kozlov and V. N. Doroshenko, *Teor. Ehksp. Khim.*, **20**, 492 (1984).
19. R. S. Davidson and S. A. Wilkinson, *J. Photochem. Photobiol. A: Chem.*, **58**, 123 (1991).
20. J. V. Crivello and V. Linzer, *Polimery*, **68**(11/12), 661 (1998).
21. J. V. Crivello and J. L. Lee, *J. Polym. Sci., Polym. Chem. Ed.*, **28**, 479 (1990).

22. J. V. Crivello, D. Bi and M. Fan, *A.C.S. Polymer Preprints*, **32**(3), 173 (1991).
23. J. V. Crivello and J. H. W. Lam, *A.C.S. Org. Ctgs. and Plastics Preprints*, **39**, 31 (1978).
24. J. V. Crivello, M. Fan, D. Bi, *J. Appl. Polym. Sci.*, **44**, 9–16 (1992).
25. J. V. Crivello, *A.C.S. Polym. Mtls. Sci. Eng. Preprints*, **65**, 319, (1991).
26. J. V. Crivello, M. Fan and D. Bi, *Proc. of the RadTech '92 North America Conference*, Boston, MA, April 26-30, 535, (1992).
27. Kopchonov, Grozdov, Krajzman, Fakin, Ogonjdov, Shik, Sidorenko, *Proc. RadTech '90 North America Conference*, Chicago, IL, March 25-29, **2**, 11, (1990).
28. J. V. Crivello, U. S. Patent 5,260,349, Nov. 9 1993.
29. R. J. Gritter and T. J. Wallace, *J. Org. Chem.*, **26**, 283 (1961).
30. M. P. Kozina, L. P. Tjimofeeva, V. A. Luk'yanova, S. M. Pimenova and L. I. Kas'yan, *Russ. J. of Phys. Chem.*, **62**(5), 609 (1988).
31. J. V. Crivello and U. Varlemann, *J. Polym. Sci., Polym. Chem. Ed.*, **33**(14), 2463 (1995).
32. A. Ledwith, *Polymer*, **19**, 1217 (1978).
33. F. A. M. Abdoul-Rasoul, A. Ledwith and Y. Yagci, *Polymer*, **19**, 1219 (1978).
34. Y. Bi and D. C. Neckers, *Macromolecules*, **27**, 3633 (1994).
35. J. V. Crivello and S. S. Liu, *J. Polym. Sci., Polym. Chem. Ed.*, 37(8), 1199 (1999).
36. J. V. Crivello and S. Liu, *Chemistry of Materials*, **10**(11), 3726 (1998).
37. B. Bednar, J. Devaty, J. Kralicek and J. Zachoval, ACS Symp. Ser., 242, 1984, T. Davidson, Editor, p. 201.
38. A. Ledwith, *A.C.S. Polymer Preprints*, **23**(1), 323 (1983).
39. J. R. Seidel, *Radiation Curing of Polymers*, edited by D. R. Randell, The Royal Society of Chemistry, London, 1987, p. 12.
40. C. B. Saunders, A. Singh, V. J. Lopata, G. D. Boyer, W. Kremers and V. A. Mason, *A.C.S. Polymer Preprints*, **31**(2), 325 (1990).
41. C. B. Saunders, A. Singh, V. J. Lopata, G. D. Boyer, W. Kremers and V. A. Mason, *A.C.S. Polymer Preprints*, **31**(2), 325 (1990).
42. C. B. Saunders, L. W. Dickson, A. Singh, A. A. Carmichael and V. J. Lopata, *Polymer Composites*, **9**(6), 389 (1988).
43. C. B. Saunders, A. A. Carmichael, W. Kremers, V. Lopata and A. Singh, *Polymer Composites*, **12**(23), 91 (1991).
44. C. B. Saunders, V. J. Lopata, W. Kremers, M. Tateishi and A. Singh, *Abstr. 37th Int'nl. SAMPE Symp.*, March 9-12, 1992, p. 944.
45. D. Beziers and B. Capdepuy, *Abstr. 35th Int'nl. SAMPE Symp.*, Apr. 2-5, 1990, p. 1220.
46. J. V. Crivello, R. Malik and T. C. Walton, *A.C.S. Polym. Preprints*, **35**(2), 890 (1994).

Complexation behavior of diazosulfonate polymers

KURT E. GECKELER,[1] OSKAR NUYKEN,[2] UTE SCHNÖLLER,[2]
ANDREAS THÜNEMANN[3] and BRIGITTE VOIT[4],*

[1]*Institute of Organic Chemistry, University of Tübingen, Auf der Morgenstelle 18, D-72076 Tübingen, Germany*
[2]*Lehrstuhl für Makromolekulare Stoffe, Technische Universität München, D-85747 Garching, Germany*
[3]*Max Planck Institute of Colloids and Interfaces, Am Mühlenberg, 14476 Golm, Germany*
[4]*Institute of Polymer Research Dresden, Hohe Strasse 6, D-01069 Dresden, Germany*

Abstract—The complexation behavior of water soluble, photo labile diazosulfonate homo- and copolymers has been studied. These polymers exhibit pK_a values of 2.0 to 2.5 which are comparable to that of poly(styrene sulfonate). Furthermore, the diazosulfonate group is able to complex metal ions with a preference for Cr(III), Fe(II), and Pb(II). Polyelectrolyte – surfactant complexes can be formed between hydrocarbon and fluorocarbon based surfactants and diazosulfonate polymers. These complexes exhibit high structural order which can be maintained and even stabilized by crosslinking after photo initiated decomposition of the diazosulfonate chromophore. Surface segregation of the fluorocarbon surfactants in the polydiazosulfonate – surfactant complexes leads to low surface energies.

Keywords: Diazosulfonate; polyelectrolyte; photo active; complexation; metal ions.

1. INTRODUCTION

Aryldiazosulfonate chromophores are known for a long time and well studied [1, 2]. These charged, water soluble azo compounds decompose upon UV irradiation with the first step being the isomerization from the stable trans-isomer of the azo group to the labile cis-form [3]. In water, the cis-isomer decomposes into the corresponding diazonium salt which can be trapped and used in dye coupling reactions. In contrast to diazonium salts, however, the diazosulfonates are thermally stable up to 200°C. Thus, low molar mass sodium or potassium diazosulfonates were applied in textile printing and reproduction techniques already in 1930's [4, 5]. However, the incorporation of these photo labile units into polymers by copolymerization of suitable monomers or by polymer analogeous reaction allowed their application as photo resins e.g. for the preparation of offset printing plates [6–9] (especially **P3**, compare Scheme 1). Irradiation of diazosulfonate copoly-

---

*To whom correspondence should be addressed. E-mail: voit@ipfdd.de

mers films leads to crosslinking under loss of nitrogen due to the formation of radicals during the decomposition process. Thus, the water soluble polymers can be converted into insoluble materials and photo imaging is possible by selective irradiation using a mask.

A wide variety of diazosulfonate containing polymers has been synthesized in the last years with the intention to optimize these polymers regarding film forming properties and especially UV sensitivity [7–11]. For this, special diazosulfonate monomers were designed, different ways for polymer preparation were explored, and the material properties of the resulting polymers were studied in detail. However, the fact that these polymers are not only photo resins but photo sensitive polyelectrolytes, too, is being explored only recently. Clearly, the sulfonate group along the polymer chain leads to a highly charged polyanion which is expected to form complexes e.g. with metal ions or with organic oppositely charged units.

Whereas the potential of coupling reactions of the polymeric diazonium ion intermediates which form upon UV-irradiation in water of polymeric diazosulfonates as well as diazosulfonate surfactants have been discussed previously [12, 13], we will now focus on the polyelectrolyte behavior and the complexation properties of the diazosulfonate polymers.

2. EXPERIMENTAL PART

The synthesis of different diazosulfonate homo- and copolymers shown in Scheme 1 is described in [9–11, 14, 15]. All polymers were prepared by free radical polymerization of 4-acryloylaminophenyl diazosulfonate **M1** or 4-methacryloylaminophenyl diazosufonate **M2** (leading to **P1** and **P2**) and their copolymerization with azo free comonomers (**P3** – **P7**). The polymers were purified carefully by reprecipitation or by dialysis in water over several days. The molar masses were determined for the copolymers (azo monomer 40 mol%) with methyl methacrylate (MMA), methyl acrylate and acrylic acid as well as for the terpolymer **P7** (azo monomer 20 mol%) with MMA and hydroxyisopropylmethacrylate by GPC (in water or DMAc/H_2O/LiCl) and varied between M_n = 7,500 to M_n = 25,000 g/mol (M_w/M_n between 2 and 3) with the highest molar masses obtained for copolymer **P3** (M_n = 25,000 g/mol) containing 40 mol% **M2** and **P7** (M_n = 16500 g/mol). Molar mass determination of the homopolymers **P1** and **P2** was difficult due to their very high polarity. GPC analysis indicated molar masses M_n around 12,000 g/mol. For comparison sodium poly(styrene sulfonate) (**P8**) has been synthesized by free radical polymerization of 4-vinylphenylsulfonic acid sodium salt **M3** in water using a standard procedure [16]. **P8** with a molar mass M_w = 110,000 g/mol (determined by solution viscosity measurements) was used for the metal ion complexation experiments. In polyelectrolyte - surfactant complexes a lower molar mass **P8** (M_n = 33,000 g/mol) was applied. Details on the formation and characterization of polyelectrolyte - surfactant complexes as shown in Scheme 2 can be found in [14, 15].

Metal ion complexation experiments were carried out with the diazosulfonate homopolymer **P1** and **P8**. For these tests a standard procedure was applied [17, 18]. The equipment consists of an ultrafiltration cell, a polysulfone membrane (molar mass exclusion 10,000 g/mol), a reservoir, a selector, and a regulator. Nitrogen served as a pressure source and the pressure was kept constant at 300 kPa during the studies. An aqueous solution of a mixture of metal cations (nitrates or chlorides, 20 mg /L of each metal ion: Cr^{3+}, Fe^{3+}, Co^{2+}, Ni^{2+}, Zn^{2+}, Sr^{2+}, Cd^{2+}, Pb^{2+}) was placed in the cell containing the polymer solution (0.5 wt%) with a total volume of 20 mL. The system was pressurized and the cell was washed with the reservoir solution. Metal ion concentrations in the retentate and in the permeate were determined by atomic absorption spectrometry. The retention R is expressed as $R = \exp.(V_p \bullet V_0)$ with V_p = volume of the permeate and V_0 = volume of the cell solution (retentate). The filtration factor Z is defined as $Z = V_p/V_0$. Z values of 10, which means the cell is flushed with 10 times its volume with pure buffer solution, is usually sufficient to decide whether an ion is complexed strongly by the polymer. The measurements were carried out at pH 7 and pH 3.

The dissociation behavior of 0.08 molar solution of **P1**, **P8**, and of the corresponding monomers **M1** and 4-vinylphenylsulfonic acid sodium salt **M3** was studied by potentiometric titrations using 0.1N NaOH and a computer controlled titration system TPC 2000 with a pH combination electrode of type N 1042A (Schott Gerätebau GmbH) at 22°C. Conductivity in mS and the pH were recorded simultaneously versus the amount of 0.1 N NaOH. The pK_a values were calculated as described elsewhere [20, 21]. Before the measurements aqueous solutions of the polymeric salts and the monomers were converted in the acid form using a cationic ion exchange resin. For comparison, the initiator 4,4'-azobiscyanovaleric acid (ABCVA), which was used in the free radical polymerizations of **P1** and **P8**, was included in the potentiometric studies.

3. RESULTS AND DISCUSSION

The complexing behavior of the diazosulfonate homopolymers **P1** and **P2** and different copolymers **P3** – **P7** (compare Scheme 1) has been studied in metal complexation experiments, potentiometric titrations or in the formation of polyelectrolyte – surfactant complexes.

For comparison, sodium poly(styrene sulfonate) **P8** has been included in these studies, not only because of the similarity of the charged units compared to poly(diazosulfonate)s but also because much is known on its dissociation behavior and its complexation ability towards metal ions e.g. in ion exchange resins [21, 22] (mostly prepared by sulfonation of polystyrene). **P8** belongs to the strongly dissociated polyelectrolytes and complexes alkali metal ions as well as heavy metals.

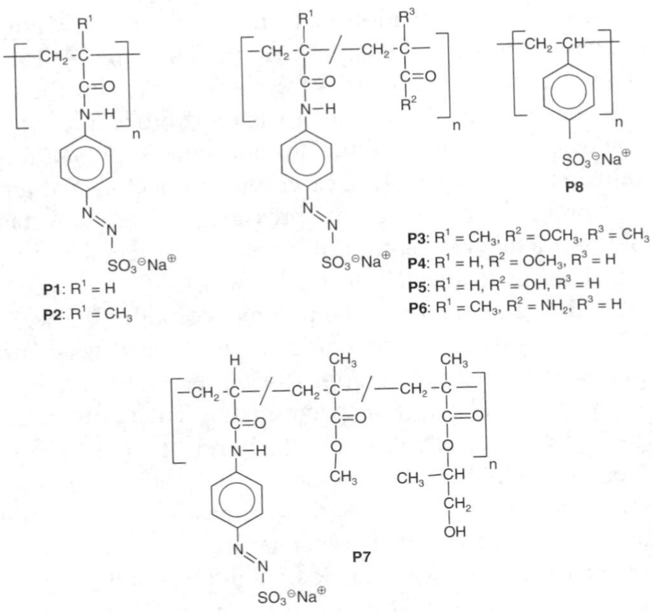

Scheme 1.

3.1. Metal complexation

In the metal complexation experiments only the homopolymers **P1** and **P8** were used to avoid any interference of comonomers with the complexation behavior of the sulfonate and the diazosulfonate group. A standard solution of a mixture of different metal salts was added to the polyelectrolyte solution which was placed in an ultrafiltration cell. A polymeric membrane kept the polymer in the cell, whereas free metal ions could pass. The retention behavior of the polymers was evaluated by measuring the metal ion concentration in the permeate with increasing dilution of the retentate in the cell. Retention R of 100% means all metal ions are complexed by the polyelectrolyte (R = exp. ($V_p \bullet V_0$) with V_p = volume of the permeate and V_0 = volume of the cell solution = retentate). The filtration factor Z is defined as $Z = V_p/V_0$ and determines the strength of the complex. High R-values at $Z=10$ is equivalent to a very strong binding tendency of the metal ion to the polymer. In our case 8 different metal ions were tested (see Table 1 and 2) with copper ions being excluded due to the known instability of diazosulfonates towards them. In Figure 1 three representative R/Z plots of the diazosulfonate polymer **P1** are shown for Cr(III), Fe(III) and Co(II). The retention behavior was measured at two different pH (pH=3, pH=7), which corresponds to the two different series of data points. Lower pH can not be applied since the diazosulfonic acid is not stable at room temperature. The results for the complexation of other metal ions (Ni (II), Cu (II), Zn (II), Sr (II), Cd (II), Pb (II)) by **P1** are summarized in Table 1. The corresponding results for **P8** can be found in Table 2.

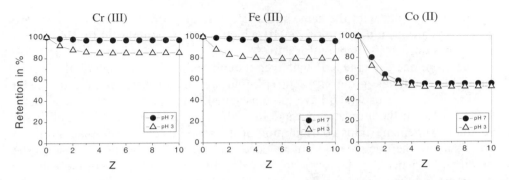

Figure 1. Retention R values in % versus filtration factor Z (V_p/V_0) for Cr(III), Fe(III) and Co(II) ions in the presence of **P1** at pH = 3 (Δ) and pH = 7 ($\bullet$) at room temperature.

Table 1.
Retention values R (in %) of diazosulfonate polymer **P1** for a series of metal ions in aqueous solution (pH3 and pH7) at filtration factor Z =1 and 10 (filtration factor $Z = V_p/V_0$, V_p = volume of permeate, V_0 = volume of retentate)

Z	Cr (III)	Fe (III)	Co (II)	Ni (II)	Zn (II)	Sr (II)	Cd (II)	Pb (II)
1 (pH=3)	92	88	80	58	52	64	68	76
10 (pH=3)	85	79	52	23	8	20	50	65
1 (pH=7)	98	99	72	77	78	76	74	95
10 (pH=7)	97	96	55	38	38	45	38	72

Table 2.
Retention values R (in %) of poly(styrene sulfonate) **P8** for a series of metal ions in aqueous solution (pH3 and pH7) at filtration factor Z =1 and 10 (filtration factor $Z = V_p/V_0$, V_p = volume of permeate, V_0 = volume of retentate)

Z	Cr (III)	Fe (III)	Co (II)	Ni (II)	Zn (II)	Sr (II)	Cd (II)	Pb (II)
1 (pH=3)	100	95	80	82	78	78	80	82
10 (pH=3)	100	92	70	75	72	68	70	75
1 (pH=7)	100	99	90	97	99	87	90	100
10 (pH=7)	100	99	88	90	99	80	86	100

In Figure 1 one can observe that at pH = 7 the retention of **P1** for Cr (III) and Fe (III) is very good, whereas the retention decreases slightly at pH = 3. This is a known effect which results from the increasing neutralization of the sulfonate group. The binding ability of **P1** towards Co (II) is lower compared to Fe (III) and Cr (II) and the effect of pH is less pronounced. Comparing the results for all metal ions (Table 1), one can conclude that the highest binding ability is found for metal ions of valence 3 (Fe (III), Cr (III)), and in between the metal ions of valence 2, Pb (II) and Co (II) gave the best results. **P8**, however, shows higher binding ability in all cases but with the same slight decrease at pH = 3. The differences are especially strong in the case of Zn (II) which is 100% complexed by **P8** (pH=7) and shows only 38% retention for **P1**.

Therefore one can conclude that even though both polymers contain the identical sodium sulfonate group in each repeating unit, major differences in the metal binding ability occur which have to be assigned to the presence of the azo function in between the aromatic unit and the sulfonate group in **P1**. It looks like the spatial arrangement (trans configuration of the azo unit) of the sulfonate group in **P1** disfavors the complexation of valence 2 metal ions but does not disturb the binding of valence 3 metal ions. One can also discuss an influence of the polymer backbone (styrene based for **P8**, acrylate based for **P1**) but this effect should be very small. The molar masses of the polyelectrolytes, however, might have a significant influence. A rather high molar mass **P8** was used for the complexation experiment (M_w = 110,000 g/mol). Unfortunately, it is not possible to obtain diazosulfonate homopolymers in a similar high molar mass due to a different behavior in the free radical polymerization. In addition, we were not able to obtain any reliable molar mass values for **P1**, since GPC and viscosity measurements do not allow determination of absolute molar mass values for these new polymers.

3.2. Potentiometric titration

The dissociation behavior of polymers **P1** and **P8** as well the corresponding monomers was studied by potentiometric titration which allows the determination of the pK_a values of the sulfonate groups. The results are summarized in Table 3, and in Figure 2 the change in conductivity and pH versus the added base (NaOH) are shown for **P1**.

The two monomers **M1** and **M3** show the characteristically shaped curves for strong acid – strong base titrations of monotropic acids. The observed pK_a values are identical for the monomers (pK_a = 2.3), however, slight differences can be observed in the polymers with **P1** having a pK_a of 2.5 and **P8** of 2.0. Still, both values are reasonable close to conclude that no major differences exist in the dissociation behavior of the two different sulfonic acid groups by titration with NaOH whereas for the complexation of metal ions with a certain spatial need the spatial arrangement of the sulfonate group had a strong influence. The reason for the observed small deviations in pK_a might be the difference in molar mass and in the flexibility of the polymer backbone of the two polyelectrolytes. This is supported by the close evaluation of the potentiometric titration of the diazosulfonate polymer **P1**, which revealed a second dissociation step corresponding to a pK_a of 4.4 (Figure 2). Determination of the pK_a of 4,4'-azobiscyanovaleric acid (ABCVA) and measurements on the mixture of **P1** and this initiator allowed to assign the second dissociation step to the carboxylic end groups of the polymer which result from the initiator used in the free radical polymerization. The determination of end groups by titration is a strong indication that **P1** is a rather low molar mass product compared to **P8**.

Table 3.
pK$_a$ values of diazosulfate and sulfonate monomers and homopolymers (determined by potentiometric titrations with 0.1N NaOH of 0.08 molar solutions, for comparison the values for the initiator are included)

	M1	P1	ABCVA (initiator)	M3	P8
pK$_a$1	2.3	2.5	–	2.3	2.0
pK$_a$2	–	4.4	4.4	–	–

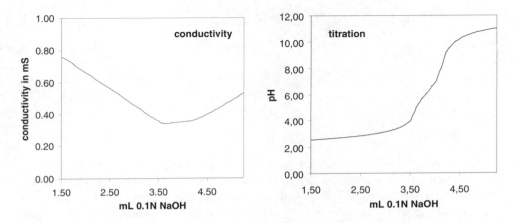

Figure 2. Changes in conductivity and pH of **P1** versus added mL 0.1N NaOH at 22°C.

The conductometric measurements for the monomers **M1** and **M3** show curves with a strong decrease followed by a strong increase as expected for strong acids. In case of the polymer **P1**, first also a strong decrease was observed in conductivity referring to the diazosulfonate groups followed by only a weak increase belonging to the carboxylic end groups of the initiator ABCVA and finally again a stronger increase which can be assigned to the remaining hydroxy ions of the NaOH titrante [21].

3.3. Polyelectrolyte surfactant complexes

An other aspect of the complexing behavior of diazosulfonate polymers is the use in polyelectrolyte – surfactant complexes. Recently we studied this using carbohydrate [14] and fluorocarbohydrate [15] surfactants and the results will be summarized here shortly. Two representative structures are shown in Scheme 2.

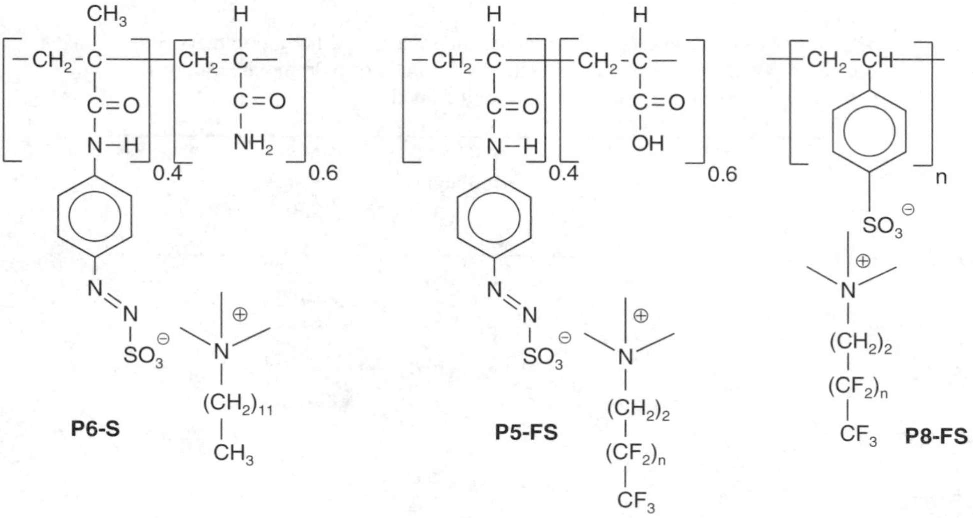

Scheme 2.

Polyelectrolyte – surfactant complexes are considered to be promising for the development of smart materials which are characterized by self-organized supramolecular structures [23, 24]. A number of polyelectrolyte – surfactants complexes and especially details on the bulk structure and the properties of these materials have been described in the last few years [23–26]. Fluorocarbon containing complexes are especially well suited for preparation of low energy surfaces [27].

The formation of polyelectrolyte – surfactant complexes has been carried out with **P2** and **P4** to **P8** under different aspects e.g. variation of the density of charged units and the flexibility of the backbone, the incorporation of additional charged units and the comparison with the azo free reference **P8**. In general, the complexation behavior of the diazosulfonate polymers did not differ from that of other strongly dissociated polyelectrolytes, meaning, the complex formation was very rapid, followed a zipper mechanism, and 1:1 complexes regarding the charged units were formed. Stable films of the complexes could be cast from organic solutions which were relatively brittle when the diazosulfonate homopolymer (**P2**) or the methacrylate derivative (**P6**) were used. In bulk the complexes exhibited the expected order, mainly lamellar structure, and this order was retained even after photolysis of the azo function [14]. The reason for this can be found in the formation of radicals upon irradiation and subsequent crosslinking reactions.

In case of the fluorohydrocarbon containing samples [15] (e.g. **P5-FS**, **P8-FS**, Scheme 2) surface segregation of the fluorinated surfactants was observed which led to low surface energies (γ_s between 11.4 and 16.9 mN/m). When copolymers with acrylic acids (e.g. **P5**) are used the charge density can be varied with the pH

(100% ionic units at pH = 9, 40% at pH = 5) and thus, the density of the packing of the fluorocarbon surfactant can be increased [15]. The fluorocarbon surfactant complex with copolymer **P5** (**P5-FS**) exhibits a surface energy γ_s of 14 mN/m at pH = 5 and of 11.4 mN/m at pH = 9. The azo free homopolymer **P8** is fully charged at pH =5 and its fluorocarbon surfactant complex **P8-FS** exhibits therefore a surface energy of 12.3 mN/m.

4. CONCLUSIONS

Diazosulfonate polymers can be considered as polyelectrolytes with a strongly dissociated sulfonate group and a similar behavior compared to sodium poly(styrene sulfonate). The density of the charged units can be varied by copolymerisation which has a significant effect on the material properties (solubility, glass transition temperature, brittleness). These photo sensitive polyelectrolytes form complexes with metal ions and with oppositely charged organic compounds. However, whereas in polyelectrolyte – surfactant complexes no major differences in the complexation behavior was observed, the binding ability towards metal ions was reduced and more selective compared to poly(styrene sulfonate). Here, the spatial arrangement of the azosulfonate groups seems to have a major influence.

Acknowledgements

We wish to thank V. Steinert and H. Hilke for their help with the titration experiments and D. Voigt for some of the GPC measurements. The financial support of the FCI and Agfa company is gratefully acknowledged.

REFERENCES

1. R. Schmitt, L. Glutz, *Ber. Dtsch. Chem. Ges.* **2**, 51 (1869).
2. R. Püttner, "Aryldiazosulfonate" in *Methoden der organischen Chemie (Houben Weyl)*, E. Müller Ed., G. Thieme Verlag, Stuttgart 1965, **Vol.10/3**, p.570ff. and references therein.
3. L. K. H. van Beek, J. Helfferich, *Rec. Trav. Chim.* **87**, 997 (1968).
4. M. S. Dinaburg, *Photosensitive Diazo compounds*, The Focal Press, London 1967, p. 18ff.
5. Ger. 560797 (1930), Ger. 560798 (1930), US Pat. 1897410 (1933), US Pat. 1909851 (1933), I. G. Farben, invs.: A. Zitscher, W. Seidenfaden; *Chem. Abstr.* **27**, 2824 and 4092 (1933).
6. US Pat. 3174860 (1965), Azoplate Corp. or Ger. 1114704 (1961), Kalle A.-G., invs.: O. Süs, K. Reiss; *Chem. Abstr.* **57**, 1792f (1962).
7. O. Nuyken, T. Knepper, B. Voit, S. D. Pask, EP 339393 A2 (2. Nov. 1989) to Bayer AG; "Polymerizable arenediazo sulfonates and their polymers".
8. P. Matusche, O. Nuyken, B. Voit, *Angew. Makromol. Chem.* **250**, 45 (1997).
9. P. Matusche, O. Nuyken, B. Voit, M. Van Damme, J. Vermeersch, W. De Winter, L. Alaerts, *Reactive Polymers* **24**, 271 (1995).
10. O. Nuyken, B. Voit, *Macromol. Chem. Phys.* **198**, 2337 (1997).
11. P. Matusche, O. Nuyken, B. Voit, M. Van Damme, *J. Macromol. Sci. A34*, 201 (1997).
12. P. Matusche, O. Nuyken, B. Voit, *Angew. Makromol. Chem.* **250**, 45 (1997).

13. O. Nuyken, K. Meindl, A. Wokaun, T. Mezger, *J. Photochem. Photobiol. A:Chem.* **81**, 45 (1994) and *ibid.* **85**, 291 (1995).
14. A. Antonietti, R. Kublickas, O. Nuyken, B. Voit, *Macromolecular Rapid Communication* **18**, 287 (1997).
15. A. F. Thünemann, U. Schnöller, O. Nuyken, and B. Voit, *Macromolecules* **32**, 7414 (1999).
16. R. A. Mock, C. A. Marshall, T. E. Slykhouse, *J. Phys. Chem.* **58**, 498 (1954).
17. K. E. Geckeler, V. M. Shkinev, B. Y. Spivakov, *Angew. Makromol. Chem.* **155**, 151 (1987).
18. K. E. Geckeler, *Polymer J.* **25**, 115 (1993).
19. S. Reinhardt, V. Steinert, K. Werner, *Eur. Polym. J.* **32**, 935 (1996).
20. "Einführung in die Theorie der quantitativen Analyse"; 7th edition E. Fluck, M. Becke-Goering, p. 201, Dr. D Steinkopff Verlag, GMbH & Co. KG; Darmstadt; 1990.
21. „Ionenaustauscher" in: Ullmanns Enzyklopädie der Technischen Chemie, 4th edition, E. Bartholomé, E. Bickert, H. Hellmann, H. Ley, M. Wickert, E. Weise (Eds.), **Vol. 13**, p.281, Verlag Chemie, Weinheim (1979).
22. "Functionalized polymers and their applications", A. Akelah und A. Moet (Eds.), Chapman & Hill Verlag, London (1990).
23. C. Ober, G. Wegner, *Adv. Mater.* **9**, 17 (1997).
24. E. D. Goddard, *Colloids Surf.* **19**, 301 (1986).
25. Y.-C. Wei, S. M. Hudson, *J. Macromol. Sci. - Rev. Macromol. Chem. Phys.* **C35**, 15 (1995) and references therein.
26. M. Antonietti, J. Conrad, *Angew. Chem. Int. Ed. Engl.* **33**, 1869 (1994).
27. A. F. Thünemann, K. H. Lochhaas, *Langmuir* **14**, 4898 (1998).

Surface modification of silica particles and glass beads

R. M. OTTENBRITE,[1] HU BIN,[1] JASON WALL[1] and J. A. SIDDIQUI[2]
[1] *Virginia Commonwealth University, Richmond, VA 23284*
[2] *Dupont Films, Hopewell, VA 23860*

Abstract—Surfaces and interfaces of particles are very important with respect to their end use. Many particles are used as fillers in polymer matrices and they usually impart properties that enhance a materials function. Therefore, the particles need to be compatible with the polymer matrix into which they are placed. To enhance the compatibility of the silica particles and silica glass beads, surface modification was carried out, such as: Surface polymerization, surface polymer grafting, and surface dendrimerization. The surface properties obtained varied, from highly hydrophilic to highly hydrophobic, from anionic and cationic to nonionic.

1. INTRODUCTION

Generally the surface properties are most important in determining the quality of the finished product [1]. Organic-inorganic composite materials are being employed more and more due to their superior properties. It is known, for example, that a solid formed from a liquid can assimilate or reject finely dispersed particles depending on the surface energy of the particles and the matrix [2]. Particle rejection from the solid leads to an inhomogeneous distribution of the particles in filler applications. Similarly, to use particles as coatings would largely determine the surface properties of the coated object. In order to obtain predictable properties from coatings and fillers, it is necessary to create well defined surfaces that can be thoroughly characterized.

Silica particles have been recognized for a variety of applications. Different types of silica, including porous silica and nonporous glass beads, have been developed and applied in environmental science, chemical processing, mineral recovery, energy production, agriculture and pharmaceutical industries. Grafting reactions [3], on porous silica [3–5] or nonporous glass beads [6] have also attracted attention and have been studied extensively. Grafting of polymers with silane end groups onto silica particles has been studied in order to improve their dispersability in organic solvents and compatibility with polymer matrices. Yoshinigawa *et al.* have synthesized polystyrenes, poly(methyl methacrylate)s, polyethylene glycols, and other polymers with terminal silane groups that are capable of reaction with surface silanol groups, thereby grafting the polymer onto the silica surface [7]. Duchet *et al.* studied the grafting of chlorosilane terminated polyethylene onto silica in hopes it would act as a connecting molecule between the silica surface and a neat polyethylene matrix [8].

Initiation of grafted polymer chains at the surface of silica has also been carried out. Suzuki et al. used covalently bound 4,4'-azobis(4-cyanovaleric acid) to initiate the polymerization of N-isopropylacrylamide at the surface of silica in an attempt to produce a thermoresponsive surface [9]. It was shown that water penetration was slower at higher temperatures for the modified poly(N-isopropylacrylamide), while there was no temperature affect for unmodified silica, indicating thermoresponsive activity. Prucker et al. reported the kinetics and mechanism of surface initiated polymerization by silane coupling agent bound azo initiators [10]. Initiator decomposition was followed by differential scanning calorimetry and volumetry. The kinetics of the polymer formation was followed by dilatometry.

Tsubokawa et al. have evaluated several methods to graft polymers onto the surface of silica particles. They have studied the cationic polymerization of several monomers at the silica surface with a benzylium perchlorate initiator [11]. The initiator system was formed from the reaction of silver perchlorate and an immobilized benzyl chloride, which was subsequently used to polymerize styrene and ε-caprolactam. They also carried out radical polymerizations at the surface with different immobilized initiators. One system was initiated by immobilized peroxides [12]. This system produced ≤70% efficiency for poly(methyl methacrylate) and ungrafted polymer was formed by free initiator fragments. The modified particles were dispersible in good solvents for poly(methyl methacrylate). Another system used an immobilized trichloroacetyl group and $Mo(CO)_6$ to form radicals to initiate the polymerization [13]. When methyl methacrylate was used as the monomer they were able to achieve 744% grafting efficiency. Tsubokawa et al. also grafted polymers to the surface of silica through bound monomers [14]. They cocondensed tetraethylorthosilicate with a variety of silane coupling agents, including methacryloyloxypropyltriethoxysilane, to produce surface functionalized silica particles. They proceeded to polymerize vinyl monomers and the polymer formed was covalently attached to the particles. Once again the particles gave stable dispersions in solvents compatible with the free polymer.

Becker et al. functionalized the surface of silica with methacryl groups in order to graft a methyl methacrylate, 2-hydroxyethyl methacrylate copolymer to the surface [15]. It was thought that the modified particles would improve the mechanical properties of the matrix at temperatures above its T_g. Beyer et al. used functionalized polymers for grafting to the surface [16]. They synthesized poly[(1-methyl vinyl isocyanate)-alt-(maleic anhydride)] and reacted it with surface immobilized amino groups. The surface obtained can be further processed to form multilayer coatings.

Surface modification of silica particles and glass beads has been studied in our group to determine the optimum grafting conditions such as time, temperature, initiator, monomer and glass beads concentration. An analysis of these grafting parameters under different reaction conditions indicated effective grafting mechanisms [17].

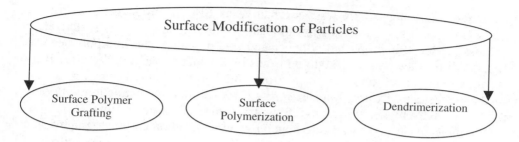

Composites containing micron-sized particles are very promising for many applications such as: chromatography, chemical sensors, ceramics and biocompatible materials. To meet the requirements of these applications the particles are generally surface modified. Due to microscopic heterogeneity of solid surfaces, it is desirable to have flexibility in constructing complex surface layers. Considering these requirements, dendritic modification of glass beads was also carried out in our laboratory [18, 19].

2. POLYMER GRAFTING OF SILICA GLASS BEADS

Polymer grafting of nonporous glass beads (2-5µm), which are used as fillers in polymer matrices to enhance polymer properties, have the potential to provide unique new materials. Compared with porous silica, which has an abundance of silanol groups on the surface as well as a larger total surface area, the glass beads that were used in this study have a very low silanol concentration (about 0.21mmol/g) and a small surface area ($\sim$2m^2/g). These surface property shortcomings, however, were overcome by surface modification via polymer grafting [17]. Potentially, a selective polymer surface layer on the glass beads can improve their compatibility and adhesion with a polymer matrix. We studied the effects of the grafting conditions such as: reaction time, temperature, initiator, monomer and the total bead surface area that can affect the graft polymerization.

There are several approaches to grafting a polymer chain onto a surface [17]. One method is free radical copolymerization of a glass surface immobilized double bond by using a single coupling agent such as methacrylpropylsiloxane or vinyl trimethoxylsilane, and another method is initiation by an immobilized free radical intiator [5]. Using these two methods, several polymers have been grafted onto silica or porous silica surface [5] while only a limited amount of data on nonporous substrates has been reported [6, 20].

P_2 glass beads were pretreated with sodium hydroxide to enhance the number of hydroxyl groups that could react with the coupling agent methacrylpropylsiloxane (MPS). The MPS modified beads were then allowed to react with acrylic acid and benzoyl peroxide (BPO) or 2,2-azobisisobutyronitrile (AIBN) to provide polymer-grafted beads. Generally, AIBN is not considered to be a suitable grafting intiator because of the resonance stabilization of the insipient radical

[21]. We found that AIBN was as effective as BPO at producing poly(acrylic acid) grafts on glass surfaces from dioxane solutions. This graft polymerization process produced both grafted polymer and free polymer. The free polymer was separated from the polymer modified beads by Soxhlet extraction. We effectively separated the free homopolymer from the grafted polymer with a 1:1 methanol/acetone mixture; Tsuzuki *et al.* [22] found that the extraction with only methanol gave a higher apparent grafting percentage than true grafting.

The molecular weight was determined by gel permeation chromatography on a μ-Styragel 10^3 column with water as the mobile phase and poly(styrene sulfonate) was used as the calibration standard. There are some differences between the molecular weight of the graft polymer and homopolymer [23], however, the concept that the two polymers have relatively similar molecular weights is commonly accepted [23–25].

Polymer grafts on the bead surface were formed either by addition of macroradical homopolymers to the double bonds on the methacrylpropylsiloxane (MPS) modified glass surface or by the primary radical initiation of these double bonds to polymerize acrylic acid. Higher grafting efficiencies and grafting percentages were obtained in dioxane in comparison with tetrahydrofuran (THF). Isopropanol, a chain transfer solvent, produced a lower molecular weight graft and grafting percent. Benzene and toluene were not good solvents because of their poor solubility for the free polymer and the grafted beads tended to form agglomerates in these solvents.

The AIBN system formed polymers on the vinyl modified glass surface by the addition of macroradicals to the vinyl groups on the surface, therefore, steric hindrance played an important role in amount of polymer grafted. In the case of BPO, primary radicals seem to be produced at the glass surface by radical addition to the immobilized vinyl group, which then reacted with the acrylic acid monomers to form the graft polymer. BPO appeared to be the better overall initiator for this graft process [26].

Temperature increases accelerated the decomposition of the initiator and molecular diffusion resulting in greater grafting percent. Increasing the initiator concentration also caused an increase in the number of grafts but a decrease in molecular weight of the polymers formed.

3. IMMOBILIZATION OF CARBOXYLIC ACID POLYMERS ON P_2 GLASS

Polymer immobilization involved the reaction of carboxylic acid polymers (CAPs) with APS (3-aminopropyl siloxane) functionalized glass beads (Scheme 1). The polymers, PAA and PMA reacted with the active APS amine groups to form ammonium carboxylate salts. These salts were heated to form stable covalent bonds between the APS modified beads and the polymers. The carboxyl and amine functional groups reacted chemically at 140°C to form amide bonds [27, 28].

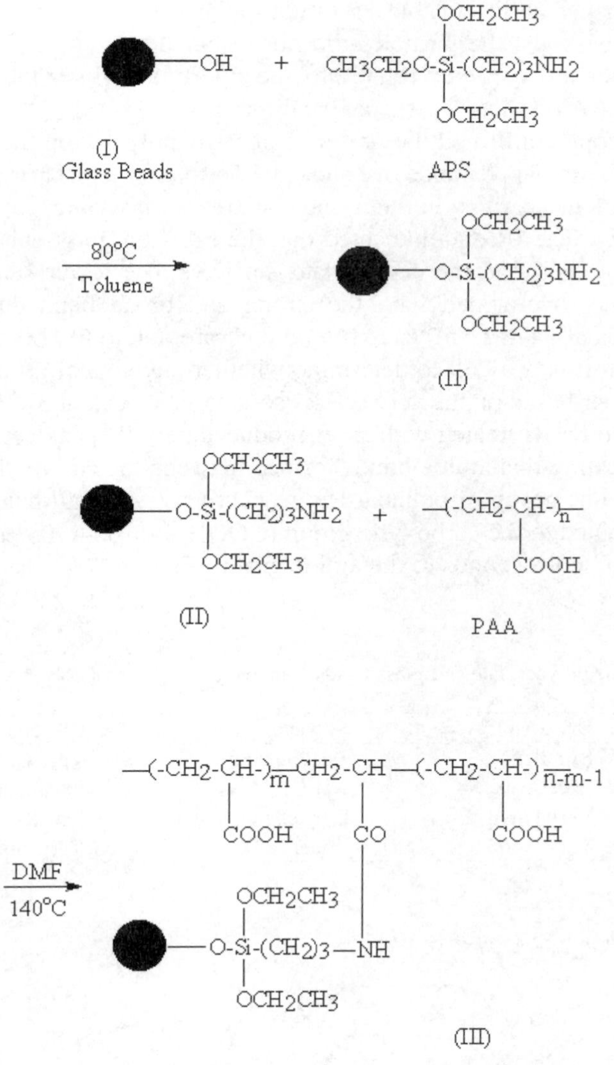

Scheme 1. Reaction for modification of APS beads with PAA by grafting.

The effect of PAA concentration on the yield of polymer graft was examined. A plateau was observed at 10 mg PAA/mL in DMF which was considered to be the optimal polymer concentration. The effects of PAA molecular weight on the degree of modification of glass beads was also determined. It was found that the higher molecular weight polymers gave slightly higher percent grafted polymer [27].

The dispersability of PAA modified glass beads composites were examined in different solvents. In polar solvents, such as methanol and acetonitrile, the particles showed good dispersability (Table 1). In non-polar solvents, such as aliphatic and aromatic hydrocarbons, the particles did not disperse very well. The dispersability of the composite particles relates to the hydrophilic/hydrophobic compatibility between PAA on the glass surface and the solvent.

Infrared spectroscopy confirmed the presence of PAA polymer on the glass surface before heating by the appearance of bands at 1701.6 cm^{-1} and 1560.6 cm^{-1} due to C=O stretching of the carboxylic acid and the free carboxylate groups of the polymer, respectively. The IR data indicated that the PAA on glass existed in two forms; the carboxylic acid and the carboxylate ion [28]. The lower IR frequency bands of the polymer overlapped with the strong and broad band due to Si-O stretching. The bands at 3448.1 cm^{-1} and 1654.6 cm^{-1} are due to O-H bond stretching and deformation. It is difficult to determine whether any silanol groups had reacted with the polymer based on the very weak characteristic peak at 3753.6 cm^{-1}.

The APS modified beads treated with PAA produced new IR peaks at 1710 cm^{-1} and 1660 cm^{-1}, indicating that amide bonds formed between the carboxyl groups of the PAA and the amine group of bound APS, respectively. A small band at 1700-1705 cm^{-1} was attributed to the carboxylic group (COOH →1700-1705 cm^{-1}), while the band at 1672 cm^{-1} corresponds to the amide group (Figure 1).

Table 1.
The dispersibility of the polyacrylate salt-glass beads composites in different solvents

Solvents	Dispersibility	Solvents	Dispersibility
Ethanol	Excellent	Acetonitrile	Very Good
Methanol	Excellent	NH$_4$OH	Excellent
DMF	Very Good	Ethylene Glycol	Excellent
DMSO	Very Good	Water	Very Good

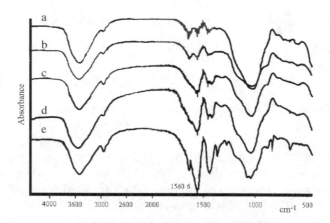

Figure 1. IR Spectra of polymer-glass composite at different reaction times a) 17 minutes; b) 30 minutes; c) 60 minutes; d) 90 minutes; e) 120 minutes.

Thermogravimetric analysis was used to determine the percent weight loss versus temperature. Modified glass beads (20 mg samples) were analyzed at a heating rate of 10°C/min in a dynamic air and helium mixture. The onset of the decomposition for all thermograms was ca. 270°C and the respective weight losses are shown in the thermograms (Figure 2). Thermogram A depicts the changes, which are minimal for the glass beads alone. Thermogram C shows a weight loss of 10 wt % after grafting PAA. Based on this work, other poly(carboxylic acid) polymers and anhydride copolymers, such as poly(styrene-co-maleic anhydride) have been grafted to the surface of P_2 beads as well [29].

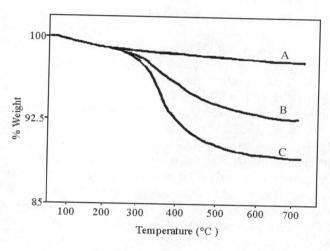

Figure 2. Thermogravimetric Curves of (A) Glass Beads (B) APS, and (C) PAA.

4. MODIFICATION OF POROUS SILICA PARTICLE SURFACES WITH POLY(ACRYLIC ACID)

Silica particles are being used for a variety of applications. Consequently grafting on porous silica [5, 30, 31] has attracted much attention and has been studied extensively. Our present concern is with the modification of porous silica particle surfaces with functional polymers. Our interest is to improve particle surface properties such as hydrophobicity and hydrophilicity. Poly(acrylic acid) was chosen as the functional polymer to provide pH-intelligent, surface-responsive particles. The PAA chains under acid conditions are usually randomly coiled, while under basic conditions the chains are extended due to electrostatic repulsion of the carboxylate ions. The surface characteristics can be tailored to respond to specific environments by controlling the pH. The pore size, size distribution and specific surface area of modified silica were evaluated from the amount of nitrogen adsorbed on the surface. The water penetration rate and porosity at different pH's were measured to determine the surface properties [6, 17, 32].

The silica surface was treated with APS and with poly(acrylic acid), following the reactions depicted in Scheme 1. The carboxyl groups on poly(acrylic acid) reacted with the APS amine groups to form salt bonds on the surface of the silica. The salt-modified beads were heated in DMF at 140°C to form the corresponding amide bonds. FTIR spectra of the APS-modified silica exhibited a peak at about 1600 cm^{-1} due to the amino group in addition to the original silica peak. After treatment with poly(acrylic acid) new peaks at 1710 cm^{-1} and 1660 cm^{-1}, indicating the carboxyl groups of poly(acrylic acid) and the amide bond formation between the poly(acrylic acid) and the bound APS, respectively [33], were observed.

Hollow, highly porous silica particles were obtained from Suzuki Oil & Fat Co. Ltd. (Japan) with particle size distribution of 2.0-2.5 μm. The silica particles were treated with a 2% sodium hydroxide solution, washed with water and dried. Surface modification of silica with APS and poly(acrylic acid) decreased the total pore volume. In addition specific surface area also decreased with each modification step. However, by controlling the amount of poly(acrylic acid) on the surface, it was possible to control the total pore volume.

Although no significant change was observed for the rate of water penetration for the APS-modified silica, the rate of water uptake by the poly(acrylic acid) coated particle changed with the pH. Similarly, the porosity of APS-modified silica was not influenced by pH, but the porosity of poly(acrylic acid)-modified silica changed directly with the pH. The poly(acrylic acid) chains can be randomly coiled at low pH (2-3) and highly extended due to electrostatic repulsion between charged carboxylate groups at higher pH. Thus the silica particles are sensitive to pH changes and the ionic strength of an aqueous solutions. Dissociation of poly(acrylic acid) at pH 5 extended the chains, while at pH 1.5 the chains remain closer to the surface. These observations are in agreement with Ito *et al.* [34], who found that poly(acrylic acid)-grafted porous polycarbonate membranes exhibited higher permeabilities at lower pH than at higher pH.

Similar to the glass beads the amount of polymer grafted increased with increasing polymer concentration. The most PAA grafted on these silica particles was 396 mg/g-silica [38].

5. DENDRITIC MODIFICATION OF GLASS BEAD SURFACES

Based on the design of divergent dendrimer and fractal structures [34–37], a new type of glass bead modification was investigated in our laboratory. We pursued this method to compensate for the low silanol (=SiOH) concentration and small surface area of the glass beads compared to the abundance of silanol groups and larger surface area of porous silica. If one visualizes the glass bead as a starburst core, then the surface SiOH groups on the glass bead could be reacted with multifunctional molecules to develop fractal-type structures on the glass surface. The dendritic procedure provides the possibility to place a great number of small molecules and reactive groups on the glass bead surface. This technique could introduce unique surface properties to the glass beads such as highly hydrophobic or hydrophilic surfaces.

Dendritic formations were initiated, according to Figure 3, by coupling APS (aminopropyltrimethoxysilane) onto the glass beads as the first generation. Propanolamine was reacted with the residual methoxyl groups to optimize the number of amine groups on the glass bead surface. The amines were reacted with trifunctional cyanuric chloride or alkyl chloride was reacted with amine groups to build the second generation. A variety of reagents were added, which offered different physical properties such as propanolamine, tris(hydroxylmethyl) aminomethane, or methyl tyrosine, were grafted to develop a third generation. The percent graft and conversion were calculated from thermogravimetric (TGA) analysis based on the weight loss by the modified glass beads during thermolysis. The temperature was scanned from 50-600°C at a rate of 2.5°C/min.

Table 2.
Reaction conversions at different generations

Generations	Reaction Conversions (%)		
	GF-1-2-3*	GF-1-2-4*	GF-1-2-5
1st	55.9	55.9	55.9
2nd	63.2	63.2	63.2
3rd	46.1	23.7	19.8

*GF: glass bead; the digits after FG represent the chemicals which reacted with glass to form different generations. 1-APS and Propanolamine; 2-Cyanuric chloride; 3-Tris; 4-Methyl tyrosine.

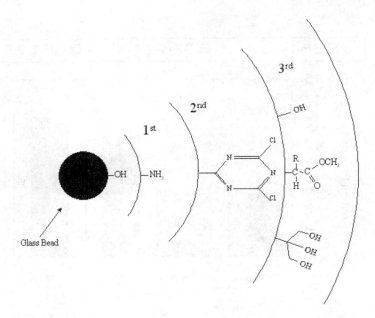

Figure 3. Dendritic Modification of glass bead.

After the coupling reaction, the IR spectra of all the samples exhibited a broad band at 1578.2 cm^{-1} due to N-H bonding vibration. This indicated that the APS was grafted on the glass surface to provide amine groups. After the propanolamine was added the IR band at 1578.2 cm^{-1} was enhanced. An anlysis of the analytical data indicated that one propanolamine was successfully added per APS to increase the number of surface amine groups. After cyanuric chloride was reacted with the amine, a new band at 1613.5 cm^{-1} appeared in the IR spectra, which was ascribed to C=N stretching vibration. Finally, three different molecules with different functional groups were attached, respectively, to provide the last generation, which were identified by a band at 2902.2 cm^{-1} due to C-H stretching vibration of the CH$_2$ groups of propanolamine and Tris and at 1689 cm^{-1} due to carbonyl vibration for –C(NHR)O. Data to support these results were obtained from TGA and titrametric analyses.

Other generations consisted of long alkyl chains of palmitoyl chloride and lauroyl chloride, respectively. These particles were highly hydrophobic with densities less than one. Consequently, these particles floated on the surface of water but were dispersable in toluene. The IR spectra showed strong C-H vibrations at 2902.2 cm^{-1} due to CH$_2$ groups. The grafting percentage increased with dendritic growth, as shown in Figure 4. Therefore, this dendritic procedure increased the number of grafted molecules on the glass bead surface [18]. The highest w/w grafting percent on the glass beads by this procedure was 21%. This is similar to the amount of polymer grafting on the same surface [38]. However, the dendritic procedure provided numerous small molecular groups that covered the glass surface with a uniform coating.

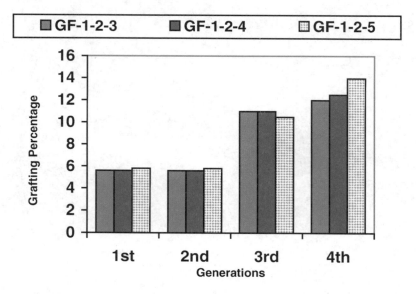

Figure 4. Grafting percentage increasing with dendritic growth.

(It was difficult to obtain 100% conversion for any generation; we found that the percent conversion changed alternately at each generation). This means that the third generation would give a low conversion if a high conversion occured in the second generation and vice versa. This phenomenon may be characteristic of the dendritic growth process. A reasonable explanation is that space hindrance plays an important role in the dendritic growth. Another factor affecting the second generation is that cyanuric chloride could react with more than one amine group in the first generation.

The hydrophobic and hydrophilic properties of dendritical modified glass beads were evaluated in different solvents for dispersibility. Low hydrophilic end groups exhibited a poor dispersibility in all solvents. Higher hydrophilic character end groups showed good dispersibility in water and methanol. When the last generations were formed by long alkyl chains good dispersibility in nonpolar solvents such as benzene and pentane was observed. Therefore, the dispersibility of the modified glass beads in various media may be tailored by the molecular structure of the last generation on the modified surface [39].

Acknowledgements

We wish to thank Dupont and High Technology Materials Center at Virginia Commonwealth University for their financial support for this project.

REFERENCES

1. E. P. Plueddermann, Silane Coupling Agent; Plenum Press: New York, 1991, Chapter 1& 2.
2. Li D.; Neumann A. W., Behavior of Particles at Solidifation Fronts, Applied Surface Thermodynamics, Marcel Dekker: New York, 1996, Chapter 12.
3. Carlier E.; Guyot A.; Revillon A., React. Polym., 1991, **16**, 115.
4. Nakatsuka T., J. Appl. Polym. Sci., 1987, **34**, 2125.
5. Boven G.; Oosterling M. L. C. M.; Challa G.; Schouten A. J., Polymer, 1990, **31**, 2377.
6. Boven G.; Folkersma R.; Challa G.; Schouten A. J., Polym. Commun., 1991, **32**, 50.
7. Yoshinigawa, K.; Horie, R.; Saigoh, F.; Kito, T., Polym. Adv. Tech., 1992, **3**, 91.
8. Duchet, J.; Chapel, J. P.; Chabert, B.; Spitz, R.; Gerard, J. F., J. Appl. Polym. Sci., 1997, **65**, 2481.
9. Suzuki, K.; Yumura, T.; Mizuguchi, M.; Tanaka, Y.; Akashi, M., Polym. Prepr., 1999, **40**, 119.
10. Prucker, O.; Ruehe, J., Macromol., 1998, **31**, 602.
11. Tsubokawa, N.; Kazahiro, S.; Yukio, S., Polym. Bull.(Berlin), 1995, **35**, 399.
12. Tsubokawa, N.; Hisanori, I., J. Polym. Sci., Part A Polym. Chem., 1992, **30**, 2241.
13. Shirai, Y.; Tsubokawa, N., React. Funct. Polym., 1997, **32**, 153.
14. Tsubokawa, N.; Kimoto, T.; Yoshikawa, S., Shikizai Kyokaishi, 1994, **67**, 231.
15. Becker, C.; Mueller, P.; Schmidt, H., Proc. SPIE-Int. Soc. Opt. Eng., 1998, **3469**, 88.
16. Beyer, D.; Bohanon, T. M.; Knoll, W.; Ringsdorf, H.; Elender, G.; Sackmann, E., Langmuir, 1998, **12**, 2514.
17. Yin R.; Ottenbrite R. M.; Siddiqui J. A., Polym. Prepr. 1994, **35**(2), 705
18. Yin R.; Ottenbrite R. M.; Siddiqui J. A., Polym. Prepr., 1995, **36**(1), 449.
19. Yin R.; Ottenbrite R. M.; Siddiqui J. A., Polym. Prepr., 1996, **37**(2), 751.
20. Hashimoto K.; Fujisawa T.; Kobayashi M.; Yosomiya R. J., Macromol. Sci., Chem., 1982, **A18**, 173.

21. Nishioka N.; Matsumoto K.; Kosai K., Polym. J., 1983, **15**(2), 153.
22. Tsuzuki M.; Hagiwara I.; Shiraishi N.; Yokota T., J. Appl. Polym. Sci., 1980, **25**, 2909.
23. Nishioka N.; Kosai K., Polym. J., 1981, **13**(2), 1125.
24. North A. M., The International Encyclopedia of Physical Chemistry and Chemical Physics, 1996, vol. 17(1), The Kinetics of Free Radial Polymerization, Pergamon Press.
25. Nishioka N.; Minami K.; Kosai K., Polym. J., 1983, **15**(8), 591.
26. Yin R.; Ottenbrite R. M.; Siddiqui J. A., Polym. Adv. Tech., 1997, **8**, 761.
27. Ottenbrite R. M.; Zengin H.; Siddiqui J. A., Polym. Prepr., 1998, **39**(2), 585.
28. Ottenbrite R. M.; Zengin H.; Siddiqui J. A., Polym. Prepr., 1998, **39**(2), 587.
29. Zengin Huseyin, Masters Thesis, Virginia Commonwealth University, 1998, May 17.
30. Naka T., J. Appl. Polm. Sci., 1987, **34**, 2125.
31. Carlier E.; Guyot A; Revillon A., React. Polym., 1991, **16**, 115.
32. Nakahara Y.; Kageyama H.; Nakahara F.; Doi Y., Daikoushi-kihou, 1989, **40**, 178.
33. Suzuki K.; Ottenbrite R. M.; Siddiqui J. A., Polym. Prepr., 1997, **38**, 335.
34. Ito Y.; Inaba M.; Chung D.; Imanishi Y., Macromolecules, 1992, **25**, 7313.
35. Tomalia D. A.; Naylor A. M.; Goddard W. A., Angew. Chem. Int. Ed. Eng., 1990, **29**, 138.
36. Newkome G. R.; Yao Z-Q.; Baker G. R.; Gupta V. K., Org. Chem., 1985, **50**, 2003.
37. Sanford E. M.; Frechet J. M. J.; Wooley K. I.; Hawker C. J., Polym. Prep., 1992, **33**(2), 654.
38. Yin R.; Ottenbrite R. M.; Siddiqui J. A., Polym. Prep., 1993, **34**(2), 506.
39. Nemeth S.; Ottenbrite R. M.; Siddiqui J. A., Polym. Prep., 1997, **38**, 365.

Molecular design of polymers containing metallo-porphyrin moieties as photoactive function

MIKIHARU KAMACHI

Department of Applied Physics and Chemistry, Fukui University of Technology, 3-6-1 Gakuen Fukui 910-8505, Japan

Abstract—Photophysical behavior of the amphiphilic polyelectrolytes with a small amount of ZnTPP moieties as a hydrophobic chromophore was studied by absorption, fluorescence, phosphorescence, and transient spectroscopies. Absorption and emission spectra of amphiphilic polyelectrolyte in aqueous solution show that the zinc(II)-tetraphenylporphyrin(ZnTPP) moieties are compartmentalized in hydrophobic microdomain which is constructed by the aggregation of hydrophobic substituents. The triplet excited lifetime of the compartmentalized ZnTPP moieties at room temperature was found to be much longer than that of ZnTPP, and emitted phosphorescence and delayed fluorescence were observed even at room temperature.

The electron transfer reactions from ZnTPP moieties to Methylviologene (MV^{2+}) and phenylmethyl-phenacylsulfonium p-toluenesulfonate(PMPS) were investigated by fluorescence spectroscopy and transient absorption spectroscopy. In the amphiphilic polyelectrolyte, $MV^{+\bullet}$ and $ZnTPP^{+\bullet}$ were clearly observed in the transient spectroscopy, although they could not be detected in polyelectrolyte with ZnTPP moieties. These results indicate that back electron transfer reaction is retarded remarkably owing to the compartmentalization of ZnTPP moieties.

Keywords: Amphiphilic polyelectrolyte; sodium 2-acrylamido-2-methylpropanesulfonate; [5-(4-acrylamidophenyl)-10,15,20-triphenylporphyrinato]zinc(II); compartmentalization; photoinduced electron transfer.

1. INTRODUCTION

Metalloporphyrins play an important role in a wide variety of biological system. For example, iron porphyrins act as catalytic centers in oxydases and catalases [1], electron carriers in cytochromes [2], and oxygen carriers in hemoglobins [3]. In these biological systems, a metalloporphyrin is situated at inner sites of protein whose microenvironment makes physicochemical properties of the metalloporphyrin differ greatly from those in homogeneous solution. Thus, the functionality of a metalloporphyrin is controlled by the microenvironment provided by a protein.

It is well-known that amphiphilic polyelectrolytes bearing hydrophobic substituents form hydrophobic domains due to an aggregation of the hydrophobic groups in aqueous solution above a critical content [4–8]. When a small fraction of hydrophobic functional groups such as tetraphenylporphyrin derivatives are covalently incorporated into amphiphilic polyelectrolytes, the functional group is encapsulated in the hydrophobic domain, in which the hydrophobic functional group is protected from aqueous phase. Photophysical and photochemical behavior of some chromophore has been found to be greatly modified when the chromophores are incorporated into self-organized microphases of amphiphilic polyelectrolytes in aqueous media [4, 5]. Furthermore, the physical properties of the microdomain formed by hydrophobic groups can be controlled by the substituents [4, 5, 9]. Accordingly, these systems are interesting in a point of view of the control of the functionality of the functional groups such as tetraphenylporphyrin derivatives buried into the microdomain [9].

We paid attention to metallotetraphenylporphyrins compartmentalized into hydrophobic microdomain of amphiphilic polyelectrolytes in their aqueous solutions as a physicochemical model for biological metalloporphyrin system. Since photophysical and photochemical behavior of zinc(II)-tetraphenylporphyrin (ZnTPP) has been investigated minutely so far [10], ZnTPP was used as a hydrophobic chromophore for an understanding of the compartmentalization effect on the photophysical and photochemical behaviors. Photo-induced electron transfer to methylviologen(MV^{2+}) from various water soluble ZnTPP derivatives has been studied [10]. Accordingly, photoelectron transfer reaction to MV^{2+} and phenyl-methyl-phenacylsulfonium p-toluenesulfonate (PMPS) [11] was used for the investigation of the effect of the microenvironment on the photochemical behavior of ZnTPP moieties incorporated covalently into amphiphilic polyelectrolytes.

Amphiphilic polymers covalently tethered with small amounts of ZnTPP were prepared by the radical copolymerization of sodium 2-acrylamido-2-methylpropane-sulfonate(AMPS), [5-(4-acrylamidophenyl)-10,15,20-triphenyl-porphyrinato]zinc(II) (ZnAATPP), and N-substituted methacrylamide containing bulky hydrophobic groups. Photochemical and photophysical behaviors of the ZnTPP moieties of the polymers were compared with those of homogeneous ZnTPP systems [12, 13].

In this chapter, it is shown that photochemical and photophysical behavior of the compartmentalized ZnTPP moieties was remarkably different from that of ZnTPP moieties in homogeneous systems [14–17].

2. AMPHIPHILIC POLYELECTROLYTE AND REFERENCED POLYELECTROLYTE

Amphiphilic polyelectrolytes with a small amount of ZnTPP moieties, as shown in the following Scheme 1, were prepared by terpolymerizations of sodium 2-acrylamido-2-methylpropane sulfonate(AMPS), N-substituted methacrylamide containing bulky hydrophobic groups[R: Lauryl(La), Cyclododecyl(Cd), and 2-naphthylmethyl(Np)], and small amounts of [5-(4-acrylamidophenyl)-10,15,20-tetraphenylporphinatozinc(II)] (ZnAATPP) by their radical terpolymerizations. For comparison, polyelectrolyte with small amounts of ZnTPP, which is described as reference copolymer, was prepared by radical copolymerizations of AMPS with small amounts of ZnAATPP.

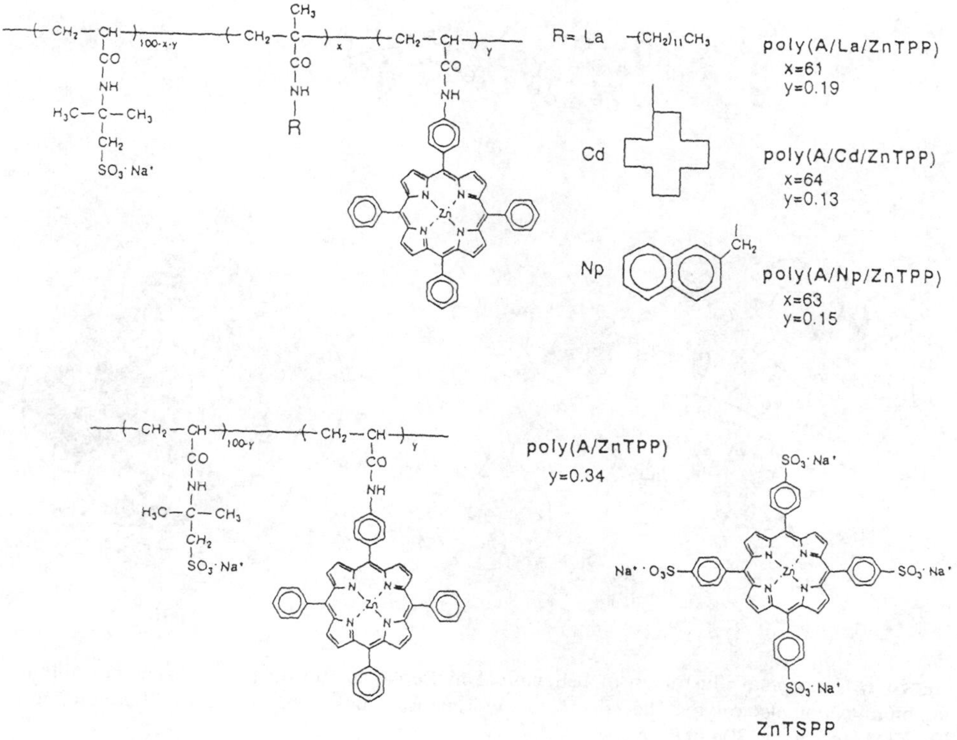

Scheme 1. Structures of amphiphilic polysulfonates and referenced polyelectrolyte with ZnTPP moieties, and their abbreviation.

3. SELF-ORGANIZATION OF HYDROPHOBICALLY MODIFIED POLYELECTROLYTE AND UNIMER MICELLE FORMATION

If the content of a hydrophobe in hydrophobically modified amphiphilic polyelectrolyte is above a certain critical content, the polymer would undergo self-organization in aqueous solution.

In general, the hydrophobic association can occur not only in an intrapolymer mode but also in intermolecular one. If interpolymer hydrophobic association occurs, then multipolymer aggregation would be formed instead of the unimers. However, whether or not the intrapolymer association predominates over the interpolymer association depends primarily on the chemical structure of the hydrophobically modified polyelectrolyte. The first step of the self-organization may be intrapolymer hydrophobic association to form micelle units along a polymer chain [18]. However, the conformation thus formed, which may be viewed as a "second order structure", may not be stable, because a significant portion of the surface of the hydrophobic cluster is exposed to water. Consequently, as schematically illustrated in Figure 1, the micelles units may congregate to form a higher order

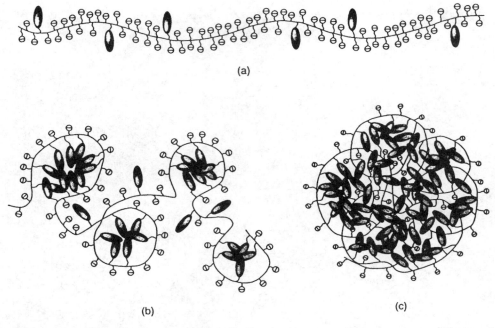

Figure 1. Conceptual illustration of behavior of hydrophobic substituents in aqueous solution of amphiphilic polyelectrolyte. The content of hydrophobic substituent: (a)below 10mol %, (b)10-30 mol %, (c) above 30mol %.

structure and thus the polymer is made up with micelle made up with a single macromolecules or aggregate intermolecularly. Recently, McCormick *et al.* [19] and Morishima *et al.* [4, 5] have independently shown that the sequence distribution of charged units and hydrophobic units along the polymer chain is an important factor to determine the mode of hydrophobic association; block sequences have a strong tendency for interpolymer association, whereas random and alternating sequences tend to an intramolecular one. Another important structural factors for unimer-forming polymers is that the hydrophobes should be bulky with cyclic structure such cyclododecane, adamantane, and naphthalene, and their contents should be higher than ca.30%. Furthermore, amido spacer bonds may be important because they form hydrogen bond networks that may contribute to retaining the compact unimer structure.

These unimers are very different from the classical micelles in that (1) all charged and hydrophobic groups are covalently linked to polymer backbone, and (2) the unimers are "static" in nature as opposite to the "dynamic" nature of the surfactant micelles which exist in equilibrium between association and dissociation [12, 18].

4. PHOTOPHYSICAL BEHAVIOR

Photophysical behavior of the amphiphilic polyelectrolytes containing ZnTPP was studied by absorption, fluorescence, phosphorescence, and transient spectroscopies [14, 15]. For comparison, photophysical behavior of a polyelectrolyte with a small amount of ZnTPP and Tetrakis(4-sulfonatophenyl)porphinatozinc (ZnTSPP) was also investigated under the same condition, respectively.

4.1. Absorption and Emission Spectra

Absorption spectra of the amphiphilic polyelectrolytes, reference copolymer, and ZnTSPP in water are shown in Figure 2 [14].

In aqueous solution, the absorption maxima for the Soret band of the amphiphilic polyelectrolytes were red-shifted by 5-9nm as compared to those of the reference copolymer and ZnTSPP, whose ZnTPP moieties are exposed to water. In DMF, the Soret band of the amphiphilic polyelectrolytes is the same as those of the reference copolymer and ZnTSPP. These observations indicate that in aqueous solution the ZnTPP moieties in the amphiphilic polyelectrolytes exist in the different environments from those of the reference copolymer and ZnTSPP. The results show that ZnTPP moieties in the amphiphilic polyelectrolytes are expected to exist in hydrophobic domain of the polymers in aqueous solution, while those in the reference copolymer and ZnTSPP are exposed to the aqueous solution.

Fluorescence spectra of amphiphilic polyelectrolyte and reference copolymer in aqueous solution at 25°C are shown in Figure 3. Fluorescence bands of amphiphilic polyelectrolytes shift to 10nm lower wave-length than those of reference copolymer, which also indicates that the ZnTPP moieties of the amphiphilic polyelectrolyte exist in the different environments from those of the reference copolymer in aqueous solution.

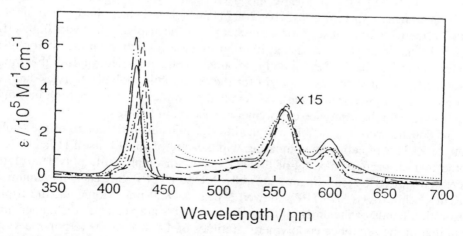

Figure 2. Absorption spectra of the amphiphilic polyelectrolytes, reference copolymer, and ZnTSPP in water. -----, poly(A/La/ZnTPP); —·—·—, poly(A/Np/ZnTPP); ············, poly(A/Cd/ZnTPP); ———, poly(A/ZnTPP); —··—··—, ZnTSPP.

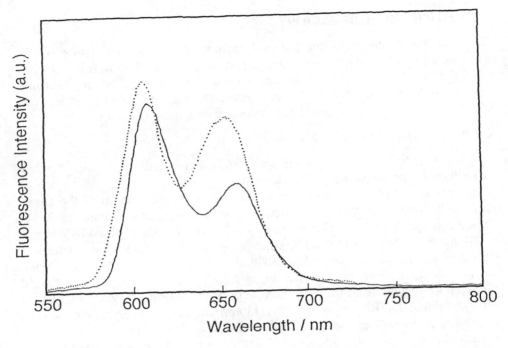

Figure 3. Fluorescence spectra in aqueous solution by exitation at the maxima of the Soret band at room temperature. ············ , poly(A/Cd/ZnTPP); ─────── , poly(A/Cd).

4.2. Transient Absorption Spectra

Photophysical behavior of ZnTPP has been minutely investigated so far on the basis of transient spectroscopy [10, 20, 21], which indicates the excited singlet state formed by light-absorption interchange to the triplet state with high efficiency by the intersystem crossing. In order to get information on the photophysical behavior of the ZnTPP moiety of amphiphilic polyelectrolytes, the time-resolved transient absorption spectra for the amphiphilic polyelectrolytes and the reference copolymer were measured. Typical examples for the amphiphilic polyelectrolytes and the reference copolymer are shown in Figure 4.

The spectra were reasonably assigned to the T-T absorption band of the triplet state of ZnTPP moieties by comparing with spectra of ZnTPP itself [10, 21]. It is noted that absorption maxima of the T-T bands for amphiphilic polyelectrolytes were red-shifted by ca. 20 nm, i.e. poly(A/Cd/ZnTPP) showed a maximum at 840 nm, while poly(A/ZnTPP) showed a peak at 820 nm. Moreover, the triplet absorption bands of the amphiphilic polyelectrolytes are persisted for longer time than that of the reference copolymer. Lifetimes of T_1-state of the terpolymers and the reference copolymers were estimated from the decay-profile of T_1 absorption at 480 nm (Figure 5).

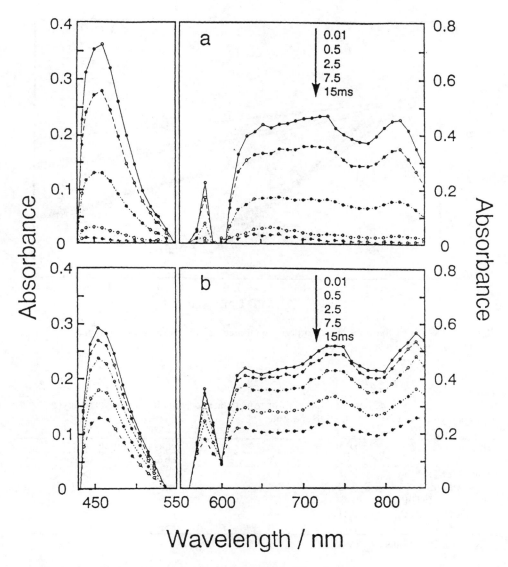

Figure 4. Comparison of the time-resolved transient absorption spectra in aqueous Solution at room temperature. a, poly(A/Cd/ZnTPP); b, poly(A/ZnTPP).

The estimated lifetimes of ZnTPP moieties for the amphiphilic polyelectrolytes and the reference copolymers in water and in DMF are shown in Table 1.

Lifetimes of the triplet states of the ZnTPP moieties in aqueous solutions of amphiphilic polyelectrolytes at room temperature were much longer than that of the reference copolymer and those of ZnTPP derivatives reported so far.

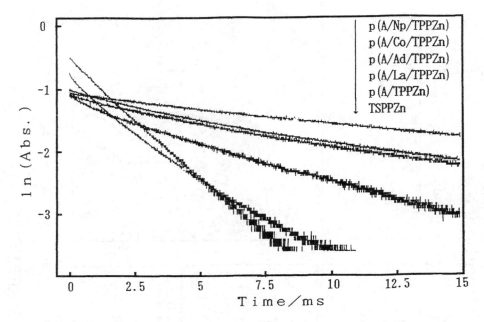

Figure 5. Decays of the transient absorption monitored at 480 nm for the terpolymers and referenced copolymer in aqueous solution.

Table 1.
Initial absorbance s(AbS$_0$) and lifetimes (τ_t) of triplet state for amphiphilic polyelectrolytes, reference copolymer, and ZnTSPP in DMF and water

sample code	DMF solution		aqueous solution	
	Abs$_0$	τ_t/ms	Abs$_0$	τ_t/ms
poly(A/ZnTPP)	0.38 [a]	3.0 [a]	0.26	3.0
poly(A/La/ZnTPP)	0.42	3.7	0.45	12
poly(A/Cd/ZnTPP)	0.42	4.3	0.25	20
poly(A/Np/ZnTPP)	0.41	3.8	0.33	19
ZnTSPP			0.53	3.5

[a]Solvent: DMF/water=9/1 (V/V).

Triplet lifetimes of ZnTPP itself have been reported to be 1.19 and 1.25 ms in ethanol and toluene solution at 300 K, respectively, while its phosphorescence lifetimes have been reported to be 25 and 26ms in ethanol and toluene rigid glasses at 77 K, respectively [20]. Lifetimes of the triplet states of the ZnTPP moieties in the amphiphilic polyelectrolytes at room temperature in aqueous solution is closer to those of ZnTPP in rigid glass than in solution, Unusually long lifetimes observed in the present study for the amphiphilic polyelectrolytes in aqueous solution at room temperature are due to the compartmentalization of ZnTPP moieties by hydrophobic substituents, because the hydrophobic residues around the compartmentalized ZnTPP moieties serve not only as protection groups for the chromophores but also as a rigid matrix for the chromophores similar to glasses at 77 K even though the amphiphilic polyelectrolytes are apparently in fluid solution.

4.3. Delayed Emission Spectra

Delayed emission spectra at 5ms after irradiation were measured for the amphiphilic polyelectrolytes, referenced copolymer, and ZnTSPP in aqueous solution at 25°C (Figure 6). The reference copolymer and ZnTSPP showed no such emission. In general, phosphorescence is not observed for organic molecules in fluid solution at ordinary temperatures mainly because the thermal deactivation of the triplet-excited state is effected by a collision mechanism. In a fluid solution, the solvent environment acts as an effective sink of thermal energy in the form of heat. Radiationless deactivation occurs by an intersystem crossing involving vibronically coupled transitions from the potential surface of the triplet-excited electronic state to the isoenergetic vibrational levels of the potential surface associated with the ground state. In the case of the ZnTPP moieties compartmentalized in the hydrophobic domain, the deactivation via collision mechanism should be precluded. In addition, the radiationless process may be suppressed bacause the ZnTPP species are firmly surrounded by the rigid cluster. It is reasonable, because the lifetime of ZnTPP moieties is shorter than 5ms. Delayed emission spectra are clearly observed for the amphiphilic polyelectrolytes as shown in Figure 6. The emission bands were assigned to the fluorescence bands and phosphores-

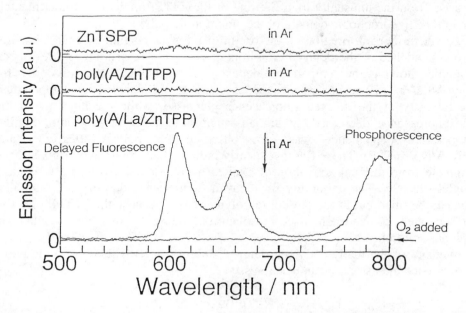

Figure 6. Delayed emission spectra of poly(A/Cd/ZnTPP), poly(A/ZnTPP), ZnTSPP in aqueous solution at 25°C.

cence band by comparison of their wave lengths with the corresponding bands, respectively. Although ZnTPP has been reported to show both fluorescence and phosphorescence in rigid glasses at 77K [22, 23], there has been no paper which has shown both fluorescence and phosphorescence at room temperature in fluid solution. Phosphorescence and delayed fluorescence in the ZnTPP moieties in the amphiphilic polyelectrolytes was found, for the first time, in aqueous solution at room temperature. These unusual phenomena are reasonably ascribable to the compartmentalization of the ZnTPP moieties by the hydrophobic domains. The delayed emission spectrum of the terpolymer was completely quenched under aerated condition as shown in Figure 6. Fluorescence was far less sensitive to oxygen. Accordingly, it is concluded that the delayed fluorescence originate from thermal activation of the triplet state back up to the excited singlet state.

Studies of thermally activated delayed fluorescence at room temperature have only been reported for zinc, magnesium, and tin protoporphyrine IX dimethylesters adsorbed on the filter paper [24, 25]. The thermally activated delayed fluorescence of metalloporphyrins in fluid solution at room temperature was for the first time found in our study. This is a remarkable feature of the compartmentalized ZnTPP system.

4.4. Origin of Delayed Fluorescence

The observation of the phophorescence and the delayed fluorescence at room temperature for the amphiphilic polyelectrolytes in aqueous solution is considered to arise from the unusually long lifetime of the ZnTPP moieties compartmentalized in the hydorophobic domain of the amphiphilic polymer.

To get further information on the origin of the delayed fluorescence, temperature dependence of the delayed emission spectra were measured. Figure 7 is an example showing the temperature dependence of the delayed emission s for poly(A/La/ZnTPP) in deaerated aqueous solution. With increasing temperature, the intensity of the delayed fluorescence increased, while the intensity of the phosphorescence decreased, giving an isoemissive point at ca740nm. Similar temperature dependence was also observed for poly(A/Cd/ZnTPP) and poly(A/Np/ZnTPP). These findings clearly indicate that the delayed fluorescence originate from thermal activation of excited triplet state(T_1) back up to excited singlet state(S_1). The possibility for delayed fluorescence due to T_1-T_1 annihilation can be ruled out in the present terpolymer cases because the ZnTPP moieties are separately compartmentalized in isolation in the rigid hydrophobic domains in aqueous solution.

Intensity of thermally activated delayed fluorescence(I_{df}) relative to the phosphorescence intensity (I_p) can be expressed as

$$I_{df} / I_p = (\phi_f k'_{isc} / k_p) \exp(-\Delta E / RT) \tag{1}$$

Where f is the quantum yield of fluorescence, k'_{isc} is the rate constant of the intersystem crossing from T_1 to S_1, kp is the radiative constant of phosphorescence, ΔE is the energy gap between T_1 and S_1, R is the gas constant, and T is the Kelvin temperature. On the basis of the emission maxima of the fluorescence and phosphorescence, ΔE values for the ZnTPP moieties in the La and Np terpolymers in aqueous solution were estimated to be 45.6 and 45.5 KJ/Mol, respectively. A ΔE value of 43.4 KJ/Mol for ZnTPP was estimated from the fluorescence and phosphorescence spectra reported in the literature. Because the ZnTPP moieties in poly(A/Cd/ZnTPP) showed unusually distorted fluorescence and blue-shifted phosphorescence bands, the ΔE value was not evaluated for the Cd terpolymer.

Figure 8 shows the Arrhenius plot of I_{df}/I_p against 1/T for the La and Np terpolymers. From the slope of the plots, ΔE values were estimated to be 44.9 and 44.2 kJ/mol, respectively. The ΔE values thus estimated for the La and Np terpolymers are in fairly good agreement with those estimated from the emission wavelengths [14].

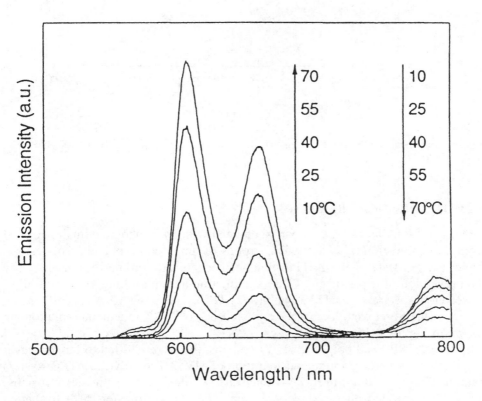

Figure 7. Delayed emission spectra of poly(A/La/ZnTPP) in deaerated aqueous solution recorded at varying temperature.

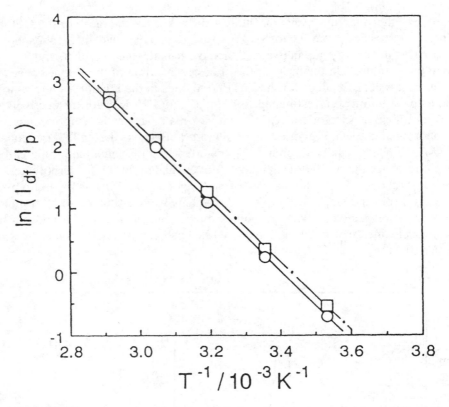

Figure 8. Arrhenius plots of I_{df}/I_p against 1/T for the terpolymers: poly(A/Cd/ZnTPP) and poly(A/Cd/ZnTPP).

5. PHOTOCHEMICAL BEHAVIOR

When ZnTPP moieties are compartmentalized with hydrophobic domains formed from the association of hydrophobic groups of amphiphilic polyelectrolytes, the triplet excited state of the ZnTPP moieties became extraordinarily long-lived by the prevention of triplet-triplet (T-T) annihilation and quenching with impurities. Absorption spectra of ZnTSPP, poly(A/ZnTPP), and poly(A/LA/ZnTPP) in the region of the Soret band in absence and presence of MV^{2+} in aqueous solution are shown in Figure 9 [15, 26, 27].

In the ZnTSPP and poly(A/ZnTPP) systems, the peaks shifted to longer wavelength with an increase in the concentration of MV^{2+}. The poly(A/ZnTPP) system exhibits the red-shift at lower MV^{2+} than ZnTSPP does, because it is electrostatistically concentrated on the polymer chain. The charge transfer (CT) formation constants (K_{CT}) for ZnTSPP, and poly(A/ZnTPP) estimated from these spectral data by use of the Nash's plot [28] are listed in Table 2. The K_{CT} value for poly(A/ZnTPP) is 1 order of magnitude larger than that for ZnTSPP.

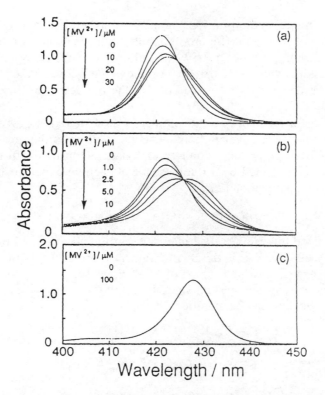

Figure 9. Soret absorption bands for (a) ZnTSPP, (b) poly(A/ZnTPP), and (c) poly(A/LA/ZnTPP) in the presence of varing amounts of MV^{2+} in aqueous solution. The residual concentration of the porphyrin moieties is 2μM.

Table 2.
Formation Constant for the CT Complex (K_{CT}), Stern-Volmer Constant for Fluorescence Quenching (K_{SV}), and Apparent Second-Order Rate Constant for Electron Transfer from Triplet Excited ZnTPP to MV^{2+} in Aqueous Solution ($k_{q,T}$)

sample	K_{CT}/M^{-1}	K_{SV}/M^{-1} ($k_{q,S}/M^{-1}s^{-1}$)^a	$k_{q,T}/M^{-1}s^{-1}$
ZnTSPP	2.5×10^4	3.2×10^4 ^{b} (1.9×10^{13})	8.2×10^9
poly(A/ZnTPP)	2.3×10^5	3.4×10^5 ^{b} (2.0×10^{14})	2.2×10^9
poly(A/LA/ZnTPP)	c	4.8×10^3 (2.8×10^{12})	2.2×10^9
poly(A/Naph/ZnTPP)	c	4.8×10^3 (2.8×10^{12})	7.4×10^8
poly(A/CD/ZnTPP)	c	2.1×10^3 (1.2×10^{12})	1.8×10^9

^{a} The apparent $k_{q,S}$ values were estimated from eq 8 by using the lifetime of singlet excited state of ZnTSPP (1.7 ns) [31]. ^{b} The apparent Stern-Volmer constant was estimated from eq 12. ^{c} No CT interaction.

In contrast, no concentration dependence on spectral profile occurred in the poly(A/LA/ZnTPP) system (Figure 9c), although the concentration of added MV^{2+} was higher than those in ZnTSPP and poly(A/ZnTPP) systems. The other terpolymer systems show that the CT complexation between the ZnTPP moiety and MV^{2+} were completely prevented due to the protection of theZnTPP moiety in the hydrophobic cluster.

Electron transfer quenching of the singlet excited state (S_1) of ZnTPP moieties (^{1}ZNTPP*) is known to occur in homogeneous solution in the presence of MV^{2+} as an acceptor(Ref). Apparent second-order constants for the ET from ^{1}ZnTPP* to MV^{2+} ($k_{q,S}$) were estimated by a fluorescence quenching experiment with the use of the Stern-Volmer equation [29]:

$$I_0/I = 1 + K_{SV}[Q] \qquad K_{SV} = k_{q,S}\tau_0 \tag{2}$$

where I_0 and I are the fluorescence intensities in the absence and presence of the quencher, respectively, Ksv is the Stern-Volmer constant, [Q] is the concentration of the quencher, and τ_0 is the lifetime of S1 in the absence of the quencher. When CT complexes exist in the system, static quenching of fluorescence occurs. Then Stern-Volmer equation can be modified [30, 31] as

$$I_0/I = (1 + K_{SV}[Q])(1 + K_{CT}[Q]) \tag{3}$$

$$= 1 + (K_{CT} + K_{SV})[Q] + K_{CT}K_{SV}[Q]^2 \tag{4}$$

if

$$[Q] \ll (1/K_{SV}) + (1/K_{CT}) \tag{5}$$

then

$$I_0/I \cong 1 + K'_{SV}[Q] \tag{6}$$

where $K'_{SV}(=K_{CT}+K_{SV})$ is the apparent Stern-Volmer constant.

The fluorescence spectra for ZnTSPP, poly(A/ZnTPP), and poly(A/CD/ZnTPP) in the absence and presence of MV^{2+} in aqueous solution are shown in Figure 10.

Fluorescence quenching was observed in all the systems containing poly-(A/La/ZnTPP) and poly(A/Naph/ZnTPP). Stern-Volmer plots on the basis of Figure 10 are shown in Figure 11. The Stern-Volmer constants obtained from the slope of the straight lines through the plots are listed in Table 2, decreasing in the follow order: poly(A/ZnTPP) > ZnTSPP > poly(A/LA/ZnTPP) ± poly(A/Nap/ZnTPP) > poly(A/CD/ZnTPP). The K'$_{SV}$ value for poly(A/ZnTPP) is larger than that for ZnTSPP because MV^{2+} is electrostatistically concentrated on the polymer chain. This suggests that most of the fluorescence quenching occurs by way of the CT complexation. In addition, K_{SV} values of all the terpolymers were 2 order of magnitude smaller than that for the reference copolymer. Since no CT complexes exists in the terpolymer systems, the fluorescence quenching can only occur by ET transfer from compartimentalized ^{1}ZnTPP* to MV^{2+}. This order shows that ZnTPP moieties in the amphiphilic polyelectrolytes are compartmentalized by the hydrophobic domain.

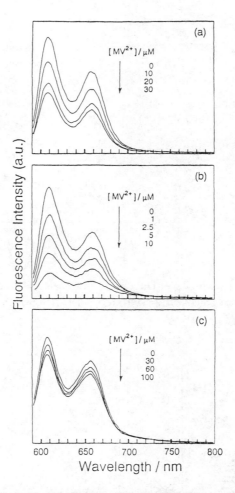

Figure 10. Fluorescence specra for (a) ZnTSPP, (b) poly(A/ZnTPP), and (c) poly(A/LA/ZnTPP) in the presence of varing amounts of MV^{2+} in aqueous solution. The residual concentration of the porphyrin moieties is 2μM. The excitation wavelength is 532nm.

Electron transfer quenching of the singlet excited state (S_1) of the ZnTPP moity is known to occur in the presence of MV^{2+} [13, 32]. The effect of the compartmentalization on photochemical behavior was investigated in photoinduced electron transfer reaction of excited ZnTPP to methylviologen(MV^{2+}). We measured time resolved transient absorption spectra of the amphiphilic polyelectrolytes and poly(A/ZnTPP) reference copolymer in the presence of MV^{2+}. Typical examples are shown in Figure12. No transient absorption spectrum was observed in the reference copolymer, which is reasonably ascribable to back electron transfer reaction. In the amphiphilic polyelectrolyte accumulation of photoproducts, $ZnTPP^{+\bullet}$ and $MV^{+\bullet}$, was found in the microsecond time region in spite of the reversible electron transfer system.

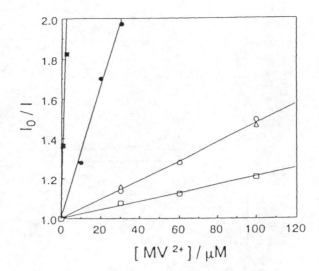

Figure 11. Stern-Volmer Plots for the ZnTPP fluorescence quenching by MV^{2+} in aqueous solution: (□) poly(A/Cd/ZnTPP), (○) poly(A/Na/ZnTPP), (△) poly(A/La/ZnTPP), (■) poly(A/ZnTPP), (●) ZnTSPP.

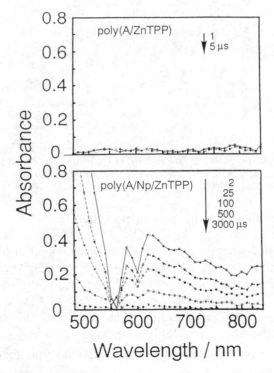

Figure 12. Time resolved transient absorption spectra of the poly(A/CD/ZnTPP) and poly(A/ZnTPP) in the presence of MV^{2+}: [ZnTPP moiety] = 100μM; [MV^{2+}] = 5 mM.

Phenylmethylphenacylsulfonium p-toluenesulfonate(PMPS) is known to be a self-destractive electron acceptor [33, 34]. When PMPS was used as an electron acceptor instead of MV^{2+}, similar trends to those found in the presence of MV^{2+} were clearly observed in absorption spectroscopy, CT-formation, fluorescence spectroscopy, rate constants for triplet quenching, and the retardation effect on back electron transfer reaction were clearly observed. As an example, fluorescence spectra of ZnTSPP, poly(A/ZnTPP), and poly(A/La/ZnTPP) in the presence of various concentration of PMPS in aqueous solution are shown in Figure 13. Fluorescences of ZnTSPP and poly(A/ZnTPP) are quenched by PMPS, while no

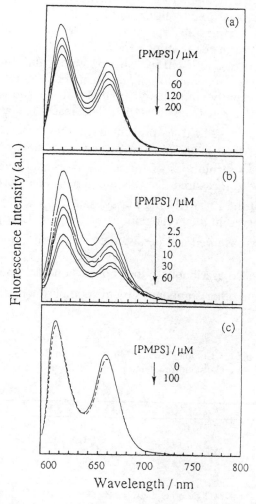

Figure 13. Fluorescence spectra for (a) ZnTSPP, (b) poly(A/ZnTPP), and (c) poly(A/LA/ZnTPP) in the presence of varing amounts of PMPS in aqueous solution. The residual concentration of the porphyrin moieties is 2μM. The excitation wavelength is 532nm.

fluorescence in poly(A/La/ZnTPP) occures even at high concentration of PMPS. The Stern-Volmer constants(K_{SV}) for ZnTSPP and poly(A/ZnTPP) could be estimated from the data in Figure 13(a) and Figure 13(b). However, K_{SV} values could not be done from the Stern-Volmer plots in amphiphilic polyelectrolytes, because no concentration effect on PMPS was observed. Although the complex formation between ZnTPP moieties and PMPS was observed in absorption spectra of ZnTSPP and poly(A/ZnTPP), no such complex formed in aqueous terpolymer-PMPS systems. This is due to the fact that the ZnTPP species are protected from CT-formation and direct contact by being buried inside the hydrophobic microdimains in amphiphilic polyelectrolytes. Time-resolved transient absorption spectra in a milisecond domain for ZnTSPP, poly(A/ZnTPP), and amphiphilic polyelectrolytes were measured by laser photolysis, and the observed spectra were assigned to the triplet-triplet absorption of ZnTPP moieties. Accordingly, rate constants for electron transfer reactions from ^{3}ZnTPP* to PMPS were estimated by using T-T absorption, and listed in Table 3. Values for amphiphilic polyelectrolytes are 2 orders of magnitude smaller than that for reference copolymer. PMPS decomposes rapidly into phenyl methyl sulfide and a phenacyl radical upon accepting an electron from ^{3}ZnTPP*. The reaction between the resulting ZnTPP$^{+\bullet}$ and phenacyl radical is prevented in the terpolymers, because the ZnTPP$^{+\bullet}$ and phenyl radical are separated inside and outside the microdomain. ESR spectra of aqueous solutions of the terpolymers and referenced copolymer containing PMPS were measured. Results are shown in Figure 14. These spectra coincide with that of the ZnTPP$^{+\bullet}$ in a glassy state at 77 K [35]. Although only a few signal was observed in the homogenious system, the ESR spectra assigned to ZnTPP$^{+\bullet}$ was clearly observed and persisted over 20 min in the terpolymers (Figure 15).

These results are also ascribable to the suppression of the reaction between the ZnTPP$^{+\bullet}$ moieties and other compounds by the compartmentalization.

Table 3.
Formation Constant for the CT Complex, Stern-Volmer Constant for Fluorescence Quenching, and Apparent Second-Order Rate Constant for Electron Transfer from Triplet Excited ZnTPP to PMPS in Aqueous Solution

sample	K_{CT}/M^{-1} [a]	K_{SV}/M^{-1} [b]	$k_{q,T}/M^{-1}s^{-1}$ [c]
ZnTSPP	3.7×10^3	1.6×10^3	1.7×10^7
poly(A/ZnTPP)	3.2×10^5	7.5×10^4	1.4×10^8
poly(A/La/ZnTPP)	[d]	[e]	3.8×10^6
poly(A/Cd/ZnTPP)	[d]	[e]	1.1×10^6

[a] Formation constant for the CT complex between ZnTPP and PMPS. [b] Stern-Volmer constant for ZnTPP fluorescence quenching by PMPS. [c] Second-order equivalent rate constant for ZnTPP triplet quenching by PMPS. [d] No CT interaction. [e] No fluorescence quenching.

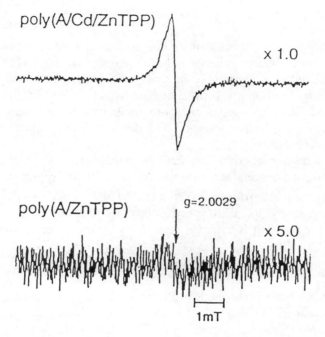

Figure 14. ESR spectra under irradiation of steady-state visible light in the presence of PMPS in aqueous solution at room temperature.

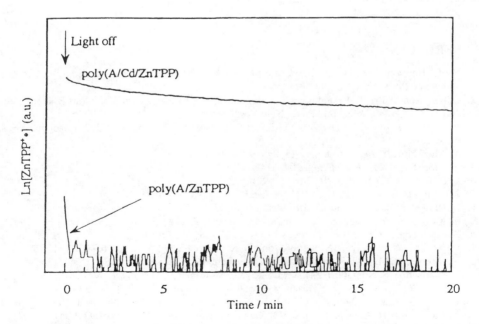

Figure 15. Decay profile for ESR the ESR absorption at 333.9 mT after irradiation with visible light for 1 min in the presence of PMPS in aqueous solution at room temperature.

4. CONCLUSION

Absorption and emission spectra of ZnTPP moieties covalently linked to an amphiphilic polyelectrolyte in aqueous solution shows that the ZnTPP moieties are compartmentalized in hydrophobic microdomain which was constructed by the aggregation of hydrophobic substituents. The triplet excited lifetime of the compartimentalized ZnTPP moieties at room temperature was found to be much longer, and emitted phosphorescence and delayed fluorescence were observed even at room temperature.

The electron tranfer reaction from ZnTPP moieties to MV^{2+} and PMPS was investigated by fluorescence spectroscopy and transient absorption spectroscopy. In the amphiphilic polyelectrolytes, $MV^{+\bullet}$ and $ZnTPP^{+\bullet}$ was clearly observed in the transient spectroscopy and ESR spectroscopy, respectively, indicating that back electron transfer reaction is retarded remarkbly owing to the compartmentalization of ZnTPP moieties.

Acknowledgment

This research has been performed by Professor Yotaro Morishima, Dr. Hiroyuki Aota, Dr. Michiko Seki, Mr. Yukio Tominaga, Mr. Katsunori Saegusa, Mr. Shigeki Nomura, Mr. Shinnichi Araki, and Mr. Takahito Ikeda in my laboratory in Department of Macromolecular Science, Faculty of Science, Osaka University from 1989 to 1996. I would like to express my sincere thanks for their efforts and excellent development.

REFERENCES

1. W. D. Hawson, L. P. Hager, in: *The Porphyrins* Vol. **7** Part B, D. Dolphin (Ed.) pp. 295–332. Academic Press, New York (1979).
2. S. Fergason-Miller, L. Braughtigan, E. Margoliash, in: *The Porphyrins* Vol.**7** Part B, D. Dolphin (Ed.) pp. 149–240, Academic Press, New York (1979).
3. Q. H. Gibson, in: *The Porphyrins* Vol.**5** Part B, D. Dolphin (Ed.) pp. 153–203, Academic Press, New York (1979).
4. Y. Morishima, *Prog. Polym. Sci.*, **15**, 949–997 (1990).
5. Y. Morishima, *Adv. Polym. Sci.*, **104**, 51–96 (1992).
6. S. E. Webber, *Chem. Rev.*, **96**, 1469 (1990).
7. J. E. Guillet, T. Wang, L. Gu, *Macromolecules* **19**, 2793 (1986).
8. M. Nawokowska, J. E. Guillet, *Macromolecules* **27**, 2151 (1994).
9. Y. Morishima, Y. Tominaga, S. Nomura, M. Kamachi, T. Okada, *J.Phys. Chem.*, **96**, 1990 (1992).
10. A. Harriman, *J. Chem. Soc., Faraday Trans.* 2, **77**,1281–1291 (1981).
11. DeVoe, M. R. V. Sahyun, N. Serpone, and D. K. Sharma, *Can. J. Chem.*, **65**, 2342 (1987).
12. Y. Morishima, M. Seki, S. Nomura, and M. Kamachi, *Proc. Japan, Acad.*, Ser. B, **69**, 83 (1993).
13. Y. Morishima, S. Nomura, T. Ikeda, M. Seki, and M. Kamachi, *Macromolecules* **28**, 2874 (1995).
14. H. Aota, Y. Morishima, M. Kamachi, *Photochem. Photobiol.* **57**, 989–995 (1993).
15. H. Aota, S. Araki, Y. Morishima, M. Kamachi, *Macromolecules*, **30**, 4090–4096(1997).
16. Y. Morishima, H. Aota, K. Saegusa, M. Kamachi, *Macromolecules*, **29**, 6509 (1996).
17. Y. Morishima, K. Saegusa, M. Kamachi, *J. Phys. Chem.*, **99**, 6505 (1996).

18. Y. Morishima, M. Seki, S. Nomura, M. Kamachi, *ACS Symposium Series* 548, American Chemical Society: Washington, DC, 1994; p. 243.
19. Y. Chang, C. L. McCormick, *Macromolecules* **26**, 5121 (1993).
20. A. Harriman, G. Porter, N. Searle, *J.Chem. Soc., Faraday Trans.* 2, **75**,1515–1521 (1981).
21. M.-C. Richoux, A. Harriman, *J.Chem. Soc., Faraday Trans. 1,* **78**,1873–1878 (1981).
22. F. R. Hopf, D. G. Whitten, in: *Polrphyrins and Metalloporphyrins,* D. Dolphin (Ed.) pp. 667–700, Elsevier, Amsterdam (1975).
23. F. R. Hopf, D. G. Whitten, in: *The Porphyrins* Vol.2 Part B, D. Dolphin (Ed.) pp. 161–195, Academic Press, New York (1979).
24. Y. Onoue, K. Hiraki, and Y. Nishikawa, *Bull. Chem. Soc., Jpn.*, **54,** 2633 (1981).
25. Y. Nishikawa, K. Hiraki, Y. Onoue, K. Nishikawa, Y. Yoshitake, and T. Shigematsu, *Bunseki Kagaku,* **32**, E115 (1983).
26. Y. Morishima, K. Saegusa, M. Kamachi, *Chem. Lett.*, 1994, 583.
27. Y. Morishima, K. Saegusa, M. Kamachi, *Macromolecules* **28**, 1203 (1995).
28. C. P. Nash, *J.Phys. Chem.*, **64**, 950 (1960).
29. N. J. Turro, Modern Molecular Photochemistry; The Benjamin/Cummings Publishing; California, (1978).
30. T. Nemzek, W. R. Ware, *J. Chem. Phys.* **62**, 477 (1975).
31. K. Kalyanasundaram, *J. Chem. Soc., Faraday Trans. 2,* **79**, 1365 (1983).
32. Y. Morishima, M. Seki, S. Nomura, M. Kamachi, *ACS Symposium Series* 548, American Chemical Society: Washington, DC, 1994; p 243.
33. R. J. DeVoe, M. R. V. Sahyum, E. Schmidt, N. Sorpane, D. K. Sharma, *Can. J. Chem.*, **66**, 319 (1998).
34. F. D. Saeva, B. P. Morgan, *J. Am. Chem. Soc.*, **106**, 4121 (1984).
35. R. H. Fetlon, D. Dolphin, D. C. Borg, J. Fajer, *J. Am. Chem. Soc.*, **91**, 196 (1964).